Springer Series in Wood Science
Editor: T. E. Timell

Springer

Berlin
Heidelberg
New York
Barcelona
Hong Kong
London
Milan
Paris
Singapore
Tokyo

Springer Series in Wood Science

Editor: T. E. Timell

M. H. Zimmermann
Xylem Structure and the Ascent of Sap (1983)
J. F. Siau
Transport Processes in Wood (1984)
R. R. Archer
Growth Stresses and Strains in Trees (1986)
W. E. Hillis
Heartwood and Tree Exudates (1987)
S. Carlquist
Comparative Wood Anatomy (1988)
L. W. Roberts/P. B. Gahan/R. Aloni
Vascular Differentiation and Plant Growth Regulators (1988)
C. Skaar
Wood-Water Relations (1988)
J. M. Harris
Spiral Grain and Wave Phenomena in Wood Formation (1989)
B. J. Zobel/J. P. van Buijtenen
Wood Variation (1989)
P. Hakkila
Utilization of Residual Forest Biomass (1989)
J. W. Rowe (Ed.)
Natural Products of Woody Plants (1989)
K.-E. L. Eriksson/R. A. Blanchette/P. Ander
Microbial and Enzymatic Degradation of Wood and Wood Components (1990)
R. A. Blanchette/A. R. Biggs (Eds.)
Defense Mechanisms of Woody Plants Against Fungi (1992)
S. Y. Lin/C. W. Dence (Eds.)
Methods in Lignin Chemistry (1992)
G. Torgovnikov
Dielectric Properties of Wood and Wood-Based Materials (1993)
F. H. Schweingruber
Trees and Wood in Dendrochronology (1993)
P. R. Larson
The Vascular Cambium: Development and Structure (1994)
M.-S. Ilvessalo-Pfäffli
Fiber Atlas: Identification of Papermaking Fibers (1995)
B. J. Zobel/J. B. Jett
Genetics of Wood Production (1995)
C. Matteck/H. Kubler
Wood – The Internal Optimization of Trees (1995)
T. Higuchi
Biochemistry and Molecular Biology of Wood (1997)
B. J. Zobel/J. R. Sprague
Juvenile Wood in Forest Trees (1998)
E. Sjöström/R. Alén (Eds.)
Analytical Methods in Wood Chemistry, Pulping, and Papermaking (1999)
R. B. Keey/T. A. G. Langrish/J. C. F. Walker
Kiln-Drying of Lumber (2000)

R. B. Keey · T. A. G. Langrish · J. C. F. Walker

Kiln-Drying of Lumber

With 97 Figures and 33 Tables

Springer

Prof. Dr. ROGER B. KEEY
Wood Technology Research Centre
University of Canterbury
Christchurch, New Zealand

Dr. TIMOTHY A. G. LANGRISH
Department of Chemical Engineering
University of Sydney
Sydney, NSW, Australia

Professor Dr. JOHN C. F. WALKER
School of Forestry
University of Canterbury
Christchurch, New Zealand

Series Editor:
T. E. TIMELL
State University of New York
College of Environmental Science
and Forestry
Syracuse, NY 13210, USA

Cover: Transverse section of *Pinus lambertiana* wood. Courtesy of Dr. Carl de Zeeuw, SUNY college of Environmental Science and Forestry, Syracuse, New York

ISSN 1431-8563
ISBN-13: 978-3-642-64071-1   e-ISBN-13: 978-3-642-59653-7
DOI: 10.1007/978-3-642-59653-7
Library of Congress Cataloging-in-Publication Data.
Keey, R. B.
　Kiln-drying of lumber / R.B. Keey, T.A.G. Langrish, J.C.F. Walker.
　　p. cm. – (Springer series in wood science)
　Includes bibliographical references (p.).
　ISBN-13: 978-3-642-64071-1
　1. Lumber–Drying. 2. Kilns. I. Langrish, T. A. G. (Timothy A. G.), 1963- .
II. Walker, J. C. F. III. Title. IV. Series.
TS837.K33 1999
674'.384–dc21   99-40864
                                                                         CIP

This work is subject to copyright. All rights reserved, whether the whole or part of the material is concerned, specifically the rights of translation, reprinting, reuse of illustrations, recitation, broadcasting, reproduction on microfilm or in any other way, and storage in data banks. Duplication of this publication or parts thereof is permitted only under the provisions of the German Copyright Law of September 9, 1965, in its current version, and permission for use must always be obtained from Springer-Verlag. Violations are liable for prosecution under the German Copyright Law.

Springer-Verlag Berlin Heidelberg 2000
Softcover reprint of the hardcover 1st edition 2000

The use of general descriptive names, registered names, trademarks, etc. in this publication does not imply, even in the absence of a specific statement, that such names are exempt from the relevant protective laws and regulations and therefore free for general use.

Product liability: The publisher cannot guarantee the accuracy of any information about dosage and application thereof contained in this book. In every individual case the user must check such information by consulting the relevant literature.

Typesetting: Best-set Typesetter Ltd., Hong Kong
Cover design: Design & Production, Heidelberg
SPIN 10645462   31/3136 – 5 4 3 2 1 0

Preface

At present, no single book adequately covers a basic understanding of wood drying in practice. This book satisfies the need for such a work. It describes the fundamental basis of kiln-drying technology, to enable forest companies to improve their drying operations as high-quality timbers become scarcer and the wasteful practices of yesteryear can no longer be tolerated. Adaptive change based on past experience is no longer good enough. Innovations require a sound understanding of the material being dried and the processes of drying. Newer techniques, such as the use of ultrahigh temperature seasoning and superheated steam under vacuum, require an even greater depth of physical understanding for these methods to be used effectively and economically.

This book provides a description of modern ideas about wood structure, moisture movement and stress development, from which models of the drying process are developed to give the kiln operator important information about the course of drying under specified conditions, and thus a means for rational process improvement. Theory is compared with practice wherever possible.

The authors have been collaborators in joint projects for over a decade, bringing together insights from material science in forestry and process understanding in engineering. This work has been sponsored in New Zealand by the Ministry of Science and Technology through grants from the Public Good Science Fund, from the Building Research Association of New Zealand and New Zealand Forest Research Ltd (Forest Research); in Australia, the work has been sponsored by the Australian Research Council, the Ian Potter and George Alexander Foundations and the Max Ralph Jacobs Fund of the Australian Academy of Science. Significant parts of this book draw upon our joint and separate activities in understanding the drying behaviour of species grown in the Australasian region.

We are conscious that there may be problems in relating our own experience with these particular species to that of others who work with other woods. We have tried to take a wider view, with the emphasis throughout on general principles governing the drying of wood in kilns, rather than on specific procedures. In this way, we trust that the absence of any familiar example will not detract unduly from what we have written. Moreover, in giving our overview of the scientific basis of kiln-seasoning technology, we have deliberately not set out to provide a manual of kiln-drying, thereby duplicating a number of excellent handbooks and guides that do that already.

We wish to express our thanks to Murray McCurdy, who has prepared most of the line diagrams to a common style. We also thank Gillian Weatherley, who has assisted us in the compilation of the references and seeking permission to publish copyrighted material. While scientific workers receive acknowledgment in the body of the text, we are conscious that our experimental efforts would have been less fruitful if it not were for the skills of our technical colleagues who have built and serviced the equipment we have used. We pay tribute to all the assistance we have had through the years from such people, namely: Paul Fuller, School of Forestry, as well as David Brown, Warwick Earl, Neville Foot, Bob Gordon and Frank Weerts of the Department of Chemical and Process Engineering at the University of Canterbury, New Zealand. When our software failed to perform to our desires, Tony Allen was always there to solve the problem. We also have a debt of gratitude to other colleagues in the Wood Technology Research Centre who share our interest in wood-related studies and have willingly provided facilities for our work. Their insights have enlarged our view.

Christchurch, New Zealand. R. B. Keey
Sydney, Australia. T. A. G. Langrish
September 1999 J. C. F. Walker

Contents

1	**The Structure of Wood**	1
1.1	The Structure of Softwoods	5
1.2	The Structure of Hardwoods	8
1.3	Cell-Wall Structure and Composition	12
1.3.1	Cellulose	13
1.3.2	Hemicellulose	14
1.3.3	Lignin	16
1.3.4	The Cell-Wall Ultrastructure of a Softwood Tracheid	18
1.3.5	Water in the Cell Wall	21
2	**Wood-Water Relationships**	23
2.1	Water in Wood	23
2.2	Moisture Sorption	24
2.3	Fibre Saturation and Maximum Hygroscopic Moisture Content	32
2.4	Theories of Sorption	34
2.5	Heat of Sorption	37
2.6	Response to Environmental Changes	40
3	**Evaporation and Humidification**	43
3.1	Moisture in Air	43
3.2	Enthalpy of Moist Air	45
3.3	Adiabatic Saturation and Wet-Bulb Temperatures	46
3.4	Humidity Charts	49
3.5	Ideal Heat Demand	52
3.6	Evaporation from a Wood Surface	54
3.7	Subsurface Evaporation	60
3.8	Mass-Transfer Coefficient Measurements	63
4	**Wood-Drying Kinetics**	65
4.1	Empirical Observations	65
4.1.1	Permeable Wood	66
4.1.2	Impermeable Wood	67
4.1.3	Empirical Models	67
4.1.4	Graphical-Analytical Methods	71

4.2	Normalisation of Drying-Rate Curves	72
4.3	Pathways for Moisture Movement in Wood	79
4.3.1	Longitudinal Movement	80
4.3.2	Transverse Movement	81
4.4	Selection of Drying Models	82

5	**Moisture Diffusion**	**85**
5.1	Driving Forces for Diffusion	86
5.1.1	The Kirschhoff Transformation	88
5.1.2	Moisture-Content Driving Force	90
5.1.3	Vapour-Pressure Driving Force	94
5.1.4	Chemical-Potential Driving Force	98
5.2	Penetration Periods and Regular-Regime Drying	100
5.3	Theoretical Modelling of Diffusion Coefficients	103
5.4	Experimental Measurements of Diffusion Coefficients	104
5.4.1	The Effect of Temperature on Diffusion Coefficients	107
5.4.2	The Effect of Moisture Content on Diffusion Coefficients	112
5.5	Conclusions	113

6	**Multiple-Mechanism Models**	**117**
6.1	Fundamental Equations	118
6.2	Experimental Observations	120
6.2.1	Temperature Profiles	120
6.2.2	Moisture-Content Profiles	121
6.2.3	Pressure Profiles	125
6.3	The Physical Process of Drying for a Softwood, *Pinus radiata*	127
6.3.1	Heartwood	128
6.3.2	Sapwood	128
6.3.3	The Moisture-Transport Equations	129
6.3.3.1	The Movement of Free Moisture	130
6.3.3.2	The Movement of Water Vapour	132
6.3.3.3	The Movement of Bound Water	133
6.3.4	The Evaporative Zone	133
6.3.5	Mass and Energy Conservation Equations	134
6.3.6	Initial Conditions	135
6.3.7	Boundary Conditions	135
6.4	Mixed-Wood Boards	136
6.5	Conclusions	137

7	**Lumber Quality**	**139**
7.1	Gross Features of Wood	139
7.1.1	Greenwood Moisture Content	139

7.1.2	Wetwood	142
7.1.3	Heartwood	143
7.1.4	Knots	143
7.1.5	Spiral Grain	146
7.1.6	Juvenile and Mature Wood	148
7.2	Intrinsic Features of Wood	152
7.2.1	Density	152
7.2.2	Collapse	157
7.2.3	Warp	158
7.2.4	Reaction Wood	162
7.2.4.1	Compression Wood	163
7.2.4.2	Tension Wood	165
7.3	Processing Implications	166
7.3.1	Sawing Strategies	166
7.3.2	Warp on Drying	166
8	**Stress and Strain Behaviour**	**171**
8.1	Mechanical Analogues	173
8.1.1	Elastic Element	173
8.1.2	Viscous Element	174
8.1.3	Maxwell Model	174
8.1.4	Kelvin Model	174
8.1.5	Burgers Model	175
8.2	Shrinkage	175
8.3	Instantaneous Strain	176
8.3.1	Linear Loading	176
8.3.2	Non-Linear Loading	181
8.3.3	Unloading	181
8.3.4	Slow-Loading Tests	182
8.4	Viscoelastic Strain	183
8.4.1	Mechanical Analogues	183
8.4.2	The Bailey-Norton Equation	185
8.5	Mechanosorptive Strain	188
8.5.1	Qualitative Observations	188
8.5.2	Quantitative Analysis	189
8.6	Relative Magnitude of Strain Components	195
8.6.1	Elastic and Other Strains	195
8.6.2	Viscoelastic and Mechanosorptive Strains	195
8.7	Solution Procedures	196
8.7.1	One-Dimensional Analysis	196
8.7.2	Two-Dimensional Analysis	197
8.8	Experimental Apparatus	198
8.9	Applications	199

9	**Airflow and Convection**	**203**
9.1	Airflow in a Batch Kiln	204
9.1.1	Velocity Distributions over a Kiln	205
9.1.2	Geometrical Considerations	206
9.1.3	Pressure Drops over Kiln Sections	207
9.1.4	Stack-Velocity Distribution	211
9.2	Flow between Boards	212
9.3	Convection in Kilns	215
9.3.1	Airflow Maldistribution	215
9.3.2	Board Irregularities	216
9.3.2.1	Uneven Thickness of Boards	217
9.3.2.2	Stack Ends	217
9.4	Bypassing	218
9.5	Kiln Economics	218
10	**Kiln Operation**	**221**
10.1	Drying under Constant External Conditions	222
10.2	Drying under Variable External Conditions	223
10.3	Practical Kiln Schedules	227
10.4	General Practical Considerations	233
10.4.1	Species-Grouped Schedules	234
10.4.2	Species-Specific Schedules	234
10.4.3	Schedule Development	236
10.4.4	Stacking	237
10.4.5	Fan Speeds and Reversals	238
10.4.6	Kiln Monitoring	239
10.4.7	Volatile Emissions and Kiln Corrosion	239
10.4.8	Equalisation	241
10.4.9	Stress Relief	242
10.4.10	Destickering	243
10.5	End-Moisture Specification	243
10.6	Handling Kiln-Dried Lumber	244
11	**Pretreatments of Green Lumber**	**247**
11.1	Protecting Wood Prior to Drying	247
11.1.1	Wet Storage	247
11.1.1.1	Control of Microorganisms	247
11.1.1.2	Relaxation of Growth Stresses	249
11.1.2	Antisapstain Treatments	250
11.1.3	Brownstain Control	255
11.2	Physical Methods to Improve Permeability	256
11.2.1	Incising	256
11.2.2	Compression Rolling	256

11.3	Low-Temperature Predrying	257
11.4	Heat Treatment	259
11.4.1	Steaming and Soaking in Hot Water	259
11.4.2	Dry Heat	261
11.5	Prefreezing	261
11.6	Antishrink Chemicals	262
11.7	Presurfacing	263
11.7.1	Problems with Surface Checking	263
11.7.2	Problems with Moulds	264
11.8	Green Finger-Jointing and Cutting Blanks	264
11.9	Precoating	266
11.10	Presorting	267
12	**Less-Common Drying Methods**	**271**
12.1	Solar Kilns	271
12.1.1	Insolation Rates and Kiln Locations	273
12.1.2	Absorbers	274
12.1.3	Glazing	275
12.1.4	Temperature Control	275
12.1.5	Humidity Control	275
12.1.6	Air Circulation	276
12.1.7	Energy Losses	277
12.1.8	Economics	278
12.1.9	Mathematical Modelling of Performance	279
12.2	Dielectric Drying	280
12.2.1	Mechanisms of Heating	280
12.2.1.1	Dipolar Rotation	280
12.2.1.2	Ionic Conduction	281
12.2.2	Interactions	281
12.2.2.1	Moisture Content	281
12.2.2.2	Density	282
12.2.2.3	Temperature	282
12.2.3	Internal Pressures	282
12.2.4	Drying Times Relative to Conventional Kilns	283
12.2.5	Economics	284
12.3	Superheated Steam Drying	285
12.4	Vacuum Drying	286
12.5	Dehumidifier Kilns	288

References	**291**
Subject Index	**315**
Species Index	**323**

1 The Structure of Wood

The woody stems of trees have a range of end-uses from large structures to small artefacts. As a material of construction, wood is easily worked with simple tools: as an object of art, the grain and texture of wood has considerable appeal.

A number of distinct roles is performed by the stem of a living tree. It must lift and support the crown above competing vegetation, conduct sap from roots to crown, and allow the counterflow of photosynthate from the foliage within the bark tissue. To provide a stable material for end-use, the moisture-containing stem is rough-sawn into lumber boards which are then dried. Normally this drying is done in a purpose-built kiln, but some predrying may be done outside the kiln should the lumber pose difficulties in drying. This book provides an insight into the kiln-drying of lumber, bridging the gap between a fundamental description of wood-water relationships on one hand and a handbook for kiln operators on the other.

The woody stem of a living tree contains both live and dead cells. The various functions of a woody stem are achieved by optimising the structure of dead, hollow conducting cells, each performing certain functions in the living tree. Optimisation means trade-offs. The tall stem of the giant sequoia (*Sequoiadendron giganteum*) represents an investment in a enormous mass of woody tissue to support the dead weight of the crown, to resist elastic buckling (height to diameter > 25 : 1) and, more critically, to withstand breaking or toppling in strong winds. Efficiencies are achieved by reducing the stem diameter roughly as the 3/2 power of the height above the ground, so that the surface stress at any point up the tapering stem is essentially constant (Mosbrugger 1990). Further, long-living forest giants face a greater chance of disaster than shorter-lived opportunistic colonising shrubs and trees. This calls for an additional investment to strengthen the stem, which is achieved partly by increasing its diameter to provide a larger factor of safety than would be needed for short-lived pioneer species. McMahon (1973) has estimated that the tallest trees are only a quarter of the height that should be achievable if buckling under self-weight were the limiting constraint.

In tall trees, a larger fraction of biomass is tied up in the stem compared with more shrubby trees, which invest relatively more in leaves, twigs, branches and roots. Thus the stem's biomass increases from around 20 to 40% of the total biomass as tree height increases from 10 to 100 m (Givnish 1995). Such biomass might have been used more productively elsewhere. There are both short- and long-term consequences of the annual investment in biomass, each season's allocation of growth, between the costs of increased stem size and branches compared with those of added foliage where the tree benefits from photosynthesis.

One way of ensuring that the tree's resources are used efficiently is to build the stem from interconnecting hollow cells. This provides strength and stiffness, while minimising the amount of wood tissue required. At the same time, these conduits perform the essential function of allowing long-distance transport of sap. The detailed strategy for accomplishing this varies between trees, which is a reflection of their evolutionary histories. Intriguingly, it appears that the earliest land plants were supported hydraulically, with water transport being of primary importance (Speck and Vogellehner 1988).

Timber is normally classed as being either a *softwood* or a *hardwood*. In softwoods, or *gymnosperms*, the seeds are "naked", whereas in hardwoods, or *angiosperms*, the seeds are enclosed in ovaries. Softwoods include the needle-leaved pines (*Pinus* spp.), the spruces (*Picae* spp.) and the firs (*Abies* spp.). These species are commonly described as evergreens, although a few, like larch (*Larix* spp.) and swamp cyprus (*Taxodium distichum*), shed their needles in the autumn or fall. Hardwoods of the temperate regions, on the other hand, are generally deciduous, shedding their foliage; examples include the oaks (*Quercus* spp.), ashes (*Fraxinus* spp.) and elms (*Ulmus* spp.). However, there are exceptions, such as the temperate, southern "evergreen" beeches (*Nothofagus* spp.), which lose their leaves as the new buds emerge in the spring, while most tropical hardwoods retain their foliage for most of the year. The terms softwood and hardwood originated in the mediaeval timber trade, which was unaware of soft hardwoods such as balsa (*Ochroma pyramidale*) and despite the relative hardness of a familiar softwood such as yew (*Taxus baccata*). This book follows well-established convention in using the terms softwoods and hardwoods for the gymnosperms and angiosperms, respectively.

Softwoods have a comparatively simple, uniform structure with few cell types. Slender, pointed cells called *tracheids* are dominant, constituting 90–95% of the stem volume. These cells provide both mechanical support and hollow conduits for sap flow (Fig. 1.1). Later, some 300 million years ago, hardwoods evolved and then different cell types became responsible for sap conduction and for structural support. The highly specialised conducting cells are known as *vessels*, while structural support is provided by short *fibres* (Fig. 1.2). Connections between cells are crucial for the flow of sap from cell to cell during its ascent to the crown.

Wood characteristics and properties are normally related to the three principal directions or axes of the stem:

- The *axial* direction is parallel to the stem of the tree itself. The majority of wood cells are elongated in this direction.
- The *radial* direction radiates horizontally from the centre of the tree out towards the bark. Varying amounts of radially-oriented tissue, known as *rays*, are aligned in this manner.

The Structure of Softwoods 3

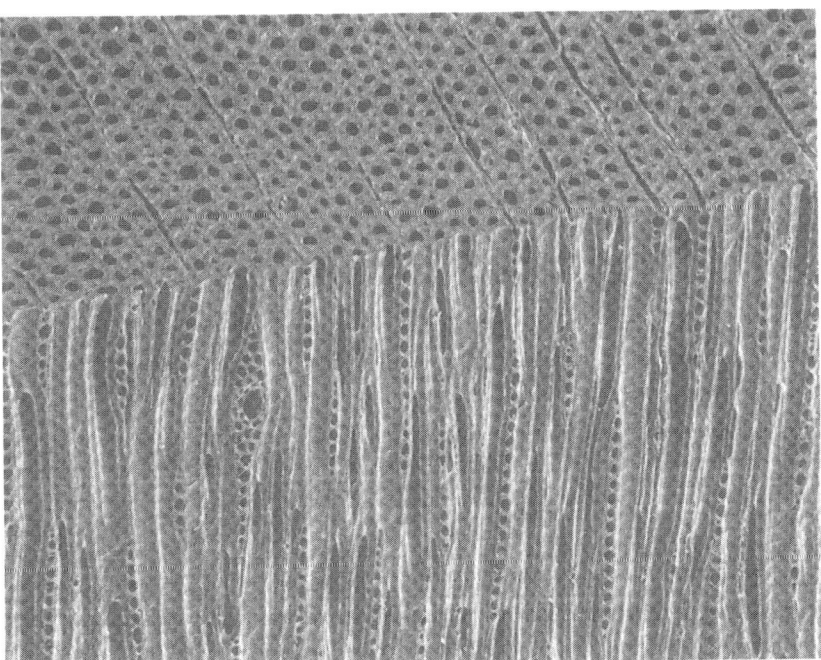

Fig. 1.1. Transverse and tangential faces of a softwood, *Larix decidua*. (Dr B.A. Meylan, pers. comm.)

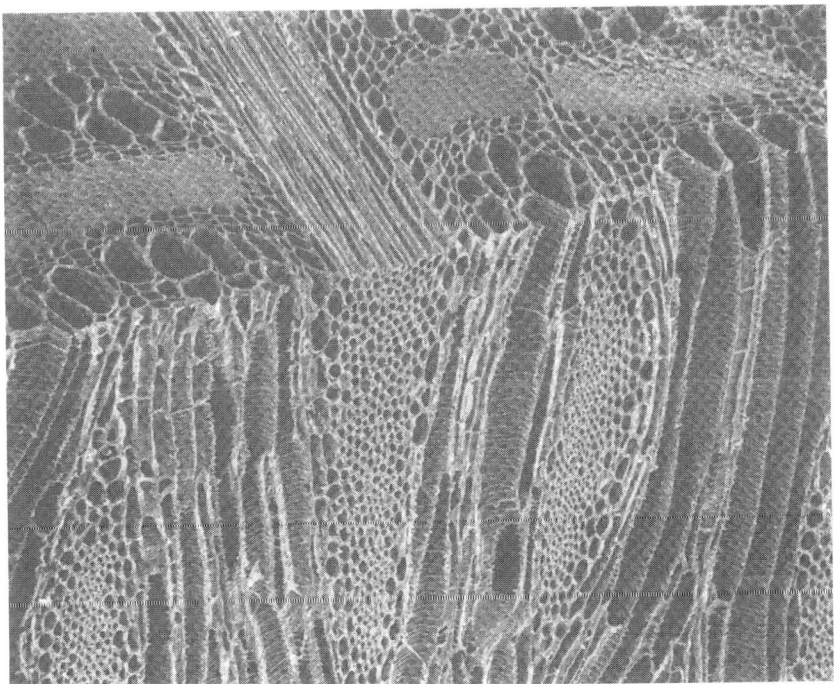

Fig. 1.2. Transverse and tangential faces of hardwood, *Plagianthus betulinus*. (Dr B.A. Meylan, pers. comm.)

- The *tangential* direction, also in the horizontal plane, is tangential to the bark. No cells are aligned in this direction, although the cellular structure permits tangential movement of sap and nutrients.

New wood is formed in a thin zone known as the *vascular cambium* lying immediately under the bark. The cambial cells divide to produce new cells both for the bark (*phloem*) and for the woody stem (*xylem*). The first few xylem cells (8–15) in the immediate vicinity of the cambium are alive and have nuclei, but at a little distance from the cambium their cytoplasms have died and the mature wood structure consists of the hollow exoskeletons of these cells through which sap can move. Thus the sequence is one of division of the cell at the cambium, either axial or radial enlargement, and differentiation from the *cambial initial* to develop into a particular cell type. In softwoods, the repeated *periclinal* division and formation of new cells at the cambium gives rise to ordered radial files of cells of uniform tangential width (Fig. 1.1). Periclinal division is unusual in biology in that the dividing cell plate grows lengthwise up and down the cell rather than taking the shortest route across the cell. Any deviation from the prescribed path results in either a short tracheid or a short cambial initial being left behind, resulting in a whole file of short cells being produced. In softwoods, repeated periclinal division of the cambial cells is accompanied by differentiation of their derivatives into wood tracheids to the inside and bark cells to the outside. Cells differentiating to the inside, the xylem, expand laterally, but increase very little in length. Initially, these cells possess only a thin primary wall which, during the next couple of weeks, is subsequently overlaid on the inside by the deposition of the three secondary wall layers. By the end of this period the cell is dead and resembles a hollow, but closed-ended, drinking straw whose internal space is called the *lumen*.

Additional cambial cells are needed as the girth of the stem increases. This is achieved by *anticlinal* (tangential) division of cambial cells, so creating additional cambial cells from which new radial files can form. In softwoods, the cell plate crosses the parent cell in a pseudotransverse direction producing two daughter cambial cells, each only slightly more than half the length of the original parent cell. These daughter cells then expand by gradual tip growth until they reach the approximate size of the former parent cell. A high frequency of pseudotransverse anticlinal division in the cambium depresses the mean length of cambial cells in the population and so reduces the length of the tracheids produced from them. As such divisions are more frequent at the end of the growing season, latewood cells tend to be shorter that those produced earlier in the season. Further, the necessity for a very high rate of pseudotransverse division in small stems contributes largely to the short cell length in the zone surrounding the pith. Repeated pseudotransverse division with the cell plate inclined in the same direction will tilt the cells with respect to the stem's axis, leading to the formation of *spiral grain*. Even a modest spiral-grain angle ($\phi < 5°$) is sufficient to cause lumber to warp on drying, which is a major problem in kiln-seasoning. The soft pithy centre of the tree consists of unenlarged *parenchyma* cells or soft, thin-walled primary xylem which has no strength.

In temperate regions, activity at the cambium is cyclic, with vigorous growth in spring and summer, and subsequent cessation of cambial division each autumn and winter. This gives rise to the familiar pattern of annual *growth rings*. In the tropics, growth rings reflect the pattern of the rainy seasons, so that several rings may be produced in a year.

In addition to axial elongated cells, trees have a system of cells, known as *rays*, aligned in the radial direction. The ray initial at the cambium has an equal size in all directions and is derived periodically from the axial elongated cambial initial. After division, ray initials elongate radially. These cells retain their cell cytoplasm and are physiologically alive in the sapwood, but lose their cell contents in heartwood, where they are often the storage sites for various extractives. The architecture of rays is relatively simple in softwoods, generally consisting of a file one cell wide, termed *uniseriate* (Fig. 1.1). Sometimes the ray is partly two cells wide (*biserate*), but only rarely is it fully *biserate* over the entire extent of the ray, as with *Cupressus macrocarpa*. In a typical ray, some 15 individual cells are stacked above one another. With hardwoods, rays are often *multiserate* (many cells wide) and the ray volume can account for up to 50% of the wood tissue in certain species (Fig. 1.2). On the other hand, some hardwoods have no ray tissue at all, e.g. *Hebe* spp. This variation is a major reason for the diversity of hardwoods and accounts for the pleasing appearance and grain of those timbers which are highly prized as furniture or finishing lumber.

As a tree grows older, a harder, darker zone begins to form at the centre of the stem. This is the *heartwood*. All its cells, even the ray tissue, are dead and physiologically inactive. The heartwood region expands as the tree increases in girth, being surrounded by an annulus of sapwood, which is typically 10 to 50 mm wide. The often dark colour of heartwood reflects the enrichment of the wood by various resins and carbohydrates known collectively as extractives. The extractives "personalise" each species and are responsible for its colour, odour and durability. These include sugars and organic acids or considerably more complex substances such as tannins and flavonoids which are driven from resin canals into the surrounding tissue to permeate the cell walls and cell spaces. Such metabolic byproducts may be inhibitory or even toxic to living cells and their accumulation probably accounts for the death of parenchyma ray tissue in heartwood. Moreover, the superior resistance to decay of some timbers is due to the presence of their particular extractives. The surrounding *sapwood* is generally perishable, but it is protected while in the living tree by the sap which saturates the tissue and excludes the oxygen needed by degrading organisms.

1.1 The Structure of Softwoods

Softwoods lack specialised conducting tissue and rely on their axially aligned tracheids for both support and water transport. Tracheids account for some

90–95% of the volume of a softwood. Individual tracheids generally range from 1 to 6 mm in length, depending on species and location within the stem, being shorter near the pith, and longer near the cambium. They are generally 15–60 μm in diameter and have a wall thickness ranging from 1 to 5 μm. Thus they are long and slender, being about 100 times longer than they are wide. Tapered overlapping ends provide strength, appearing sharply tipped when viewed tangentially, but more rounded in radial view.

Growth rings are visible because the size and wall thickness of tracheid cells vary systematically across each ring. In temperate species, the *earlywood*, formed in the spring and start of summer, has large-diameter, thin-walled cells as these are principally involved in sap conduction: by contrast, *latewood* cells, formed in the late summer and autumn, are thicker-walled and rectangular, with a much reduced radial width, and thus contribute more to structural support. The transition from earlywood to latewood is abrupt in European larch (*Larix decidua*) and Douglas-fir (*Pseudotsuga menziesii*), but quite imperceptible for the spruces (*Picea* spp.).

Tracheids have abundant pitting in their radial walls, as illustrated in Fig. 1.3. These *bordered pits* allow sap to pass through the cell walls of adjacent tracheids and so ascend the stem. Also, the presence of pits in the radial walls allows circumferential redistribution of sap. The lack of pitting on the tangential walls, except in the last few cells at the end of a growing season and the first few cells at the beginning of the next season, severely limits radial translocation of water. In the living tree, active sap conduction is restricted to a few growth rings adjacent to the cambium. This zone, known as the sapwood region, is characterised by a very high sap content, although the last growth ring or so adjacent to the central heartwood zone surrounding the pith may be as dry as the heartwood itself. The sapwood annulus ranges in width from a few millimetres to a maximum of about 50 mm, depending on species, tree age and height up the stem, with a general tendency for the width to be smaller in maturer trees, especially towards the base.

In many of the following chapters, the role of pits is discussed, when water flows in wood in contrast to diffusion of water across cell walls, and where the mechanism of pit closure or aspiration is considered, both in sapwood during drying and in heartwood while the tree is standing still alive.

The bordered pits are specialised valves designed to seal and isolate individual tracheids if they become damaged or embolised by cavitation. This is needed because the sap stream in a tree is under capillary tension as the sap is "pulled up" to the crown. Any break in the capillary column would kill the tree unless it had a mechanism for confining and isolating the failure.

While tracheids dominate the composition of softwoods, other axial cell types are present in small amounts including axial parenchyma, strand tracheids and resin canals, although resin canals are strictly "ducts" that form between cells. Parenchyma cells are thinner-walled than tracheids and may retain their cell contents for several years. Strand tracheids probably represent an intermediate evolutionary stage between tracheids and parenchyma: they

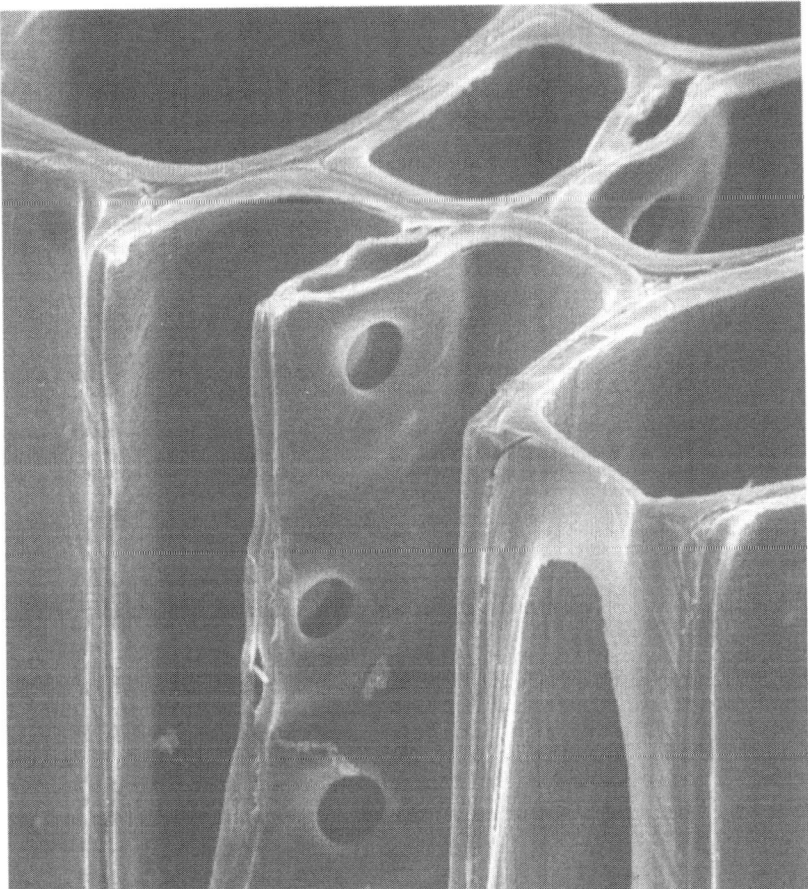

Fig. 1.3. Softwood pitting. (Dr B.G. Butterfield, pers. comm.)

arise from the transverse subdivision of an axially elongated cell that otherwise might have developed into a tracheid. They differ from parenchyma cells in having bordered pits. They are often associated with resin canals.

Resin canals, or resin ducts, are a normal feature of some pines (*Pinus* spp.), but are not found in some firs (*Abies* spp.) and hemlocks (*Tsuga* spp.). Indeed, there are more genera without resin canals than with. The axial resin canals mostly occur singularly, although in Douglas-fir (*Pseudotsuga menziesii*) they often occur in pairs. Resin canals are surrounded by thin-walled, epithelial parenchyma cells. Epithelial cells in the sapwood actively secrete resin into the canal and, with time, exudation pressure drives this resin into the heartwood, so enriching it with extractives. The epithelial cells in spruce (*Picea* spp.), Douglas-fir (*Pseudotsuga menziesii*) and larch (*Larix* spp.) are thick-walled and lignified. In pines (*Pinus* spp.), the axial resin canals are particularly large (100–200 μm in diameter), abundant and surrounded by thin-walled, unlignified epithelial cells. This difference in wall thickness and lignification is significant, as the thin, unlignified tissue can collapse, providing a new network of paths for an enhanced mass flow in already-dried wood on preservation and on redrying. Resin in sapwood of all these species is fairly fluid and can bleed from dried and dressed millwork, unless dried at temperatures above 80 °C.

Physiological or physical stress arising from insect or fungal attack, mechanical injury or wind damage can result in *traumatic resin canals*. Furthermore, these canals can be found in *Abies, Cedrus, Sequoia* and *Tsuga* spp. which do not form normal resin canals. Traumatic resin canals tend to be large and form an arc within the weaker earlywood. In some softwoods, *resin pockets* form as a result of extensive separation between tracheids. In *Pinus radiata*, these pockets can be up to 40 mm wide in the tangential direction, 100 mm along the grain and 5 mm thick. They are most prevalent in trees found on windy sites where the cambium is periodically subject to considerable flexure. In old trees, such resin pockets are most abundant some distance from the cambium, presumably having been formed many years earlier when the thinner stem was subject to larger strains.

There is another system of cells known as *rays* which lie in the radial direction. Usually they are composed only of parenchyma cells, but many of the pine family incorporate *ray tracheids*, which are devoid of protoplasts and have secondary walls with bordered pits. Ray tracheids tend to occur at the upper and lower margins of the ray. A longitudinal-tangential view of a wood reveals the height and width of the ray. A large resin canal in a uniserate ray results in a central bulge in an otherwise single row of cells in the ray. The radial transport within ray parenchyma has an important function of delivering carbohydrates such as starch to the cambium, especially at the start of a period of active growth.

1.2 The Structure of Hardwoods

The specialised hardwood vessel comprises a number of individual cells called *vessel elements* stacked on top of one another. Their end walls are partly or completely dissolved during the final stages of cell maturation, thus allowing the short vessel elements to combine to form a single long capillary, ranging in length from a few millimetres to many metres. The end-wall openings linking the individual vessel elements are termed *perforation plates*. These can be broadly categorised as *simple perforation plates*, comprising a single large opening between two adjacent cells, while *multiple perforation plates* have a number of openings variously arranged.

Viewed in cross-section, the vessels of some hardwoods are of relatively uniform diameter and are distributed evenly throughout the growth ring. These are described as *diffuse-porous* hardwoods. Such vessel distribution is found in a number of genera in the temperate zone, including *Acer*, *Aesculus*, *Populus* and *Salix* spp., and is typical of most tropical trees. By way of contrast, woods with very large vessels in the earlywood and much smaller vessels in the latewood are described as *ring-porous* hardwoods, of which *Castanea*, *Fraxinus* and *Quercus* spp. are well-known examples. Ring-porous hardwoods are in the minority.

Further, vessels can be solitary or grouped together in various patterns. Vessels need to interconnect with one other in order to create a continuous conducting system up the stem. The paths of vessels are irregular and of varying length. They can be followed by infusing a specimen with a water-soluble paint containing micron-sized pigment particles which are confined to the vessels, being too large to migrate into adjoining tissue. The network of vessels for some American hardwoods is illustrated in Fig. 1.4. With ring-porous species the large-diameter (ca. 300 μm) earlywood vessels can extend for many metres, occasionally running the full height of the stem, whilst the narrower vessels (ca. 50 μm) in the latewood are between 0.1 and 1.0 m in length. The vessels of diffuse-porous, temperate hardwoods are invariably narrow, ca. 75 μm, and short, the majority being no more than 0.1 m long, in contrast with diffuse-porous, tropical hardwoods which often have wide vessels greater than 200 μm in size. The principal constriction to sap flow is offered by the pitting between adjoining vessels. Inhibition to sap flow is minimised by the vessels presenting a very large area of pitting on the lateral walls (Fig. 1.5). These intervessel pit membranes are very homogeneous with innumerable "invisible" openings (generally $\ll 0.1\,\mu$m) for water transfer that are not large enough for the smallest particulate matter. Intervessel pit membranes lack the sealing mechanism found in softwoods, so hardwoods have developed alternative mechanisms to protect against cavitation and embolisms within vessels.

In hardwoods, wide, long vessels are individually far more efficient for conduction, but they risk embolisms after freezing and from low water potentials. Hardwoods occupying droughty, cool environments tend to have high-conductivity plumbing, wide earlywood vessels with simple perforation plates, when water is available in the spring, but narrow latewood vessels during dry, warmer spells to withstand low water potentials without developing embolisms. Hardwoods protect their conducting network in one of two ways, utilising *tyloses* or gums. In general, species with large-diameter vessels rely on tyloses, which are ingrowths from neighbouring cells through pits into the vessel. Where cavitation has occurred, activity in the live parenchyma cells surrounding the vessel results in numerous such intrusive growths of a special layer entering through pit apertures and blocking the vessels in the timber. Species with small-diameter vessels normally rely on resin exudates from the adjacent live parenchyma cells to block the vessel.

As there are no living parenchyma cells in heartwood, any occlusions must have preceded the formation of heartwood. While the durability of heartwood arises primarily from the presence of appropriate extractives, hardwoods can

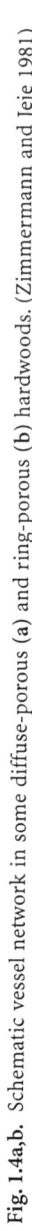

Fig. 1.4a,b. Schematic vessel network in some diffuse-porous (a) and ring-porous (b) hardwoods. (Zimmermann and Jeje 1981)

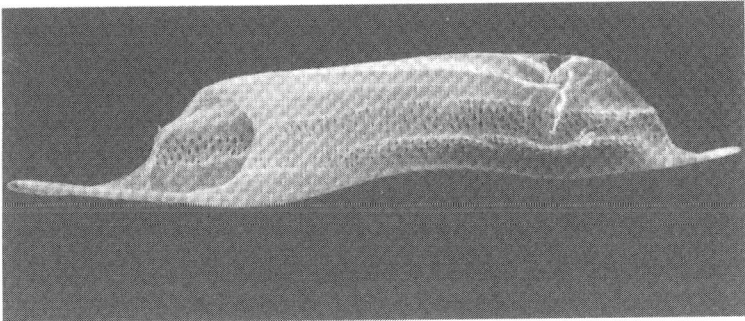

Fig. 1.5. Intervessel pitting in hardwoods. (Dr E. Wheeler, pers. comm.)

anticipate attack by microorganisms by blocking vessels in the inner sapwood zone as a precaution.

Structural support is provided by the fibres, which differ from softwood tracheids in being comparatively short (0.25–1.5 µm), more rounded in outline and thick-walled with restricted lumens. Fibres can be classified as fibre tracheids, libriform fibres and septate fibres. Although *fibre tracheids* have bordered pits, these and the other fibres in hardwoods can be considered as high-density blocks of poorly conducting tissue. *Libriform fibres* are a little thicker-walled and longer than fibre tracheids, and have simple pits. Either fibre type can be septate, where the cell is partitioned across internally into a number of septa which retain their protoplasts. These thin transverse walls form after the cell's secondary wall has been laid down. In most technical work, the individual characteristics of the distinctive fibre types can be ignored, referring to all indiscriminately as fibres. The only fibres that have a self-evident role in sap conduction are the vessel-centred or *vasicentric tracheids*, which are present in some hardwoods, often in close proximity to vessels. In some woods, vasicentric tracheids resemble vessels in that they are stacked above one another, but differ in that they retain their end walls and have abundant small, bordered pits on all side walls.

Of the other cell types, parenchyma cells are generally very abundant and are mostly associated with vessels (*paratracheal axial parenchyma*) in about half of the hardwoods, as illustrated in Fig. 1.2. They have relatively thin walls interrupted by irregularly arranged, small, circular simple pits. Paratracheal parenchyma may consist of either a broad band of tangentially aligned parenchymous tissue embedding many vessels or a ring of parenchyma cells around a vessel in which case parenchyma tissue is clearly not abundant. By contrast, *apotracheal parenchyma*, which are found in about one-third of all hardwoods, can occur in similar bands of tissue but avoiding the vessels, or as a tangential strand which is a cell or so wide, or as isolated cells. Hardwoods may show both types of parenchyma or neither. The distribution of parenchyma helps in identifying timbers. Large volumes of parenchyma result in hardwoods of low density. While it is clear that live parenchyma cells are able to play varied roles, in healing wounds, formation of adventitious shoots and food storage, these functions do not yield a complete explanation in view of the large quantitative variation in the number of parenchyma cells in various hardwoods.

Hardwood rays are exclusively composed of parenchyma cells, but display an enormous variation in size and shape: from uniserate rays only a few cells tall in willows (*Salix* spp.) to massive multiserate structures, which can be over 30 mm tall, as found in most oaks (*Quercus* spp.), that frequently taper to uniserate margins. The largest rays are aggregates of small rays barely separated by a few vertical fibres. The majority of ray cells are *procumbent*, being elongated and radially oriented. Nevertheless, especially in the larger ray systems, there may be some axially elongated cells, composed of upright parenchyma cells, which are either confined to the upper and lower extremities of the ray, as in holly (*Ilex aquifolium*), or form a sheath around the ray. Upright cells, however, are present even in some woods with narrow rays such as willows (*Salix* spp.). The rays cells are a store of food materials and often contain starch grains. In heartwood, food reserves like starch have disappeared, with the entire zone being infused with gums and resins.

The proportion of the various cell types in hardwoods is extremely varied. In North American hardwoods, for example, the vessel volume ranges from 6 to 55%, fibre volume from 29 to 76%, ray volume from 6 to 31% and axial parenchyma volume from 0 to 23% (French 1923). The density of hardwoods is largely determined by the proportion of fibres to other cell types. In low-density hardwoods, the fibres occupy a minor proportion of the volume, whereas in high-density hardwoods the fibres occupy a large proportion of the volume. Furthermore, vessel and parenchyma cells are thin-walled, while with fibres the wall thickness can vary greatly. In low-density hardwoods, the fibres are thin-walled with large lumens, whereas in denser woods the fibres are smaller and more numerous, the walls are thicker and the lumens contracted.

The variety of wood cells found in hardwoods is a major reason why most fundamental studies and much applied research have been undertaken on softwoods such as pines having a simpler anatomy.

1.3 Cell-Wall Structure and Composition

Wood is a composite material, with cellulose constituting the *skeletal* substance, hemicelluloses the *matrix* and lignin the *encrusting* substance. Without lignin, a tree could not grow taller than about 3 m; it would collapse under it own weight. Cellulose is the substance responsible for the tensile strength of wood, while lignin provides most of its compressive strength.

1.3.1 Cellulose

Cellulose, the most abundant organic chemical on earth, is also the most important constituent of wood, for which it accounts for 40–45% of the cell wall in normal wood. It is formed in a condensation reaction, in which a glucose ($C_6H_{12}O_6$) residue is added to the polymer while a water molecule is removed. All three hydroxyl groups in the glucose residues, as well as the glucosidic bonds connecting them, are equatorial; that is, they are located in the plain of the glucan chain which, unlike that in starch is linear, as shown in Fig. 1.6.

Wood cellulose is a large, linear polymer with an average degree of polymerisation of about 10 000. The basic functional unit of cellulose, wherever it occurs naturally, is the *microfibril*, which in the secondary cell wall of wood is about 10 nm wide. In the alga *Valonia* sp., it is twice as wide, but in the primary wall in wood only 1–2 nm (Preston 1974). The degree of crystallinity of wood cellulose is considerably less than it is in cotton or in the G-layer of tension wood. The differences vary according to the experimental procedure used, such as NMR or X-ray diffractometry, and it is necessary to recognise differences that arise from material with the crystals' interiors and the surface cellulose (Newman and Hemmingson 1994). The cellulose crystals cannot be penetrated by water because of their strong hydrogen-bonding, but the less ordered, amorphous regions are hygroscopic.

The cellulose chains are held together in the chain direction by strong covalent bonds. In the other two directions, there are relatively strong hydrogen bonds and weak van der Waals forces. Bond energies lie in the ranges:

- Covalent bonding 200–300 kJ mol^{-1}
- Hydrogen-bonding 10–40 kJ mol^{-1}
- Van der Waals forces 1–10 kJ mol^{-1}

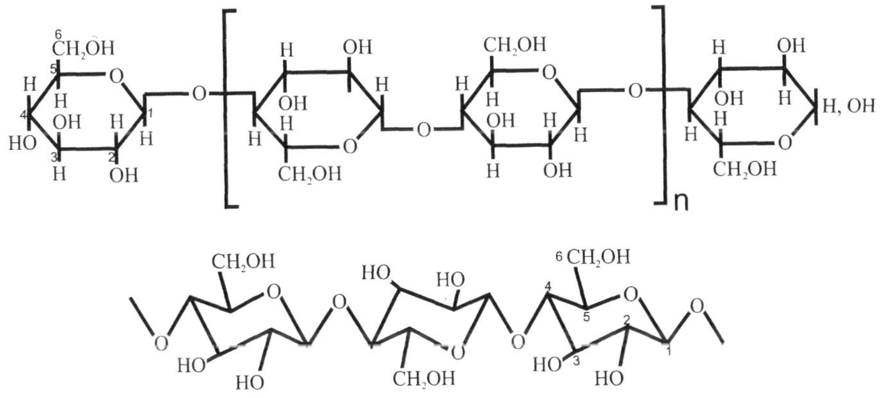

Fig. 1.6. Haworth formula (above) and conformational formula (below) of cellulose

Cellulose has the characteristic of being both a chain and a sheet lattice. This is the main reason for the anisotropic behaviour of fibres and thus of wood. For example, the strength and shrinkage are vastly different when comparing properties parallel and perpendicular to the cellulose microfibrils.

In the S_2-layer, which is the dominant layer of the secondary wall, the cellulose microfibrils are oriented at an angle of 5–30°, a fact responsible for the small longitudinal shrinkage of wood on drying. In juvenile wood and compression wood, this angle is as large as 40–50°, resulting in excessive shrinkage and swelling in the longitudinal direction.

1.3.2 Hemicellulose

Hemicellulose is a collective name for a number of closely related polysaccarides, which are present in most plant-cell walls where they are invariably associated with cellulose and lignin (Timell 1967). The sugar units are linked together to form linear chains in the same manner as cellulose, but the chains are much shorter, and they consist not only of glucose, but also of other sugar residues, mannose and xylose being the most common. In addition, the main chain of the hemicelluloses in most cases carries single-unit side chains of other sugar units. Unlike cellulose, the hemicelluloses are amorphous as they exist in the cell wall.

Hardwoods and softwoods contain different, albeit related hemicelluloses. The dominant hemicellulose in hardwoods (20–30% by wood weight) is a xylan (Fig. 1.7a). It carries side chains, one per ten xylose units, consisting of a single residue of a methylated glucuronic acid. Unlike glucose, xylose has no primary hydroxyl groups, and many of the secondary hydroxyls are acetylated, which reduce the ability of the xylan to develop hydrogen bonds. However, in certain algae that lack cellulose, xylan is present in the form of highly crystalline microfibrils.

The dominant hemicellulose in the softwoods is a galactoglucomannan (20–25%) (Fig. 1.7b). The main chain consists of partly acetylated glucose and mannose residues at a ratio of 1 : 3. There are single-unit side chains of galactose, where one of the hydroxyl groups is axial. A second hemicellulose in softwood is a xylan (8–12%). Unlike the xylan in hardwoods, it has no acetyl groups and also contains twice as many methylglucuronic acid side chains (one per five xyloses). A second side chain consists of a single arabinose residue, which, like mannose and galactose, has one hydroxyl group.

The hardwood xylan contains 200 xylose residues on average. The other wood hemicelluloses cannot be isolated without some degradation, and their original molecular weight is therefore not yet known. There are, however, indications that they are probably of the same size as the hardwood xylan. The wood hemicelluloses are located in the cell wall around and between the cellulose microfibrils. Hemicelloses and lignin are linked covalently.

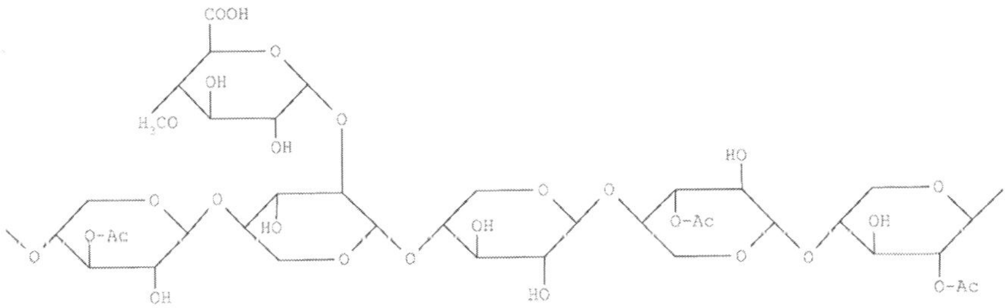

Fig. 1.7a,b. Partial chemical structures of a) galactoglucomannan, abundant in softwoods, and of b) methylglucuronoxylan, a dominant hemicellulose in hardwoods. (Reproduced from Walker 1993, Primary Wood Processing, with kind permission of Kluwer Academic Publishers)

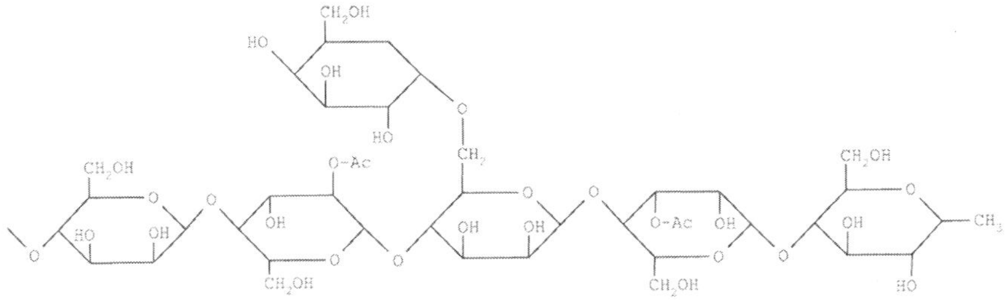

The acetyl groups and bulky side chains of hemicelluloses prevent them as they exist in wood from developing ordered, strong hydrogen bonds and crystallising. However, if the acetyl groups are eliminated and if the side-chains are reduced in frequency (though not completely removed), all wood hemicelluloses can form ordered hydrogen bonds and crystallise. In this process, they can often become firmly attached to cellulose. In fact, two wood hemicelluloses were the first polymers ever observed in the form of a single crystal (Timell, pers. comm.).

The function of the hemicelluloses in wood is uncertain. They may be needed to form connections between the hydrophilic cellulose and the hydrophobic lignin (Fujita et al. 1983), so permitting effective transfer of shear stresses. Another function, suggested by recent research on the biosynthesis of bacterial cellulose, could be the prevention of cellulose in wood from becoming more highly crystalline than it is now. A crystallinity as high as that found in cotton is likely to weaken the strength characteristics of wood.

In summary, the hemicelluloses are formed from polysaccharides just like cellulose, but differ in having a relatively short main chain/backbone with a few, short side branches. They are unable to pack tightly together. The hydroxyl groups are not perfectly oriented and spaced to form strong hydrogen bonds with one another, and therefore they are accessible to intracellular water molecules, with which they bond in the living cell wall. This possibility has

16   The Structure of Wood

important implications for the rate of drying and the energy needed at low moisture levels.

1.3.3 Lignin

The third principal component of the cell wall is lignin. This is an amorphous polymer linked in a variety of ways to produce a substance that is almost totally insoluble in most solvents. Softwood lignin is thought to incorporate roughly 500 phenylpropane units. However, the variety of linkages between individual units means that representations of the structure of lignin are only "model" structures that include the variety of cross-links and their relative proportions. One of the earliest and simplest merely shows 16 phenylpropane units (Fig. 1.8).

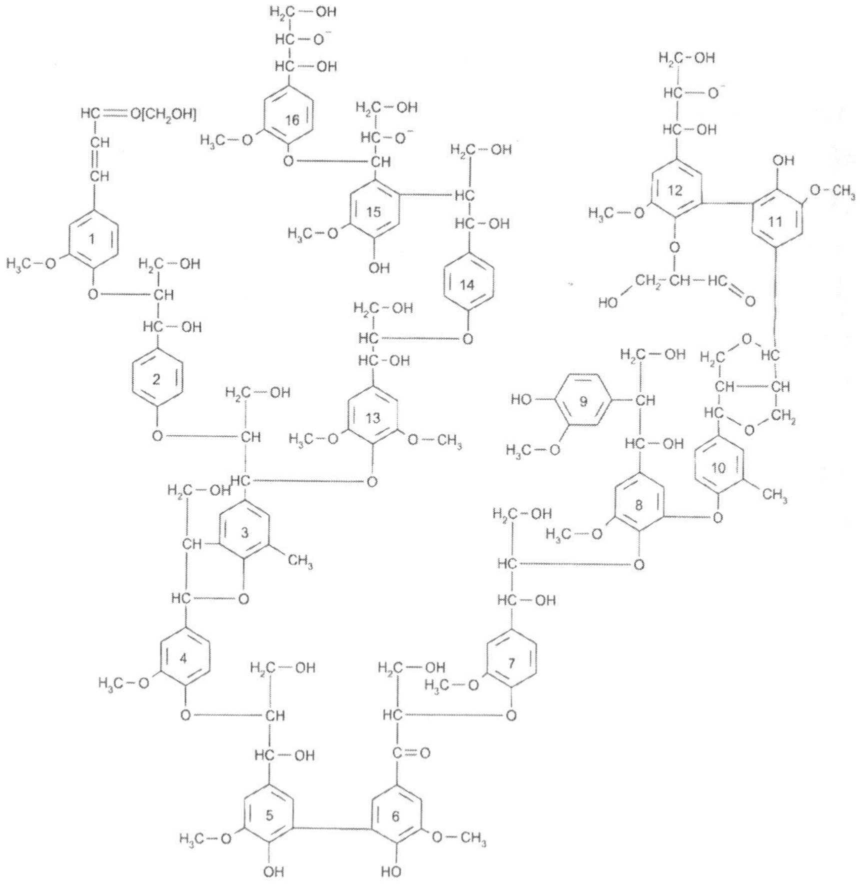

Fig. 1.8. Model structure of lignin. (After Adler 1977)

In examining the structure, one observes two kinds of cross-linking. There are C-O-C ether linkages, frequently between the aromatic head of the monomer and a point on the three-carbon aliphatic tail of an adjacent monomer. Second, there are direct carbon-carbon bonds linking monomers.

Hardwood lignin is a little different since it is a mixture of two monomer units, as illustrated in Fig. 1.9: the guaiacylpropane unit, which is found in softwoods, and the syringylpropane unit. The relative proportions of guaiacyl- and syringylpropane in hardwoods vary from 4:1 to 1:2. Syringylpropane has two methoxyl groups, -OCH_3, attached to the aromatic head of the molecule, whereas there is only one in guaiacylpropane. As shown in Fig. 1.9, some cross-links are only possible because there is a single methoxyl group adjacent to the phenolic group (-OH) in the aromatic ring: these are the carbon-carbon cross-links formed between units 3 and 4, units 5 and 6, units 11 and 12, and units 14 and 15, as well as the C-O-C ether bridge between units 8 and 10. Such links are not possible with syringylpropane units, as they are blocked by the presence of the second methoxyl group. This probably explains why the lignin in hardwood is of lower molecular weight than that in softwood.

In wood biosynthesis, the cellulose framework of the cell wall is laid down first and only subsequently are the intervening spaces filled by the amorphous matrix of hemicelluloses and lignin. In fact, lignin deposition within the cell wall does more than just fill the gaps between the microfibrils. The wall swells in thickness to accommodate the lignin and there is a corresponding decrease in the length of the cell, generating growth stresses that are locked into the living trees only to be released when the timber is milled. This is a particular problem with *Eucalyptus* spp. In Europe, such growth stresses are most noticeable in beech (*Fagus sylvatica*). The timber warps as it comes off the saw, presenting problems in achieving dry, straight timber.

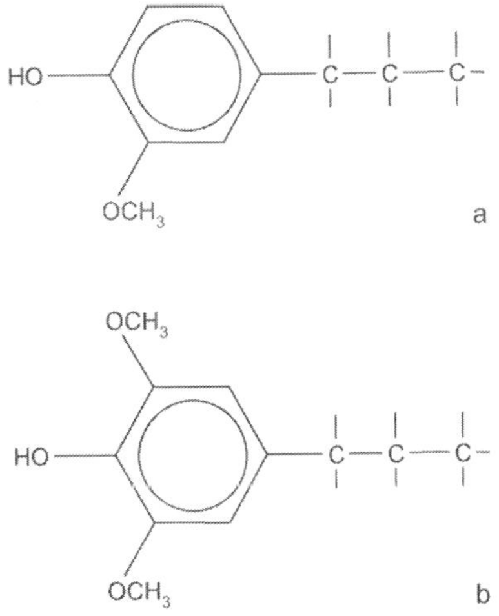

Fig. 1.9. a Guaiacylpropane monomer which dominates the lignin structure of softwoods; b Syringylpropane monomer which is present in varying amounts in hardwoods. (Reproduced from Walker 1993, Primary Wood Processing, with kind permission from Kluwer Academic Publishers)

1.3.4 The Cell-Wall Ultrastructure of a Softwood Tracheid

The description of the cell wall is framed in terms of the orientation of the cellulose microfibrils. The main features are the middle lamella (ML), a thin primary wall (P) and a compound secondary wall (S). The general orientation of the microfibrils in the various layers is shown in Fig. 1.10.

Strictly, the middle lamella is not part of the cell wall and it contains no cellulose microfibrils. The middle lamella occupies the intercellular region and consists of material that holds adjacent cells together. At the cambium, where the cells are dividing and expanding, the intercellular material is largely pectic and only later does the middle lamella become highly lignified.

Initially, the cell wall of the enlarging cell is very thin (0.1–0.2 μm) and consists of only the primary wall. At this stage, the wall is a colloidal membrane of protein, pectinaceous and related hemicellulosic material lightly reinforced by cellulose microfibrils. The wall is both elastic (allowing the cell to grow axially) and plastic (permanently extendable) during its early growth. Longitudinal extension of cells occurs prior to laying down the secondary wall layers and prior to lignification. The multinet theory of wall growth envisages that during the first instances of primary wall formation the relatively small amount of cellulose present becomes reoriented from a flat transverse orientation to a steep helix as the cell is extended. As more of the primary wall is laid down, this is stretched and reoriented to a smaller degree. Thus the primary wall becomes a sequence of microfibril "nets" each experiencing less and less rotation. Overall, the orientation of the microfibril network is random

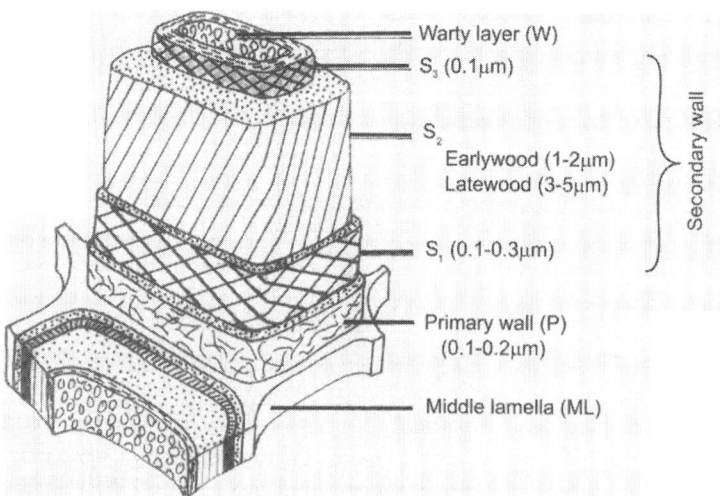

Fig. 1.10. Orientation of the cellulose microfibrils across the cell wall. in a softwood tracheid. (Reproduced from Walker 1993, Primary Wood Processing, by kind permission of Kluwer Academic Publishers)

except near the cell corners where it is aligned parallel to the length of the cell. The primary wall only begins to lignify after the deposition of the microfibrils in the secondary wall. Frequently, it is not easy to distinguish or physically separate the primary wall from the middle lamella, so where both are examined simultaneously the combined region (ML + P) is called the compound middle lamella (CML).

The secondary wall is laid down after the cell extention is complete. Three distinct layers are recognised in the compound secondary layer, the S_1, S_2 and S_3 layers, by virtue of the differing orientation of the microfibrils in each layer (Fig. 1.10). The outermost layer, the S_1, is only 0.1–0.3 μm wide, and here the microfibrils wind round the cell wall in a lazy helix at an angle between 50° and 75° to the cell axis. There are a number of very thin concentric lamellae within the S_1 layer which distinguish themselves from one another by slight differences in alignment of the cellulose microfibrils between each lamella or by an abrupt reversal of direction in which microfibrils in adjacent lamellae wind round the cell. By convention, the direction or sense of the microfibrils is described as an S- or Z-helix. They form an S-helix where they wind up from the lower right to the upper left and they form a Z-helix where they wind up from the lower left to the upper right, when viewing the cell wall from the outside (Fig. 1.10). The S or Z orientation of the microfibrils is deduced from the middle stroke of the letters Z or S (the latter being a stylised mirror image of the letter Z). In the S_1 layer, the S-helix lamellae are more strongly developed.

After the few lamellae of the S_1 layer have been laid down, the orientation of the microfibrils changes rapidly to that found in the S_2 layer. The S_2 layer is the dominant feature of the cell-wall ultrastructure. Even in the thin-walled earlywood it is 1–2 μm thick, while in the latewood it is 2–5 μm thick. In the S_2 layer the microfibrils are densely packed and steeply inclined, making an angle of 10° to 30° with the tracheid axis. All are similarly oriented (within 5° or so) and wind round the cell in the Z-sense. There are a large number of lamellae in the S_2 layer ranging from 30 in the earlywood to 150 in the latewood.

The last part of the cell to be formed is the S_3 layer lying adjacent to the lumen. The S_3 layer is thin, 0.1 μm, with the microfibrils inclined in an S-helix at an angle between 60° and 90° to the cell axis. Finally, in most species, the inner surface of the S_3 layer is overlaid with a thin non-cellulosic film thought to be composed of remnants of the protoplast, with lumps, which gives rise to its name, the warty layer.

Thus far, the cell-wall ultrastructure has been interpreted in terms of the orientation of the cellulose microfibrils. Once the microfibrils in the primary wall have being laid down and formation of secondary wall cellulose has begun, the initial deposition of lignin also starts at the junctions or intercellular corners between cells and then extends to the whole middle lamella. Lignification of the cell wall commences during later stages of S_2-cellulose formation and proceeds more rapidly after the entire microfibrillar network has been formed (Fujita et al. 1983). Lignin monomers diffuse into the cell wall and polymerise in situ. One widely accepted description of the final structure is provided by Kerr and Goring (1975). They stained thin sections of black spruce (*Picea mariana*) and, using the electron microscope, observed regular striations with a repeat distance of 7 nm in the radial direction: both the stained, lignin-rich bands and unstained cellulose microfibrils were roughly 3.5 nm wide. On scanning in the tangential direction, they found the repeat distance was approximately 15 nm. From this, and other evidence, they proposed their ultrastructure model of the cell wall which is reproduced in Fig. 1.11a.

There is sufficient evidence to permit a broad analysis of microfibril dispersal within the cell wall. According to Kerr and Goring (1975), the microfibrils are approximately 3.5×15 nm in cross section. In an earlywood tracheid, the cell wall is, say, 1.5 µm thick, while the cell diameter is, say, 30 µm. Further, cellulose occupies 40% of the volume of the cell wall, on the

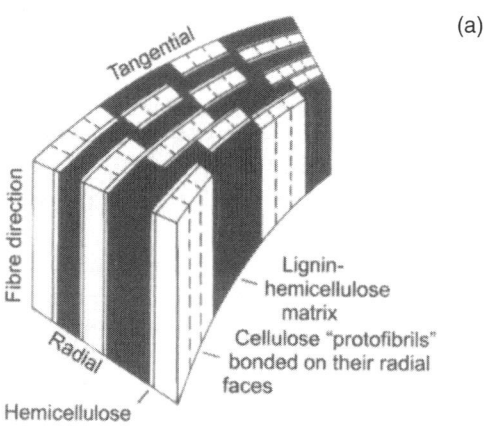

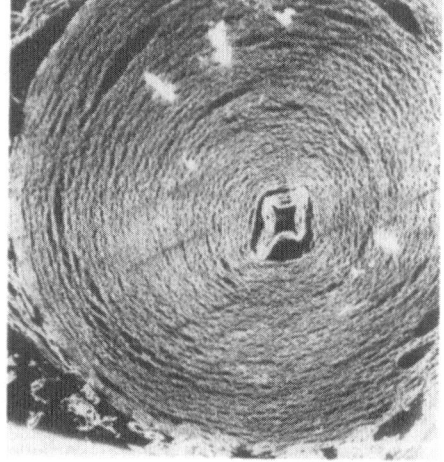

Fig. 1.11a,b. Ultrastructural detail. a Microfibrils embedded in the lignin-rich matrix (After Kerr and Goring 1975); b Cell-wall layering after extensive delignification. (Reproduced from Young (1986), Cellulose Structure Modification and Hydrolysis, by permission of John Wiley & Sons, Inc)

assumption that crystalline cellulose is a little denser than the amorphous cell-wall constituents. Thus, there are about 3 million microfibrils in an earlywood tracheid. The fine dispersal of microfibrils amongst the numerous lamellae is well demonstrated in the ultrastructure after removing most of the lignin, as shown in Fig. 1.11b.

Wood is the ultimate reinforced-fibre composite. The cellulose microfibrils give timber its tensile strength, the lignin provides good compressive strength and prevents the slender microfibrils from buckling, while the hemicelluloses, which are formed contemporaneously with both the cellulose and the lignin, link the two. A modest *microfibril angle* (the Z-helical angle in the S_2 layer) is an excellent compromise between axial stiffness (which increases as the microfibril angle gets smaller) and toughness or the ability to absorb energy (which gets larger with increasing microfibril angle) as the fibres shear apart and buckle.

Just as the microfibrils are finely dispersed throughout the cell wall, so in turn is the amorphous matrix of lignin and hemicelluloses. Although the middle lamella is extremely lignin-rich, with the lignin accounting for 70–75% of the material in the compound middle lamella, the bulk of the lignin is found in the S_2 layer, by virtue of the volume of material in this layer. Approximately three-quarters of all the lignin is found in the secondary wall.

1.3.5 Water in the Cell Wall

The cells differentiate and enlarge while immersed in the sap stream. Consequently, the cell walls are saturated with water. In the cell wall, the water hydrogen-bonds with accessible hydroxyls on the non-crystalline hemicelluloses and perhaps on the lignin, while being unable to penetrate the crystalline microfibrils. The fine dispersal of water within the cell wall and its thermodynamic characteristics strongly influence the drying of timber. The physical nature of the wood-water relationship is explored in the following chapter.

2 Wood-Water Relationships

2.1 Water in Wood

Water profoundly affects wood. Perfectly dry wood swells as the cell walls take up moisture until the fibres are saturated. The gain in water content transforms a relatively brittle material into one that is more plastic and deformable under load. Wood above a certain moisture content is also biodegradable.

A living tree contains a considerable quantity of sap, mainly water. Although a felled log remains stable while the fibres remain saturated, as soon as moisture is removed from the cell-wall material, the wood shrinks essentially in proportion to the water lost, and the bulk material may distort in response to the shrinkage-related stresses. Cracks or checks may appear where the logs are exposed to the atmosphere.

The maximum amount of moisture that a species may contain can be found from the *basic density* ρ_b, which is the oven-dry mass of a weighed specimen relative to its green, swollen volume. This value is species-dependent. By comparison, the density (ρ_c) of oven-dry cell tissue of all woody plants varies little with species, and a value of $1500 \, kg \, m^{-3}$ is a reasonable approximation for oven-dry woody tissue (Walker 1993). The fraction of woody tissue in the swollen material is ρ_b/ρ_c; therefore the greatest mass of moisture which can be held in unit volume of swollen wood is $(1 - \rho_b/\rho_c) \rho_w$, where ρ_w is the density of water. Whence the maximum moisture content (on a dry-matter basis) becomes

$$X_{max} = (1 - \rho_b/\rho_c)(\rho_w/\rho_b). \tag{2.1}$$

Basic wood densities can vary from $50 \, kg \, m^{-3}$ for some balsawood (*Ochroma lagopus*) to $1400 \, kg \, m^{-3}$ for lignum vitae (*Guaiacum offinale*), corresponding to maximum moisture contents of $19 \, kg \, kg^{-1}$ and $0.05 \, kg \, kg^{-1}$, respectively, when estimated from the above expression. However, most commercial species have densities in a narrower span from 350 to $800 \, kg \, m^{-3}$ (Walker 1993), with a corresponding maximum moisture-content range of 2.2 to $0.58 \, kg \, kg^{-1}$.

In general, the green heartwood of all species has a moisture content well below this limit. In hardwood timber, both the heart and sapwood have similar moisture-content levels, but the sapwood of softwoods normally has a significantly higher moisture content than their heartwood. For example, the average moisture content for the green sapwood of *Pinus ponderosa* is about $1.5 \, kg \, kg^{-1}$ which is nearly four times that for its heartwood (Hoadley 1980).

Almost all species show seasonal variations in moisture levels, but the effect is particularly noticeable in some deciduous species such as birch (*Betula* sp.). A rise in moisture content accompanies leaf emergence, dwindling in late summer, with a further rise at the time that leaves fall. In other species, like American beech (*Fagus grandifolia*), there is little year-round variation.

The effect of tree age on wood density and moisture content for the softwood *Pinus radiata* has been investigated by Cown and McConchie (1982). An increase of nearly 30% in tree-mean basic density and 40% in tracheid length was noted in stems aged from 12 to 52 years, with the whole-tree moisture content falling from 2.15 to 0.95 kg kg^{-1}. This study also emphasises that there will be considerable variation in wood basic density and moisture content within a given tree of this species, with less dense wood occurring near the pith and denser wood being found with distance away from the younger wood.

Normally, one would expect from Eq. 2.1 that the greater the basic density the lower the moisture saturation. However, Zobel and van Buijtenen (1989, p 31) report a contrary trend for *Pinus taeda*, where the moisture content is 1.29 kg kg^{-1} for wood of basic density 420 kg m^{-3}, 1.08 kg kg^{-1} at 450 kg m^{-3} and 0.93 kg kg^{-1} at 470 kg m^{-3}. In this case, the trend appears to be primarily driven by the proportion of heartwood to sapwood, as the percentage saturation is estimated to be 75, 69 and 63% in these three instances.

2.2 Moisture Sorption

Oven-dry wood will take up water from moist air. At a given temperature, the wood will reach an equilibrium moisture content that depends upon the moisture-vapour pressure p_v relative to the vapour-saturation value p_v^o at that temperature. This relative vapour pressure p_v/p_v^o is commonly called the *relative humidity*. From thermodynamic considerations, the relative humidity is almost equal to the *activity* of pure water at equilibrium. The adsorption of water is accompanied by swelling of the cell wall and the gross dimensions of the wood. Conversely, if wet wood is exposed to dry air, water is lost until equilibrium is reached with the relative humidity of the atmosphere, and shrinkage takes place.

A plot of the water content as a function of equilibrium relative humidity at constant temperature yields the moisture *isotherm* which is sigmoid in shape. As shown in Fig. 2.1, the sorption processes of wetting and drying are not reversible. There is a hysteresis, with a higher equilibrium moisture content being reached when the equilibrium is approached by desorption (or drying) than when it is attained by adsorption (or wetting). Further, once wood has been dried, the original moisture content is not approached should the material be rewetted. Drying produces an irreversible loss of capacity for water or hygroscopicity (Spalt 1958). A closed hysteresis loop is found only whenever a

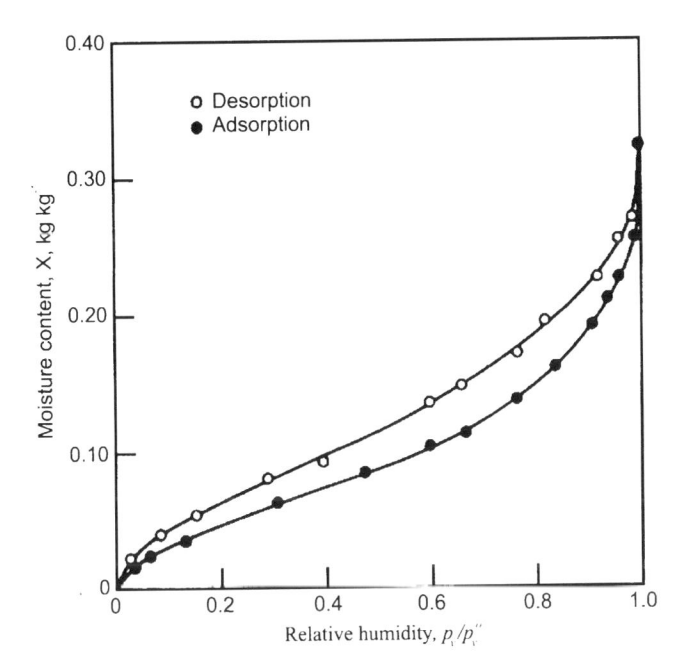

Fig. 2.1. Sorption of water by Klinki pine (*Araucaria klinkii*). (After Kelsey 1957)

desorption isotherm is determined after an adsorption one, as illustrated in Fig. 2.2. Normally, only the initial desorption isotherm obtained by drying green (saturated) wood is of interest in kiln-seasoning lumber. The hysteresis loop for wood spans practically over the whole range of equilibrium relative humidities. A number of explanations have been put forward to explain this behaviour.

Cohan (1944) proposed that the contact angle for wetting differs from that for desorption. The attractive forces between the sorbed moisture and the surfaces compete with those between the moisture and permanent gases in the void spaces of the material, to result in a smaller amount of moisture being held on wetting compared with that on drying to the same equilibrium relative humidity. This theory is supported by Luikov's claim (1966) that the hysteresis of "some typical capillary-porous bodies" disappears when experiments are carried out in high vacuum when permanent gases are almost absent. Other possibilities are based on considering the behaviour of pores with bottlenecks and those open at both ends (Cohen 1944).

An alternative theory is based on the idea that the sorption sites are less available after wood has been dried from green (Stamm 1964; Walker 1993). In green wood, hydroxyl groups in the carbohydrate/polysacharide matrix are thought to be hydrogen-bonded to water molecules. As the wood dries out, some of these hydroxyl groups are freed of attached water molecules and are able to bond with each other as the cell walls shrink. Such interlinked hydroxyl

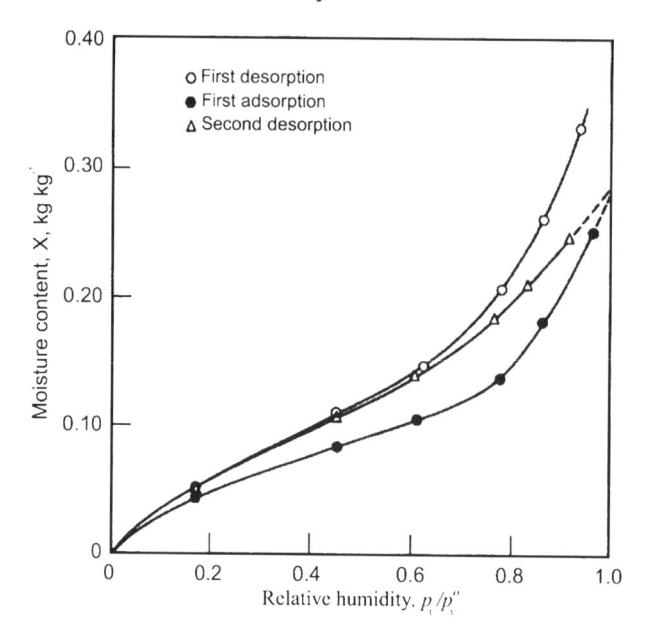

Fig. 2.2. Sorption hysteresis for Douglas-fir (*Pseudotsuga menziesii*) at 32.2 °C. (After Spalt 1958)

groups are not available for sorption when the wood is rewetted. Since wood is not a simple porous body, an explanation for hysteresis based solely on capillary phenomena seems unlikely to be the whole story, and the availability of sorption sites would appear to be more determinative of the observed behaviour.

The adsorption and desorption isotherms for a wood diverge with increasing equilibrium relative humidity, the corresponding equilibrium moisture content being higher on drying (desorption) of never-dried wood than wetting (adsorption) from complete dryness, with the values differing by their greatest amount under near-saturation conditions. The ratio of the value for adsorption compared with that for desorption normally ranges between 0.75 and 0.85 (Schniewind 1989). As the extent of a relative-humidity swing between two levels becomes smaller, so the moisture-content difference in the hysteresis loop becomes smaller. Orman (1955) notes, for example, when various New Zealand woods were subject to environmental cycles between 50 and 80% relative humidity, the isotherms for adsorption and desorption were, at the most, apart by 1 to 2% moisture content at a given relative humidity.

The hygroscopicity of wood derives mainly from the hemicelluloses and cellulose, with a lesser contribution from the lignin. Although there are marked changes in relative distribution of these components throughout the cell wall, the overall distribution does not vary greatly with common species, as illustrated in Table 2.1. The lignin content is somewhat higher in softwoods than in hardwoods, but the carbohydrate content, which dominates the

Table 2.1. Typical wood composition (Reproduced from Walker 1993, Primary Wood Processing, by kind permission of Kluwer Academic Publishers)

Components	Softwood (%)	Hardwood (%)
Cellulose	42 ± 2	45 ± 2
Hemicelluloses	27 ± 2	30 ± 5
Lignin	28 ± 3	20 ± 4
Extractives	3 ± 2	5 ± 3

Table 2.2. Adsorption of water by a hardwood

Components	Sorptive capacity relative to wood[a]	Fractional contribution to total sorption[b]
Hemicelluloses	1.56	0.46
Cellulose	0.94	0.42
Lignin	0.60	0.12
Wood	1.00	1.00

[a] Estimated by Christensen and Kelsey (1958) for chemically isolated components.
[b] Composition assumed from Table 2.1.

hygroscopic properties (Table 2.2), is fairly uniform at about 70% of the total. Therefore, large differences in sorptive behaviour between species are not expected.

The data in Tables 2.1 and 2.2 represent broad generalisations for genera in common commercial use. A comprehensive study reveals somewhat larger variations in chemical analyses than indicated in Table 2.1. A 1975 summary of data for 153 softwood and hardwood species reproduced by Fengel and Wegener (1984) shows that the cellulose content of softwoods can vary from 38.4% for Japanese thuja (*Thijopsis dolabrata*) to 61.6% for yellow pine (*Pinus strobus*), while the lignin content of softwoods can vary from 25.6% for Serbian spruce (*Picea omorika*) to 39.4% for pencil cedar (*Librocedus decurrens*). There is a similar large spread of chemical analyses for hardwoods from temperate zones.

Nevertheless, the differences in chemical composition of various common species are sufficiently small that these woods hold roughly the same amount of moisture at a given relative humidity. Browning (1963) notes that the percentage moisture content of most woods in an almost completely saturated atmosphere at room temperatures reaches 28 to 32% (dry basis). At a relative humidity of 0.5, the range is from 7.5 to 9.5%. Shubin (1990) is able to correlate with a single curve the desorptive behaviour of five common European timbers, both hard and softwoods, covering a range in basic densities from 350 to 656 kg m^{-3}, as illustrated in Fig. 2.3.

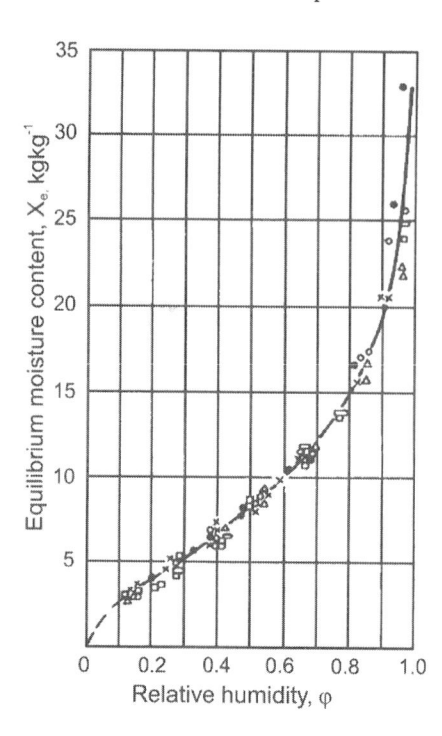

Fig. 2.3. Combined isotherm for various species at 42.4 °C. (After Shubin 1990)

Careful study of Fig. 2.3, however, does reveal a wide range of equilibrium moisture contents at the highest relative humidities determined. Percentage moisture-content values range from 33% for oak (*Quercus* sp.) to 22% for a pine (*Pinus* sp.) at 95% relative humidity, while the trend of the combined data suggests that values might reach 35% at complete saturation. Several explanations may be adduced for such a spread.

All woods contain water-soluble extractive material to a greater or lesser extent. The concentration of polar extractive compounds in the *fusca* subgroup of *Nothofagus* spp. in New Zealand can be as high as $9 \pm 3\%$ (Uprichard et al. 1975). Hoadley (1980) notes that, in species that have a high extractives content, such as redwood (*Sequoia sempervirens*) and mahogany (*Swietenia mahogoni*), the fibres remain saturated up to about 22 to 24% moisture content, whereas birch (*Betula* spp.) may have a moisture content up to 35% at fibre saturation. The presence of water-soluble organic material will enhance the thermodynamic activity of the adsorbed moisture in the cell wall, and thus the equilibrium relative humidity is higher at a given moisture content than if the moisture were pure water.

Shubin (1990) produces an alternative, and possible additional explanation based on the ultrastructure of vessels and tracheids. He notes that the fractional void space ε or air capacity of various woods must differ in conformity with their basic wood density:

$$\varepsilon = [1 - \rho_b(1/\rho_c + X/\rho_w)], \quad (2.2)$$

where X is the moisture content (dry basis) and ρ_w is the sap density. For the sapwood of pine (*Pinus* sp.), the void fraction ε varies from about 0.6 in the air-dried state to 0 when totally saturated, but oak (*Quercus* sp.) has a value of about 0.4 when air-dry, while the very dense *Guiaiacum offinale* has maximum fractional air capacity of only 0.25. Thus the nature of the cell walls' inner surfaces has an important bearing on the hygroscopic nature of the wood itself at very high relative humdities approaching saturation. Microcavities such as pits in the cell wall can induce a vapour-pressure lowering. For a cylindrical cavity, this reduction is given by the well-known Kelvin equation,

$$\varphi \doteq \frac{p_v}{p_v^o} = \exp\left[-\frac{2\sigma\rho_v}{\rho_w p_v^o r}\right], \quad (2.3)$$

in which σ is the surface tension, ρ_v is the moisture-vapour density and r is the radius of the cavity. When this radius is 1 μm, the relative humidity is 0.999; for a radius of 0.1 μm the relative humidity is 0.99. Any microcavities, surface depressions or crevices are unlikely to be perfectly cylindrical, so the Kelvin equation is only indicative of the magnitude of the possible vapour-pressure lowering. By assuming that hygroscopicity is caused by both adsorption within the cell wall and by filling the various microcavities, Shubin (1990) is able to calculate the maximum hygroscopic moisture content. For pine this limit is 32.3%; for oak it is 41.3%. While these estimates involve certain assumptions regarding the limits of sorptive behaviour and the influence of the microcavities themselves, the calculations do demonstrate the possibility of large variations in hygroscopicity at high relative humidities.

A further cause of variation is suggested by Feist and Tarkow (1967), who measured a fibre-saturation value of 52% for the very light balsa (*Ochroma lagopus*). They suggested that, with a wood of very low basic density (in this case, 250 kg m^{-3}), the cell walls have less resistance to swelling than woods of average density, thus enabling the walls to take up more moisture.

Finally, we observe that the measurement and control of relative humidity close to complete saturation is extremely difficult, and thus some of the reported differences are undoubtedly due to experimental variation.

Notwithstanding these foregoing uncertainties, the hygroscopic behaviour of common commercial woods is sufficiently uniform at relative humidities of normal interest for tables of equilibrium moisture contents as a function of relative humidity and temperature to be published. At 65% relative humidity and 26°C, the equilibrium moisture content for 15 coniferous species (including both heart and sapwood) has been found to vary from 10.7 to 13.5%, with a mean of 12.2% and a standard deviation of 0.6% (Harris 1961).

Bramhall and Wellwood (1976), for example, present averaged data to 115°C for the Canadian lumber industry, while almost identical data to 120°C are given by the German kiln manufacturer, Hilderbrand GmbH (1979). At 70°C and 65% relative humidity, for instance, both sets of tables give the same equilibrium moisture-content value of 9.0%. A more extensive chart of moisture-equilibria data is reported by Shubin (1990), presumably for common northern European species. These data cover a range of temperatures to 180 °C and total pressures to 0.3 MPa.

Raising the temperature displaces the equilibrium in favour of less adsorbed moisture. Desorption isotherms presented by Stamm (1964) for never-dried Sitka spruce (*Picea sitchensis*) are illustrated in Fig. 2.4 for temperatures from 25 to 100 °C. At this latter temperature, the maximum hygroscopic moisture content is shown to fall from about 31 to 23%. This shift implies that, when cold lumber is placed in a heated kiln, the relatively small thermal expansion of woody material is overwhelmed by the larger shrinkage accompanying the loss of cell-wall moisture.

At temperatures above 100 °C the relative humidity is calculated by reference to the saturated-steam pressure p_s^o at the temperature being considered:

$$\varphi = p_v / p_s^o, \tag{2.4}$$

where p_s^o is always greater than 0.1 MPa. In those cases when the gas is only water vapour, as in superheated-steam drying, the corresponding relative humidity is given by

$$\vartheta_{max} = P / p_s^o, \tag{2.5}$$

where P is the total pressure of the system. Contours of constant equilibrium moisture content ascend when the associated relative humidity is plotted against moisture content. As shown in Fig. 2.5, these contours extend into the superheated-steam region. At a total pressure of 0.1 MPa (atmospheric), the

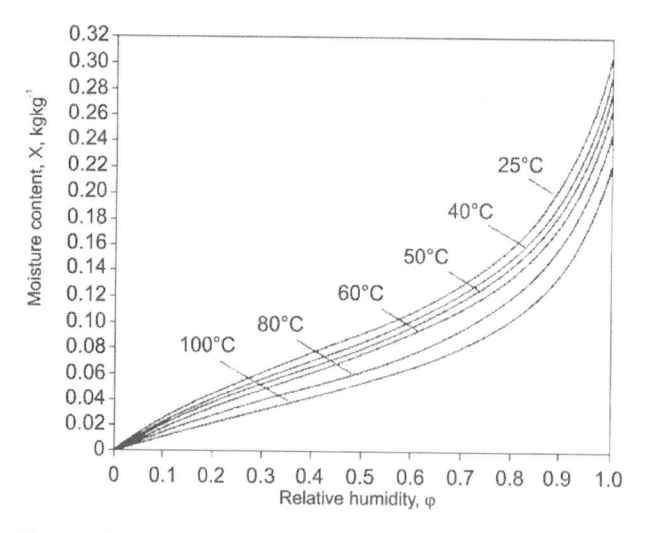

Fig. 2.4. Desorption isotherms for green Sitka spruce (*Picea sitchensis*). (Reproduced from Stamm (1964), Wood and Cellulose Science, by permission of John Wiley & Sons, Inc)

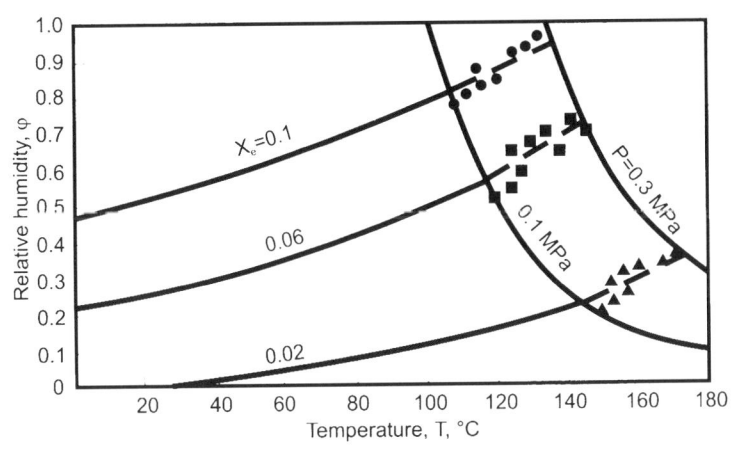

Fig. 2.5. Extrapolated contours for equilibrium moisture contents of wood to a total pressure of 0.3 MPa. (After Shubin 1990)

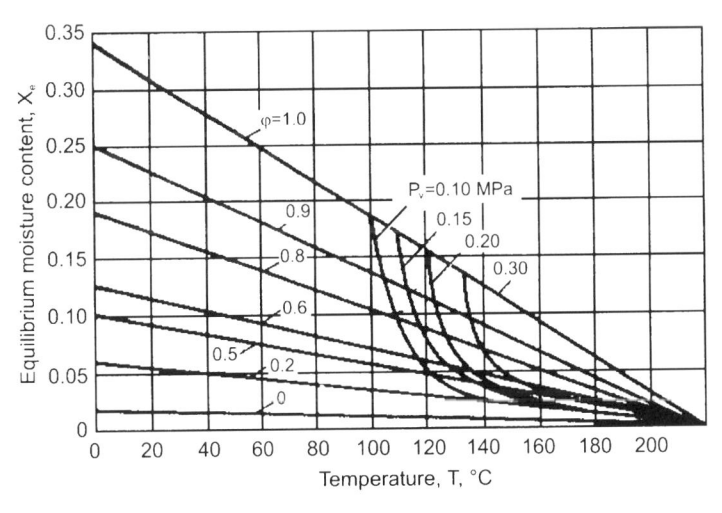

Fig. 2.6. Variation of equilibrium moisture content with temperature, with contours of total pressure to 0.3 MPa. (After Shubin 1990)

equilibrium moisture content falls rapidly with temperature, being only about 5% at 120°C and under 2% at 150°C.

Shubin (1990) produces another interesting, transformed diagram by plotting the equilibrium moisture content as a function of temperature with the relative humidity as parameter, as shown in Fig. 2.6. The contours of relative humidity appear to converge towards a virtual point at a temperature just above 200°C on the limiting envelope for zero relative humidity.

2.3 Fibre Saturation and Maximum Hygroscopic Moisture Content

In general, moisture in wood is distinguished between the sap or *free moisture* within the lumens of tracheids and vessels and *bound moisture* which is held in the cell walls. It is the bound moisture that is associated with the hygroscopic nature of the material, the extent of bonding being reflected in the lowering of the moisture-vapour pressure.

As noted in the previous sections, there is some uncertainty about the limits of hygroscopic behaviour. Feist and Tarkow (1967) used a solution of a water-soluble polymer, polyethylene glycol, having a molecular weight of 9000, to determine the amount of water that would be contained only in the lumens and pits. They found that the percentage moisture content was in the range of 35 to 40% in the saturated fibres of virgin green softwoods with a basic density of $300\,kg\,m^{-3}$ or more. These values are consistent with Shubin's calculations (1990) in which allowance was made for capillary-condensation effects in microcavities in the cell wall.

However, it is useful to define the *fibre-saturation point* in a more arbitrary way. If capillary-condensation effects in pores less than $0.1\,\mu m$ in equivalent cylindrical diameter are neglected, then the fibre-saturation point can be defined precisely as the equilibrium moisture content of a wood sample in an environment of 99% relative humidity. At room temperature, this definition would yield a fibre-saturation point for most common commercial species between 30 and 32% (dry basis).

The variation of fibre-saturation point with temperature may be found from the envelope of asymptotic values of the moisture isotherms at relative humidities approaching full saturation. For example, the sorption data for sitka spruce (*Picea sitchensis*) plotted in Fig. 2.4 reveal that the fibre-saturation point falls from about 31% at 25°C to 23% at 100°C. The moisture becomes less strongly bound with rising temperature.

For temperatures above 100°C, Shubin's chart may be used to determine the fibre-saturation point by tracing the equilibrium conditions along the 0.1 MPa contour. As noted earlier, at 120°C, the value of the fibre-saturation point so obtained has fallen to 5%, and at 150°C it is below 2%. Direct experimental measurement of fibre saturation above 130°C is difficult, however, because of the onset of thermal degradation with some species (Noack 1959). However, highly permeable softwoods, such as *Pinus radiata*, are often dried commercially at temperatures of 120°C and above. Clearly, one advantage of high-temperature drying is that very little bound moisture is left in the lumber at these temperatures.

Shubin (1990) also presents data for the change in fibre saturation as the temperature falls below 0°C, reflecting harsh winter conditions in the northern hemisphere. The data for four species fall from a mean of 32% at -2°C to 21% at -30°C. At these temperatures, the saturation-vapour pressure over ice is very low. Moisture from the frozen cell wall can sublime, so the remainder will reach a new, and lower steady-state value below that measured at room temperature when maximum sorption can occur.

The ultimate hygroscopic moisture content is higher than the normally-acknowledged fibre-saturation point. Some idea of the possible limit is given by the experiments of Feist and Tarkow (1967) already mentioned, but another estimate can be made by invoking Luikov's correlation of data for higher relative humidities, as explained by Keey (1992). Luikov (1968) was able to correlate a large body of experimental data, mainly for building materials including wood, by an expression of the form

$$\frac{1}{X} = \frac{1}{X_{max}} - B\ln\varphi, \tag{2.6}$$

where X_{max} is the maximum hygroscopic moisture content when vapour-pressure lowering is negligible. This equation essentially fits an exponential expression to the isotherm at the higher relative humidities. When Spalt's data points for Douglas-fir (*Pseudotsuga menzesii*) are fitted to Luikov's equation, as shown in Fig. 2.7, a value of 33% is obtained for the limiting moisture content.

This method is only an extrapolation and thus is insensitive to the effect of very small irregularities in the cell-wall surfaces and the presence of polar compounds in the sap, which separately may cause a virtually immeasurable lowering of the moisture-vapour pressure at relatively high moisture contents.

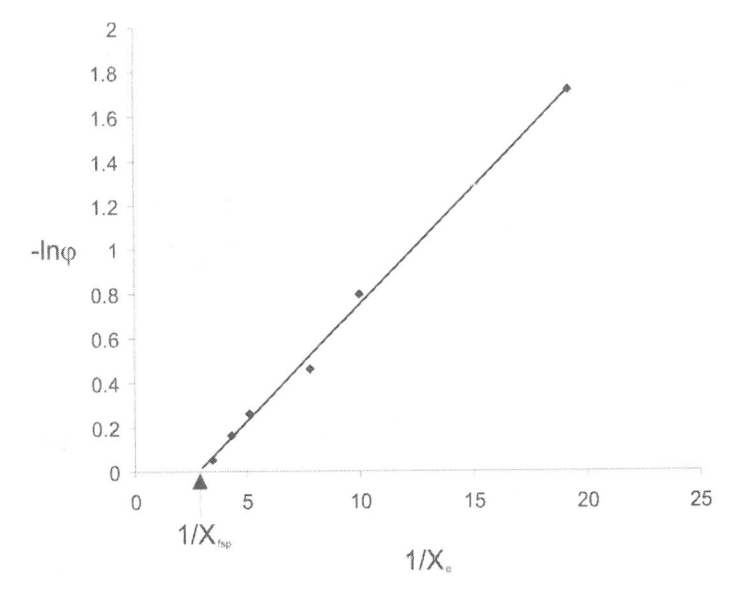

Fig. 2.7. Spalt's sorption data (1958) for Douglas-fir (*Pseudotsuga menziesii*) fitted to Luikov's equation

Indeed, Shubin (1990) postulates that moisture in wood is always bound to some degree. While a sharp transition between hygroscopic and any non-hygroscopic behaviour does not occur, the concept of fibre saturation is a useful one, indicating the transition from behaviour which is dominated by what happens within the cell wall to that within the lumens of the vessels and tracheids themselves.

2.4 Theories of Sorption

A vapour may be attracted to condense on a bare, dry surface. Under steady-state conditions, the rate of condensation of the vapour on the dry areas must be equal to the rate of evaporation from the wet regions. If x is the fraction of the surface covered, then the rate of evaporation is equal to $k_1 x$, where k_1 is a constant of proportionality. The rate of condensation is proportional to the vapour pressure p and may be equated to $k_2(1-x)p$, where k_2 is another constant. These considerations lead to the expression first derived by Langmuir (1918) for the fraction of surface covered:

$$x = k_2 p / (k_1 + k_2). \tag{2.7}$$

The quantity x is also the ratio of the equilibrium moisture content X to the value X_1 when the surface is completely covered with a single layer of molecules, while the vapour pressure p is directly proportional to the relative humidity φ. Thus Langmuir's equation can be rewritten as

$$x = \frac{X}{X_1} = \frac{c_2 \varphi}{(c_1 + c_2)}, \tag{2.8}$$

in which c_1 and c_2 are new coefficients.

This expression describes the moisture-sorption behaviour at low relative humidities below about 0.2. In wood, this monomolecular adsorption can be considered to correspond to the hydrogen bonding of water molecules to accessible hydroxyl groups in non-crystalline regions and on the surface of the microfibril crystallites (Browning 1963). The adsorption is associated with swelling of the cell-wall structure.

Brunauer, Emmett and Teller [BET] (1938), recognising that more than one layer of molecules could be sorbed on a surface, have generalised Langmuir's equation by assuming that the rate of condensation on top of one layer is equal to the rate of evaporation from the layer above:

$$a\, p\, x_{i-1} = b\, x_i \exp[-E_i / RT], \tag{2.9}$$

where a and b are coefficients and E_i is the energy of activation for the ith layer. In an extension of the idea to adsorption in capillaries, Brunauer and colleagues (1940) assumed that $(2n-1)$ layers could fit into a slit-like space between two parallel plane walls. The equations for the evaporation-condensation equilibria remain the same, except the one for the last layer in the middle:

$$a\, p\, x_{n-1} = b\, x_n \exp[-(E_i + Q)/RT], \tag{2.10}$$

where $(E_i + Q)$ is the heat of adsorption of the last layer. However, if one assumes that the heat of adsorption for the second and subsequent layers differs from the heat of condensation on a bare surface by a constant value, which may be positive or negative, as Jaafar and Michalowski (1990) propose, then the modified BET equation

$$x = \frac{X}{X_1} = \frac{C\varphi}{(1-k\varphi)\{1+(C-k)\varphi\}} \tag{2.11}$$

results. Here, the parameter x is the effective number of sorptive layers, with X_1 being moisture content for a complete monolayer. The coefficient k is equivalent to the term $\exp(\pm\Delta H/RT)$, where ΔH is the constant enthalpy difference in adsorptive heats between the first and successive layers. When k is zero, this equation reduces to Langmuir's isotherm for monomolecular adsorption. The value $k = 1$ corresponds to the limit when ΔH is infinitesimally small and the number of layers correspondingly infinite. Over the range, $0 < k < 1$, the equation predicts a finite moisture content at saturation (100% relative humidity) which may be associated with the fibre-saturation point.

This theory leads to some surprising results. Stamm (1964) has used the BET equation to estimate the average number of molecular layers in the adsorbate and the surface area involved in the sorption of water vapour on two softwoods, as shown in Table 2.3. The theory predicts surface areas many times greater than the internal surfaces of the lumens. At fibre saturation for sitka spruce (*Picea sitchensis*), the adsorbate thickness between adjacent surfaces is about 5 nm for the dual layer about 13 molecules thick, compared with the tracheid diameters which are at least 1000 times larger.

Jaafar and Michlowski (1990) have tested the modified BET equation for both adsorption and desorption of water vapour on 29 materials at room temperature over a wide relative humidity range from 0.07 to 0.97, and for some of these materials at temperatures between 45 and 75 °C within the narrower range, $0.06 < \varphi < 0.6$. In most cases, experimental data could be fitted to within

Table 2.3. Analysis of water-vapour adsorption on sitka spruce (*Picea sitchensis*) and sugar maple (*Acer saccharum*)

Species	Average no. of layers at saturation	Relative humidity for monolayer	Adsorption heat, kJ kg^{-1}	Internal surface area, m^2 g^{-1}
Acer	7.5	0.21	350	210
Picea	6.5	0.22	360	250

±8% up to a relative humidity of 0.7, and in some instances over the whole relative humidity range, with markedly better agreement than the unmodified BET equation does. However, the BET equation and its derivatives are generally satisfactory for rigid capillary-porous materials, but less satisfactory with materials, such as wood, that swell with increasing moisture content.

There have been other attempts to modify the BET equation. Hartley and Avramidis (1994) proposed that the formation of clusters of water molecules can occur when vapour is adsorbed at moisture contents above 15%. Newly condensing molecules would tend to form bridges with already-sorbed molecules of water rather than attach to the hydroxyls in the cellulosic matrix. Beyond a moisture content of 20% these clusters would become larger, and may have more than ten molecules in any one cluster. This behaviour can be modelled, as shown by Le and Ly (1992), who modified the original BET model by allowing only a fraction of the first-sorbed layers to be covered with succesive layers. This model leads to the equation

$$x = \frac{X}{X_1} = \frac{C_\varphi}{(1-C\varphi)} \frac{\{1 - \alpha\varphi^n - (1-\alpha)\varphi^m\}}{1-\varphi}, \qquad (2.12)$$

where C is a coefficient, α is the fraction of adsorption sites that contain up to n layers, and the remaining sites have m layers ($m > n$). Although this equation was developed for the adsorption of moisture in wool, the expression also serves to correlate the sorption behaviour of wood (Hartley and Avramidis 1994).

Hailwood and Horrobin (1946) considered that the uptake of water vapour in a polymer could be described in terms of equilibrium between the polymer, polymer hydrates and dissolved water. Such hydrates could form in wood by the linking of one water molecule to two adjacent hydroxyl groups in the carbohydrate/polysacharide matrix. In the case of a single hydrate layer, Hailwood and Horrobin's model leads to the equation

$$x = \frac{X}{X_1} = \left[\frac{K_1 K_2 \varphi}{1 + K_1 K_2 \varphi} + \frac{K_2 \varphi}{1 - K_2 \varphi} \right], \qquad (2.13)$$

where K_1 is the equilibrium constant between the dissolved water and the hydrate, K_2 is that between the dissolved water and the surrounding water vapour. This equation provides a good empirical fit of sorption data on wood (Simpson 1973).

The difficulty of extending the BET-type equations to describe behaviour at very high relative humidities has led workers to add a "correction" for presumed adsorption in microcavities by invoking the Kelvin equation for vapour-pressure lowering in capillaries. As shown in Fig. 2.8, this procedure can yield a remarkably good fit of the experimental data.

The various adsorption theories, although based on plausible biophysical pictures of the ways in which water can be taken up by wood, are all used essentially as correlating tools. Any three-coefficient equation can normally be

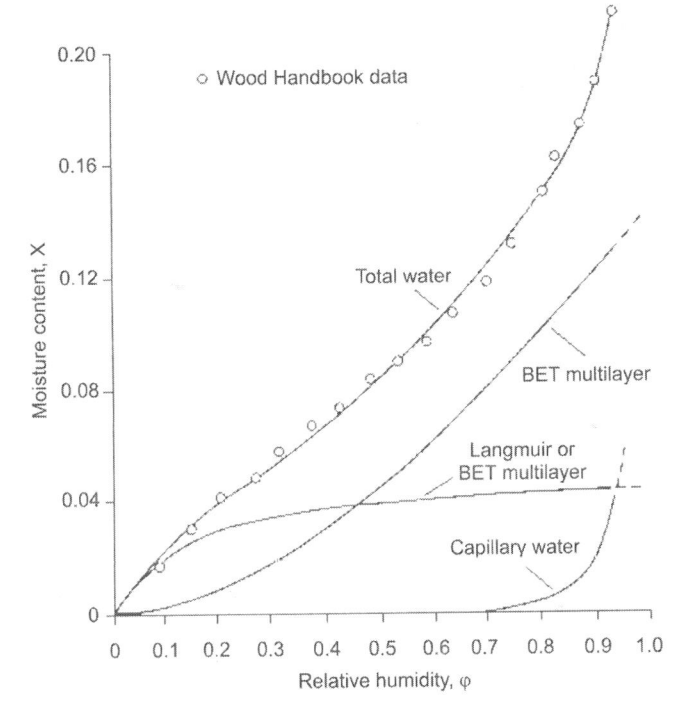

Fig. 2.8. Moisture-sorption data taken from the *Wood Handbook* fitted with the BET and Kelvin equations. (After Simpson 1980)

employed to fit a sigmoid isotherm. In many cases, it is equally convenient to interpolate tabulated data.

2.5 Heat of Sorption

Adsorbed water is held tenaciously in the cell walls. From thermodynamic considerations, the energy required to sever the bonds of a unit quantity of moisture from its host material is given by

$$-\Delta G = -RT \ln p_v/p_v^o = -RT \ln \varphi, \tag{2.14}$$

where R is the universal gas constant and T is the absolute temperature. Bond energies calculated from this equation vary from less than $100\,\mathrm{J\,mol^{-1}}$ for the largest pores to as much as $5000\,\mathrm{J\,mol^{-1}}$ for molecules adsorbed on a monolayer.

The bonding of condensed moisture to its host material is also reflected in a difference in heat content or enthalpy between that of the bound moisture (H_w) and that of free moisture (H_w^o) at the same temperature and total pressure. This difference $(H_w - H_w^o)$ may be calculated from a form of the Clausius-Clapeyon equation, which is derived by assuming that the vapour phase behaves as an ideal gas and that the molal volume of the condensed phase is negligible compared with that of the vapour (Smith and Van Ness 1975):

$$\left[\frac{\partial \ln(p_v/p_v^o)}{\partial T}\right]_X = -\frac{1}{R}\left[\frac{H_w - H_w^o}{T^2}\right]. \tag{2.15}$$

The vapour-pressure ratio, p_v/p_v^o, is the relative humidity, R is the universal gas constant and T is the absolute temperature. The enthalpy difference $(H_w - H_w^o)$ is called the *differential heat of sorption*, $-\Delta H_w$, which, being dependent upon the relative humidity, is a function of the equilibrium moisture content. Rearrangement of the above equation yields an explicit expression for this heat of sorption:

$$\Delta H_2 = RT^2\left[\frac{\partial \ln \varphi}{\partial T}\right]_X = R\left[\frac{\partial \ln \varphi}{\partial (1/T)}\right]_X. \tag{2.16}$$

It follows that the heat of sorption can be found from the moisture isotherms by evaluating the slope of a graph of $\ln \varphi$ plotted against the reciprocal absolute temperature, $1/T$. We call this estimate the *point* value for the differential heat of sorption at the given moisture content X. Over small temperature intervals, it may be assumed that the heat of sorption is a constant. Thus, if data at two, fairly close temperatures (say 19 and 21 °C) be known, then values of the relative humidities for these temperatures at the specified moisture content X can be read off the isotherms, and the heat of wetting is calculated as

$$-\Delta H_w = \frac{RT_1T_2 \ln(\varphi_1/\varphi_2)}{T_1 - T_2}, \tag{2.17}$$

with $\varphi_1 < \varphi_2$ and $T_1 > T_2$. This estimate may be termed the *finite-difference* value.

Skaar (1988) has given correlations for the heat of sorption based on fitting the data of Stamm and Loughborough (1935). At temperatures near 50 °C, Skaar gives a simple exponential relationship for ΔH_w:

$$\Delta H_w = 1172 \exp(-1400\,X). \tag{2.18}$$

Pang and coworkers (1993) compare this equation with both the point and finite-difference derivations from the Clausius-Clapeyron equation. The comparison is shown in Fig. 2.9. The correlation deviates from these derived equations at moisture contents below 0.07 kg kg^{-1} or relative humidities below 0.35. The calculated reduction in the heat of sorption at low relative humidities is unlikely, since the moisture bonding is expected to strengthen monotonically as the moisture content falls as indicated by the free energy change $-\Delta G$. Rees (1948), however, notes that the sorption relationship becomes unreliable at low moisture contents, its unreliability being masked by forcing the equation

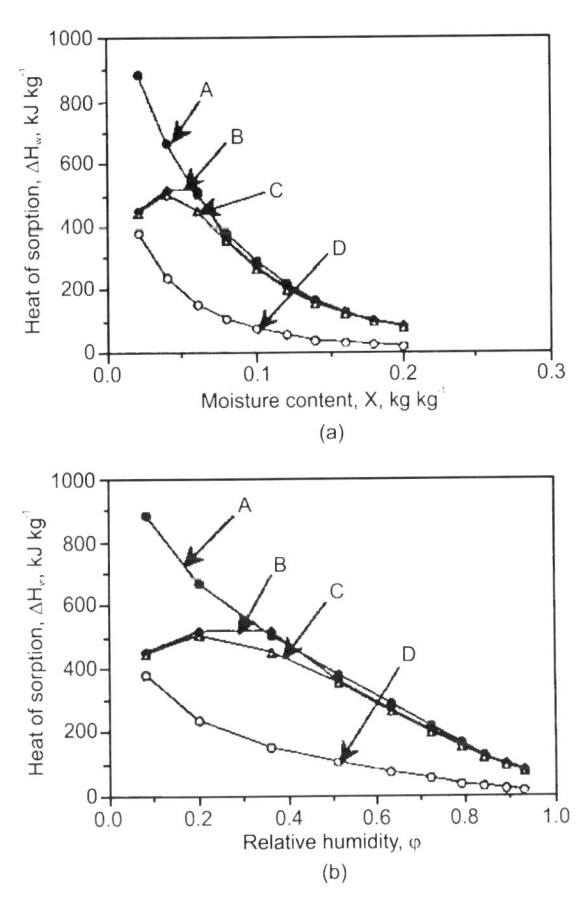

Fig. 2.9a,b. The heat of sorption of wood. **a** Variation with moisture content; **b** Variation with relative humidity. *A* Skaar's correlation (1988); *B* Differential form of the Clausius-Clapeyon equation; *C* Integral form of the Clausius-Clapeyron equation; *D* Free energy change. (Reprinted from Pang et al. 1993, Drying Technol 11: 1076, Courtesy of Marcel Dekker Inc)

through the origin, (0,0). Heats of sorption at low moisture contents (or relative humidities) are best estimated in another way.

Calorimetrically, heats of sorption can be obtained by immersing oven-dry, groundwood particles in a known quantity of water and measuring the temperature rise (Stamm 1964). The heat released is known as the *heat of wetting*, which progressively becomes smaller as the groundwood is preconditioned at increasingly higher moisture contents.

For most woods, the initial differential heat of sorption is approximately 1250 kJ kg^{-1}, about half the latent heat of vaporisation of water, reflecting the strong bonds of the first-sorbed moisture. By the time a complete monolayer has formed by about 5% moisture content, this heat of sorption has fallen to one-half to one-quarter of the latent heat.

On drying wood, the enthalpy value of interest is the integral heat of sorption, which is the averaged value over a specified moisture-content range:

$$\overline{\Delta H_w} = \frac{1}{X_1 - X_2} \int_{X_2}^{X_1} \Delta H_w dX. \tag{2.19}$$

In kiln-seasoning, the integral heat of sorption is rarely a significant component of the energy demand, as ΔH_w only becomes significant compared with the latent when the equilibrium relative humidity is less than 50% and the moisture content is less than about 10%.

Figure 2.9 shows that the free energy change, $-\Delta G$, is always less than the enthalpy difference, $-\Delta H_w$. For an isothermal process, the two thermodynamic quantities are related through the entropy, ΔS:

$$\Delta S = (\Delta H_w - \Delta G)/T. \tag{2.20}$$

The variation of entropy with moisture content follows the same trend as that for the differential heat of sorption, but a discontinuity in the entropy curve for wood has been reported by Kelsey and Clarke (1956) at about 8 to 9% moisture content. This has been attributed to changes in the relative contributions of the inherent ordering of the adsorptive process and the disorganisation of the cellulosic structure accompanying adsorption of water. Earlier, Stamm and Loughborough (1935) had attributed the entropy change to the dimensional changes in wood on swelling. Their values for the entropy of sorption as function of moisture content show a monotonic decrease with increasing moisture content, with no negative values, and only a slight variation with temperature.

2.6 Response to Environmental Changes

The response of wood to a sudden change in its environmental conditions depends upon its past history. The moisture in wood tends to equilibrate with that in the air. The equilibrium moisture content of never-dried wood at a given relative humidity is higher than that of very dry wood which has taken up moisture; it is also higher than that of wood that has been re-wetted to a higher moisture content, and then dried out again. At a given temperature, the absorption and secondary desorption isotherms describe equilibrium boundaries for the wood's response to changes in the environmental relative humidity. The moisture content of the wood oscillates within these limiting envelopes in response to changes in relative humidity (Perralta 1995). These intermediate isotherms are sometimes called *scanning curves*. The various sorption isotherms are illustrated in Fig. 2.10.

The gradient of the intermediate isotherms, $dX/d\varphi$, is essentially linear. However, Time (1998), on reviewing data for these isotherms reported by

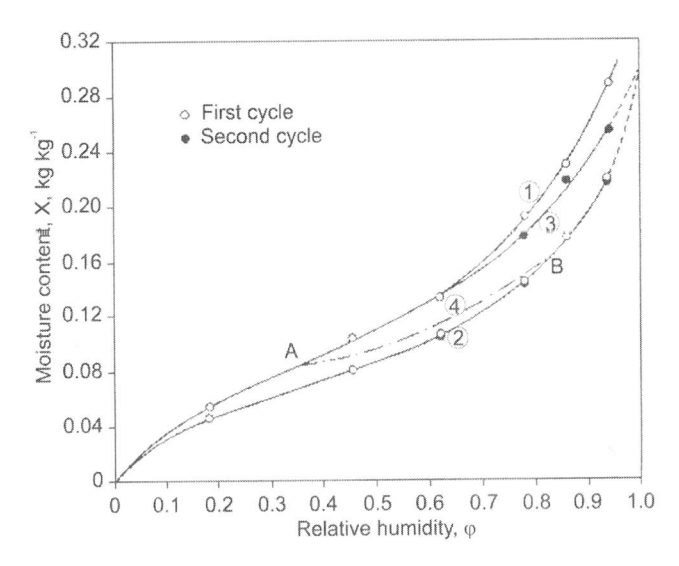

Fig. 2.10. Sorption isotherms for wood. *1* Primary desorption isotherm; *2* Adsorption isotherm; *3* Secondary desorption isotherm; *4* An intermediate isotherm. (After Peralta 1995)

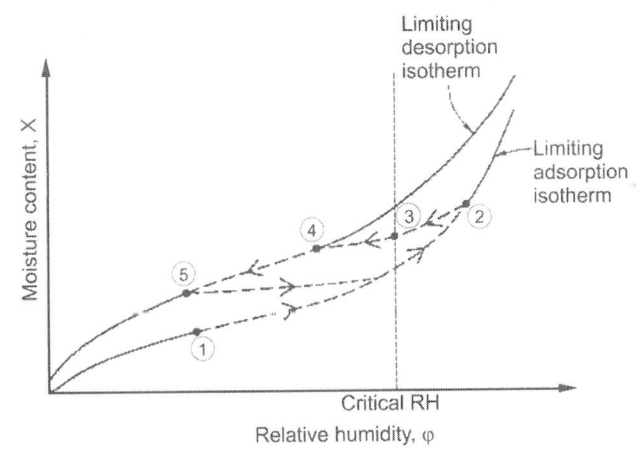

Fig. 2.11. Sorption paths above and below the critical relative humidity. Curve *1-2* is the limiting absorption isotherm; curve *2-3-4* is an intermediate desorption isotherm through the critical relative humidity; curve *4-5* is the limiting desorption isotherm; curve *5-6* is an intermediate absorption isotherm below the critical relative humidity. (After Time 1998)

Algren (1972) and Peralta (1995), defines a *critical relative humidity*, which separates the regions where the slopes of intermediate curves differ, as shown in Fig. 2.11. For the data considered, this critical relative humidity is found at 75% of the saturation vapour pressure.

The response to step changes in relative humidity is far from instantaneous, even with thin samples. Wadsö (1993) reports the fractional weight change in European spruce (*Picea abies*) samples of 2- and 4-mm thickness when exposed to an atmosphere of 75% relative humidity from one of 54%. About 90% of the expected change occurred after about 70 h exposure, with a much slower change thereafter. Similar findings are reported by Time (1998) for specimens of the same species up to 10 mm in thickness. These results seem to suggest that the sorption is a two-step process, with an initial uptake into the cell-wall matrix being followed by a slower redistribution within it.

The temperature at which wood is dried appears to alter its responsiveness to environmental changes (Kininmonth and Whitehouse 1991). Both the equilibrium moisture content of untreated *Pinus radiata* after a year's exposure and rate of response to climatic fluctuations decreased with the temperature at which the wood had been dried. Heartwood appeared more susceptible than sapwood. The mean equilibrium moisture content of air-dried heartwood, for example, was found to be $0.188\,kg\,kg^{-1}$, whereas that for lumber dried at 115 °C it had fallen to $0.124\,kg\,kg^{-1}$, a reduction of over one-third. The mean weekly change in moisture content was $0.0066\,kg\,kg^{-1}$ and $0.0043\,kg\,kg^{-1}$ respectively, a fall of similar relative magnitude. Similar studies were obtained in an earlier study by Kininmonth (1976) involving smaller specimens of *Pinus radiata* and a hardwood, *Beilschmiedra tawa*. This loss of hygroscopicity is probably related to loss of some volatile components and other chemical changes at the kiln temperatures.

3 Evaporation and Humidification

3.1 Moisture in Air

The amount of moisture in the surrounding air has a crucial bearing on the rate at which a piece of wet wood will dry out. The ratio of water vapour to dry air, on a mass basis, is called the *humidity*. This quantity is dimensionless, but at low moisture-vapour levels it is often convenient to record humidities in grams water vapour per kilogram of perfectly dry air. Sometimes values of the absolute humidity are reported on a volume basis. However, it is easier to follow changes in humidity throughout a kiln on a dry air basis, since the mass of circulating moisture-free air is not altered by the drying process.

Air itself is a mixture of nitrogen, oxygen, some inert gases and carbon dioxide. The nitrogen content is substantially uniform over the earth's surface, but the small amount of carbon dioxide is variable and mean values have been slowly increasing over the past century from the burning of carbonaceous fuels. The composition of dry air has been defined by Goff (1949) in his final report to an international committee for psychrometric data as being the mixture set out in Table 3.1. Air is heavier than water vapour at the same temperature (the molar-mass ratio of air to water vapour being 0.0290/0.0180 = 1.611). Thus, steam would tend to concentrate at the top of the kiln and the air to sink to the bottom at the start if the circulating fans were not switched on. To maximise production from high-temperature kilns for softwood, the final steam conditioning is generally done in a separate chamber which does not have fans. Some stratification of the kiln environment may occur, depending upon the manner of steaming.

The pressure-temperature behaviour of air can be expressed by the equation of state

$$Z = \frac{PV}{RT}, \qquad (3.1)$$

where V is the specific molar volume and R is the universal gas constant which takes a value of 8.314 J mol^{-1}. At pressures close to ambient and at kiln temperatures the compressibility factor Z is close to 1, and air behaves as an ideal gas. The same equation may be used to represent the behaviour of water vapour. At atmospheric pressure, the value of Z is 0.986 at 380 K [107 °C] (Hilsenrath et al. 1955), and thus steam, too, may be regarded as behaving essentially as an ideal gas.

Table 3.1. Composition of dry air (Goff 1949)

Substance	Molar mass kg mol^{-1}	Composition Mole fraction
Nitrogen	0.028 016	0.7809
Oxygen	0.032 000	0.2095
Argon	0.039 944	0.0093
Carbon dioxide	0.044 01	0.0003

With air-steam mixtures, various interactions between like and unlike molecules are possible, but such effects are relatively very small, and again for most practical purposes moist air may be treated as an ideal-gas mixture when following state changes within a kiln. Since the total pressure is the sum of the partial pressures of the air and water vapour, it follows that the humidity Y is related to the partial pressure p_W of the water vapour through the expression

$$Y = \left[\frac{M_W}{M_G}\right]\frac{p_W}{P - p_W}, \qquad (3.2)$$

where the molar-mass ratio of water vapour to air, $[M_W/M_G]$ is 0.622. The maximum value that p_W can take is the saturation value $p^o{}_W$, which is often known simply as the *vapour pressure*. Tables of water-vapour pressure, e.g. Krischer and Kast (1978) and Keey (1978), show that the vapour pressure rises rapidly with temperature, varying from about one-thousandth of the atmospheric pressure over ice at −20 °C, to 2.34 kPa at +20 °C and 47.4 kPa at 80 °C. It is equal to the standard atmospheric pressure (101.325 kPa) at the boiling point of 100 °C. Thus the humidity of air saturated with moisture vapour correspondingly rises rapidly from 0.0149 at 20 °C, to 0.559 at 80 °C and is infinite at the boiling point when the fraction of dry air vanishes. Clearly, the moisture-holding capacity of air is markedly enhanced as the temperature is raised by only a few degrees. This is one economic reason among others for choosing the highest possible temperature in fixing a kiln schedule.

The *relative humidity* φ of a damp gas is a measure of its fractional saturation with moisture. At thermodynamic equilibrium, the activity of the moisture in wood is closely equal to the relative vapour pressure $p_W/p^o{}_W$ of the environment. Thus for convenience, the relative humidity is defined on a similar basis, namely

$$\varphi = p_W/p_W^o \qquad (3.3)$$

rather than as a fractional humidity on a mass basis. An explicit expression for the relative humidity in terms of the absolute humidity Y and the vapour pressure $p^o{}_W$ can be obtained by an algebraic rearrangement of Eq. (3.2) and (3.3):

$$\varphi = \frac{PY}{(Y + M_W/M_G)p_W^o}. \qquad (3.4)$$

Only when the moisture-vapour pressures are very much less than the total pressure P will the relative humidity take closely similar numerical values to the fractional saturation humidity. This equation shows that the relative humidity is directly proportional to the total pressure. The pressure of the atmosphere falls with elevation above sea level, being about 88 kPa at 1000 m. Thus, with increasing height, the relative humidity at given temperature and moisture-vapour content will progressively be less than that in a standard atmosphere having a pressure of 101.325 kPa. The equilibrium moisture content will also be slightly smaller in consequence. A similar effect is observed in vacuum drying in which the operating pressure is deliberately reduced to enhance drying under relatively low temperatures.

3.2 Enthalpy of Moist Air

All substances have internal energy, which is identified with the energies of molecular motion. For convenience, this internal energy is lumped with flow energies in forcing a substance into a system against pressure and forcing it out. The composite energy is known as the *enthalpy H*. Absolute values of the enthalpy cannot be fixed. One convenient and commonly used reference state for zero enthalpy is liquid water under its own vapour pressure of 611.2 Pa at the triple-point temperature of 273.16 K.

The change of enthalpy with temperature at constant pressure is known as the *heat capacity C_P*:

$$C_P = \left[\frac{\partial H}{\partial T}\right]_P. \tag{3.5}$$

Heat capacities of air and water vapour increase linearly with temperature, so that it is possible to evaluate enthalpy differences of these gases from the heat capacity at the arithmetic mean temperature $[(T - T_o)/2]$ above the datum T_o:

$$\Delta H = [C_P]_{av} \cdot (T - T_o). \tag{3.6}$$

The heat capacity of dry air is approximately equal to 1 kJ kg^{-1} K^{-1} at kiln temperatures. Thus, the numerical value of the specific enthalpy (in kJ kg^{-1}) of dry air is almost equal to the temperature (in degrees Celsius) when 0 °C is taken to be the datum temperature. The heat capacity of water vapour is significantly higher, varying from 1.87 kJ kg^{-1} K^{-1} at 20 °C to 2.01 kJ kg^{-1} K^{-1} at 100 °C. The specific enthalpy of water vapour is also very much higher than that of dry air because there is the additional and substantial contribution of vaporising moisture from its datum state.

Since enthalpy is an extensive property, the enthalpy of a moist gas may be estimated from the partial enthalpies of the various constituents. Like humidity, it is convenient to compute enthalpies of moist air on a dry-gas basis. Thus the specific enthalpy of unit mass of dry air plus its associated moisture vapour is given by:

$$I_G = H_G + H_W Y \quad (3.7)$$

where H_G is the specific enthalpy of dry air, H_W is that for water vapour and Y, the humidity, is the mass of moisture vapour per unit mass of dry air. This expression ignores any residual enthalpy due to heats of mixing and other interactions which are relatively very small (List 1958). The quantity I_G is known as the *humid enthalpy*. As shown by Keey (1978), the humid enthalpy at low humidities can be calculated by

$$I_G = C_{PG}T + [C_{PW}T + \Delta H_{VO}] Y \quad (3.8)$$

where the heat capacities are evaluated at a temperature $T/2$, the temperatures being measured in degrees Celsius, and ΔH_{VO} is the latent heat of vaporisation of water at the datum temperature of 0 °C which takes a value of 2501 kJ kg^{-1}. This equation becomes inaccurate for precise calculations at humidities greater than 0.05 kg kg^{-1}. The alternative expression

$$I_G = C_{PG}T + [C'_{PW}(T - T_S) + C_{LW}T_S + \Delta H_{VS}] Y \quad (3.9)$$

may be used if greater precision is needed, where C'_{PW} is the mean heat capacity between temperatures T and T_S, the temperature at which the air would be saturated at the given humidity, C_{LW} is the mean heat capacity of liquid water between zero and T_S, and ΔH_{VS} is the latent heat of vaporisation of water at the temperature T_S.

Krischer and Kast (1978) present tabular values of the humid enthalpy of saturated air over the temperature range from −20 to +100 °C. These rise rapidly with saturation temperatures, being 57.9 kJ kg^{-1} at 20 °C, 277 kJ kg^{-1} at 50 °C and infinite at the boiling point. Thus most of the enthalpy of humid air is tied to the moisture vapour, and heat-recovery schemes at kiln temperatures are effectively worthless unless this latent heat of the vapour can be captured, as in dehumidifier-kiln arrangements.

3.3 Adiabatic Saturation and Wet-Bulb Temperatures

A large kiln may be regarded as an adiabatic chamber to a first approximation. Consider the situation where air at temperature T_G and humidity Y_G is being humidified with copious amounts of water at a lower temperature T_S (Fig. 3.1). The air will leave the chamber saturated at this temperature T_S, there being so much water that its temperature changes imperceptibly. The corresponding humidity of the saturated outlet air is then Y_S.

An enthalpy balance over the system yields the expression

$$\frac{Y_S - Y_G}{T_S - T_G} = -\frac{C_{PY}}{\Delta H_{VS}}, \quad (3.10)$$

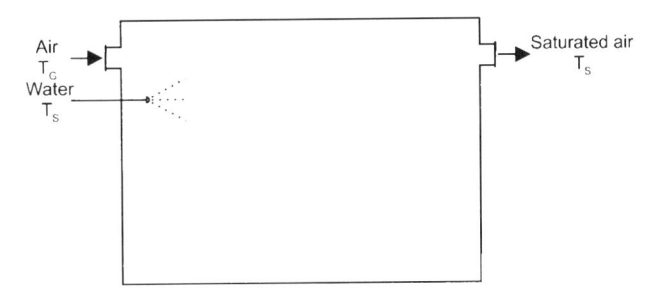

Fig. 3.1. Adiabatic-saturation chamber

where $\hat{C}_{PY}$ is an effective heat capacity, known as the *humid heat*, and is the quantity $(C_{PG} + C_{PW}Y)$ (e.g. see Keey 1978,1992). The ratio of the humid heat to the latent heat of vaporisation is virtually independent of the starting conditions, and thus there is a common adiabatic-saturation path for a given end temperature T_S. This is known as the *cool-limit* or *adiabatic-saturation temperature*. The temperature at which dew would form in air of the given inlet humidity, the *dewpoint*, is always less than this adiabatic-saturation value, as Fig. 3.2 shows.

If the air had been exposed to lumber below its fibre-saturation point rather than to water, then the equilibrium vapour pressure of moisture would be less than the saturation vapour pressure of water at the same temperature. The corresponding cooling of the air is less than in the case of humidifying with water, and the thermodynamic end-point temperature is consequently higher. If the heat of wetting of the lumber is known, then this higher temperature can be calculated (Keey 1992).

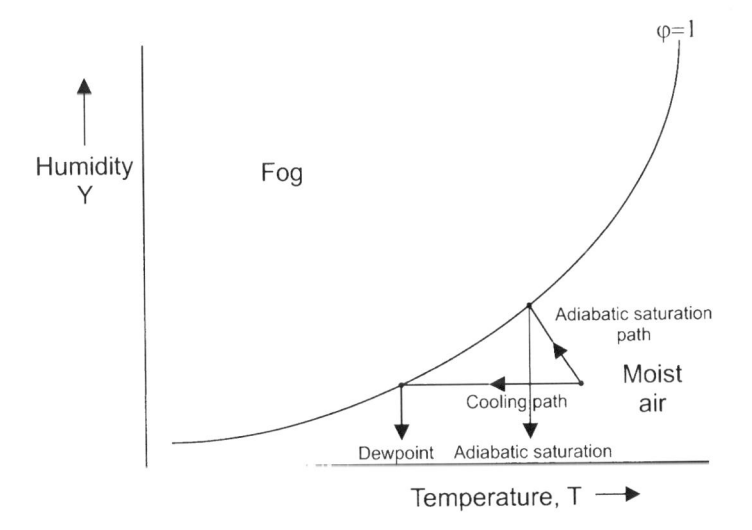

Fig. 3.2. Adiabatic-saturation and dewpoint temperatures

There is another commonly observed equilibrium temperature which is the steady-state value reached by a fully saturated surface under certain conditions. When the surface is heated convectively (with no heating by conduction through the material), the temperature difference between that T_G of the flowing gas and that T_W of the wet surface becomes

$$T_G - T_W = \gamma \frac{\Delta H_{VS}}{C_{PY}}(Y_G - Y_W), \qquad (3.11)$$

where Y_G is the bulk-gas humidity and Y_W is the saturated humidity at the surface. The coefficient γ is a function of the thermophysical properties of the moist-gas system, the bulk-gas humidity and the extent to which the surface is heated by radiation (Keey 1992). This coefficient takes a value close to 1 under most kiln conditions with steamy air. For example, γ is 1.06 when the humidity is 0.05 and the thermal radiation received is 10% of the convected amount. Thus, it is often assumed that the coefficient is 1 for the situation of a water-saturated surface exposed to warmer air. The temperature of the surface so calculated is called the *wet-bulb temperature,* which is the temperature reached by the surface of the wood for a period in the early part of the drying schedule. The temperature T_G of the bulk of the air in contact with the board is known as the *dry-bulb temperature.*

The equation for the wet-bulb temperature under these conditions is exactly the same as that for the adiabatic-saturation temperature [Eq. (3.10)]. This equivalence has been useful in estimating air humidities by means of *psychrometric measurements* in which temperatures have been recorded by thermometers with wet and dry bulbs, respectively. The difference in temperatures, T_G–T_W, is called the *wet-bulb depression,* and is sometimes used as a control parameter in drying.

The apparent simplicity of the method is deceptive, however, if accurate measurements are wanted. The air velocity must be at least $3\,\mathrm{m\,s^{-1}}$ over the surface which needs to be shielded from excessive radiation and receive no conduction. At relative humidities below 20%, it is difficult to maintain even a small surface fully wetted. Although psychrometric measurements are convenient, it is difficult in industrial practice to measure a relative humidity (RH) to an accuracy greater than ±3% RH. Even under ideal laboratory conditions, the uncertainty with an aspirated psychrometer with a built-in fan is ±1% RH at a temperature of 25 °C and a humidity of $0.01\,\mathrm{kg\,kg^{-1}}$ (Wiederhold 1995). By contrast, at a dry-bulb temperature of 80 °C and a wet-bulb value of 50 °C, a ±1 °C uncertainty in the wet-bulb depression, which is not unlikely with standard industrial resistance thermometers, corresponds to a ±3% RH uncertainty in the relative humidity. Lumber-kiln control, then, probably relies more on the dry-bulb temperature settings than on maintaining an environment of a specified relative humidity. In particular, at the start of a schedule when very small wet-bulb depressions may be specified, the temperature rise of the kiln stack becomes the determinative factor in controlling the rate of drying.

3.4 Humidity Charts

The properties of moist gas systems can be conveniently graphed on thermodynamic charts. These diagrams, once used extensively for process-design calculations, still retain a usefulness in displaying the air conditions in and the performance limits of kilns.

The earliest diagram still in use is that proposed by Grosvenor (1907). He suggested that the extent to which air humidifies in contact with a solid containing water is given with sufficient accuracy by the expression

$$\frac{dY}{dT} = \frac{aC_P}{b - T_E}, \qquad (3.12)$$

in which a and b are coefficients, and T_E is the temperature at which the humidification is stopped. This process is thus represented by a set of straight lines for each end temperature on a graph of temperature against humidity as shown in Fig. 3.3. The various humidification lines could be plotted over contours of constant relative humidity. Most modern Grosvenor charts, as reproduced in handbooks, often display other thermophysical data as well.

A more elaborate chart is that often attributed to Mollier (1923), although there are claims that Ramzin introduced the idea some five years earlier (Solukhin and Kuts 1983). Instead of temperature, the Mollier-Ramzin chart uses a modified enthalpy which is linearly related to humidity, thereby having important characteristics in the ease of estimating the properties of mixed gas streams. The so-called Lever Rule holds, in which the mixed-gas property

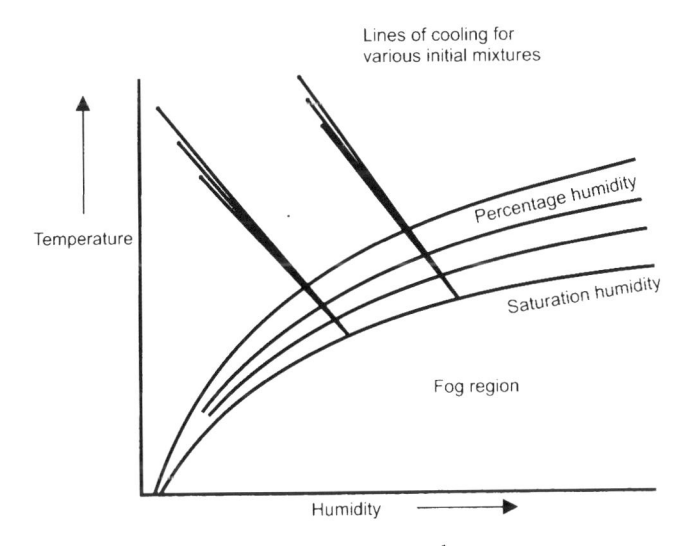

Fig. 3.3. Grosvenor's humidity-temperature chart

(enthalpy or humidity) is the mass-weighted mean of the property of each individual stream (see Keey 1978, 1992).

The humid enthalpy can be expanded as

$$I_G = [C_{PG}T_G + C_{PW}T_G Y_G] + \Delta H_{VO} Y_G = I'_G + \Delta H_{VO}. \tag{3.13}$$

The use of the modified humid enthalpy I'_G as ordinate opens up the working area of the diagram for unsaturated moist-air mixtures, but there are other important properties. If the humid enthalpy is taken to be zero at 0 °C, then it follows from Eq. (3.13) that

$$\left[\frac{dI'_G}{dY_G}\right]_{I_G=0} = -\Delta H_{VO} \tag{3.14}$$

It therefore follows that all the isenthalpic contours on the chart will have this same slope, $-\Delta H_{VO}$. Equation (3.13) also shows that the isothermal lines

$$\left[\frac{dI'_G}{dY_G}\right]_{T_G} = C_{PW}T_G \tag{3.15}$$

have positive slopes equal to the heat capacity of water vapour. The intercept of the isotherm on the zero-humidity axis reveals the heat capacity of the air. The interpretation of these various thermophysical parameters on the chart is illustrated in Fig. 3.4.

The various air-state changes in a lumber kiln are sketched on this kind of chart in Fig. 3.5. Air (A) is drawn through the open vent in the roof and mixes with the humid air (Z) leaving the outlet face of the stack. The mixed air (M) is drawn through the fan; some is expelled to the outside, but the majority passes through the heating coils to the inlet face of the stack at condition (O). The line segment OZ represents the cooling path of the air as it passes through the stack of lumber, picking up moisture. The ratio of line segments AM/AZ is the ratio r (on a dry-air basis) of the more humid leaving air that goes back to the fan to be recirculated, compared with the total mass of air which passes through the stack. By mass balance, (1 − r) is the fraction of fresh air that is drawn in. The vertical line segment MO is the amount of heat (per unit mass of air throughput) added by the heating coils.

The state conditions at the wood surface during the drying process can also be displayed, as shown in Fig. 3.6. In this example, the kiln's air state is shown held at condition O. The state path represents the change in equilibrium humidity at the surface as the wood dries out. This state path follows the moisture-saturation contour, $\varphi = 1$, from the time of loading the charge (at condition C) until the fibre-saturation point is reached at F. The state path then follows a curve, tracing points of dwindling relative humidity and rising temperature, to point E which represents the surface condition at the end of the drying schedule. The curve FE follows closely (in reverse) the adiabatic-saturation path OF, but is not coincident with it because of the hygroscopic nature of the wood's surface.

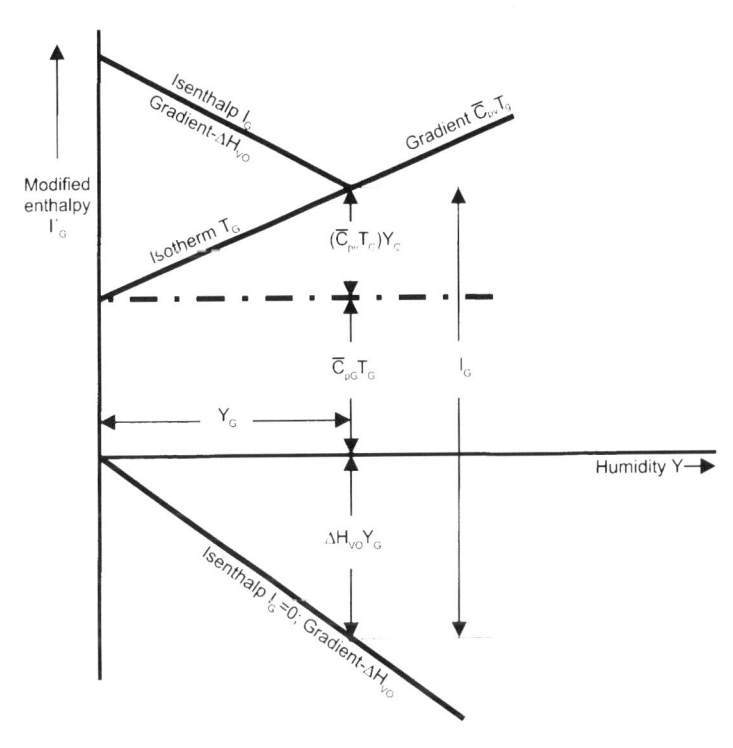

Fig. 3.4. Basic construction of the Mollier-Ramzin chart for moist air. (Reproduced from Keey 1992, Drying Principles and Practice, by kind permission of Taylor & Francis)

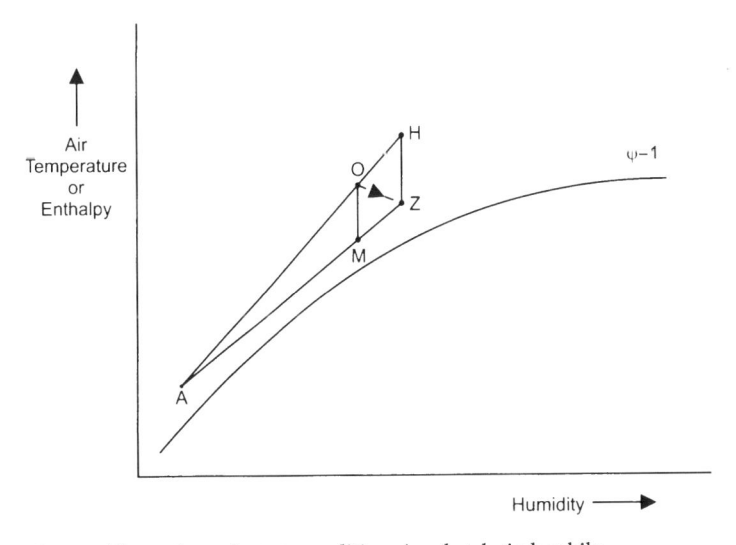

Fig. 3.5. The various air-state conditions in a batch timber kiln

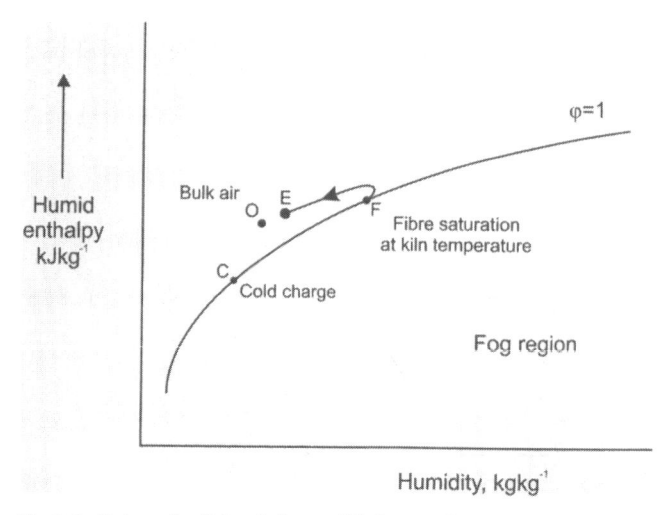

Fig. 3.6. State path of the air in equilibrium with the wood's surface during kiln-drying

This diagram also illustrates an important aspect of the kiln-drying process. When a cold charge is introduced into a warm, humid kiln, the equilibrium humidity of the cold wood will be much less than that of the bulk of the warm, humid air. Therefore, condensation on the lumber boards will occur until the wood has warmed up sufficiently so that the surface-air humidity is greater than that in the bulk of the air. For example, in the accelerated conventional schedule for *Pinus radiata*, with a dry-bulb temperature of 90 °C and a wet-bulb temperature of 60 °C, the bulk-air humidity is 0.014, nearly ten times greater than the saturation humidity at 20 °C. Only when the wood has reached 58 °C will drying begin.

3.5 Ideal Heat Demand

The *ideal heat demand*, which relates solely to the heating requirements of the air in maintaining kiln conditions without any heat losses though the kiln walls or any sensible heating of the wood, is a useful benchmark in assessing the thermal economy of kilns. This demand depends primarily on the dry and wet-bulb temperature settings and the extent to which the vents are open to maintain them, but there is also a small influence of the location of the heaters with respect to the overhead fans.

Consider the case where the heaters are placed at one side of the kiln, as shown in Fig. 3.7. Suppose the vents are partially open, so that a fraction r (on a dry-gas basis) is recirculated and the remainder $(1 - r)$ is expelled to the outside. It follows that a corresponding amount of fresh air is drawn in.

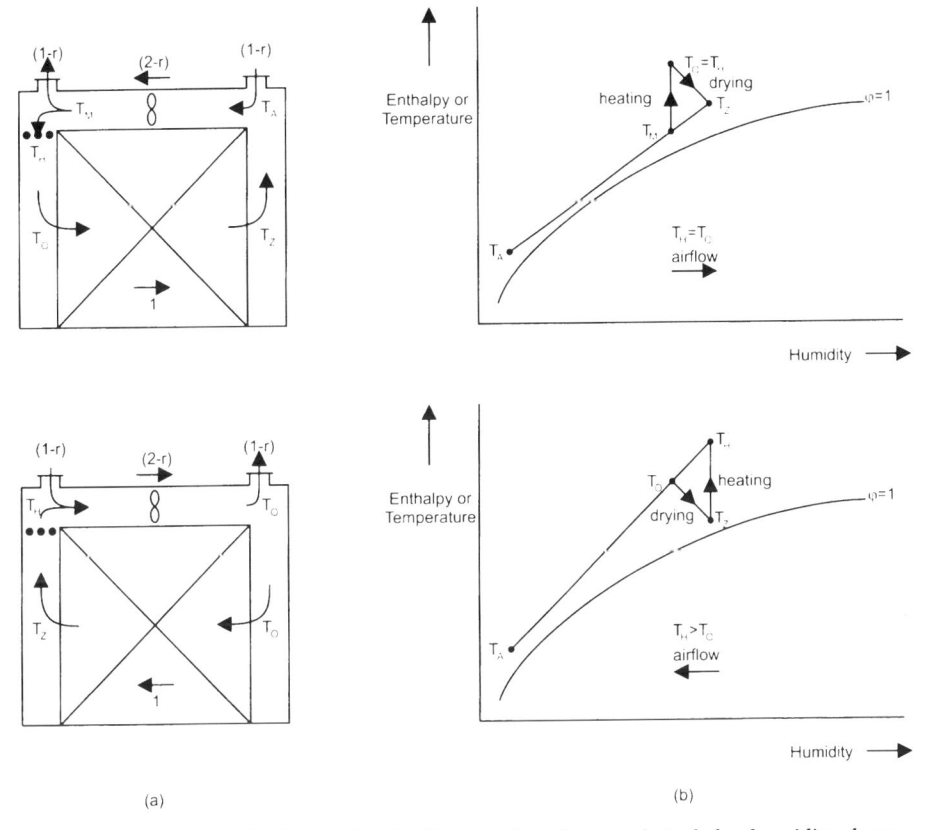

Fig. 3.7a,b. Temperature levels in a batch kiln. **a** Airflow diagram, **b** Enthalpy-humidity chart showing air-state changes

When the fan pushes the air towards the heater, fresh air at temperature T_A mixes with the more humid air from the outlet face of the stack at temperature T_Z to give a mixed airstream at T_M which is then heated to T_O, the temperature of the air at the inlet face of the stack. When the air flows in the reverse direction, with the fan pulling the air from the heater, the heater now accepts air at temperature T_Z and raises it to a value T_H so that on mixing with fresh air the inlet-face temperature of T_O is maintained. In this case, slightly more heat is required as effectively the exhausted fraction of the air has to be heated through an extra temperature difference $(T_O - T_M)$. This can also be seen from the enthalpy-humidity diagram, Fig. 3.7b. The temperature T_H must also be higher than the the inlet temperature T_O.

When the air is being pushed towards the heater, the heat supplied is the sum of two quantities: that to heat a fraction $(1 - r)$ from T_A to T_O and a fraction r from T_Z to T_O. The ideal heat demand is then given by

$$Q^* = \frac{C_{PY}(1-r)(T_O - T_A) + C_{PY}r(T_O - T_Z)}{(Y_Z - Y_O)}, \tag{3.16}$$

since $(Y_Z - Y_O)$, the humidity change over the stack, is the moisture pick-up per unit mass of dry air passing through. On the assumption that the humid heats are everywhere the same, and the humidification is adiabatic, Eq. (3.16) reduces to

$$Q^* = \left[1 + \frac{(1-r)(T_Z - T_A)}{T_O - T_Z}\right]\Delta H_{VS}, \tag{3.17}$$

where ΔH_{VS} is the latent heat of vaporisation at the adiabatic-saturation temperature, which may be taken as the wet-bulb temperature for the air-steam system. This expression shows that the ideal heat demand is always greater than the heat of vaporisation of moisture when air is being expelled from the kiln even if there are neither heat losses through the walls nor warming of the wood.

The equivalent expression for the case when the air is being pulled away from the heater is given by

$$Q^* = \left[r + \frac{(1-r)(T_H - T_A)}{T_O - T_Z}\right]\Delta H_{VS}. \tag{3.18}$$

In the case where the dry-bulb temperature is 90 °C, the wet-bulb temperature is 60 °C and the outlet-face temperature falls to 75 °C, the recirculation ratio r is 0.956 (just over 4% of the through-air being expelled per cycle); the foregoing equations suggest that the ideal heat demand varies between 18 and 20% more than the vaporisation load, depending upon the rotational direction of the fan.

Rosen (1995) notes that the drying of solid wood products requires at least 50% more energy than the heat of vaporisation, with southern yellow pine (*Pinis taeda*) requiring between 3.7 and 5.1 MJ per kg moisture evaporated, Douglas-fir (*Pseudotsuga menziesii*) from 4.7 to 7.0 MJ kg^{-1}, while refractory lumber such as 50-mm-thick white oak (*Quercus alba*) can require more than 14 MJ kg^{-1}. Low-temperature dryers can be fitted with dehumidifiers to reduce the energy consumption, but the associated compressor draws electrical energy, normally more costly than the thermal energy that may be derived from the combustion of wood residues of little commercial value. Dehumidifier kilns are briefly considered in Chapter 12.

3.6 Evaporation from a Wood Surface

The evaporation from a wood surface into the air flowing through a stack in a kiln is determined by the convective heat transfer to the exposed surfaces of the boards. The heat-transfer rate to unit surface is given by Newton's law (1701):

$$q_c = \alpha(T_G - T_S), \tag{3.19}$$

in which T_G is the bulk air temperature, T_S is the surface temperature and α is the heat-transfer coefficient. This coefficient depends upon the geometry of the surface and the airspeed over it. Strictly, this equation does not apply to a wet surface without modification. The moisture vapour leaving the surface conveys enthalpy with it, thus reducing the net heat transfer. However, at normal kiln-drying rates, the effect is very small and may be ignored. Thus, the heat transfer to the surface can be examined separately from the mass transfer from it, although the two transfer processes are directly related and described by analogous equations.

The rate of evaporation itself depends upon the difference in moisture-vapour pressure between that at the surface (p_S) and that in the bulk of the air (p_G) (Dalton 1802). Unlike the heat transfer, the mass transfer is non-linear in its driving potential because of the additional bulk-flow effect hindering the transport of moisture vapour. In the case where an airstream flows over a flat, solid surface, the rate of evaporation over unit surface becomes

$$N_V = \tilde{\beta} \ln\left[\frac{P - p_G}{P - p_S}\right] = \tilde{\beta} \ln\left[\frac{1 - y_G}{1 - y_S}\right], \tag{3.20}$$

where the coefficient $\tilde{\beta}$ has units of flux (mol m^{-2} s^{-1}), y_G is the mole fraction of moisture vapour in the bulk air ($y_G = p_G/P$) and y_S is the mole fraction adjacent to the surface ($y_S = p_S/P$).

It is more useful to rewrite this expression in terms of humidities. We have:

$$Y \equiv \frac{Dy}{1 - y}, \tag{3.21}$$

where D is the density ratio or the molar-mass ratio of water vapour relative to dry air ($D = 0.622$). It thus follows that

$$N_V = \beta D \ln\left[\frac{D + Y_S}{D + Y_G}\right], \tag{3.22}$$

where β is a mass-transfer coefficient and equal to $\tilde{\beta} M_V$.

Since the heat-transfer rate to the surface is directly proportional to the temperature difference, it is often assumed that, likewise, the evaporation rate should be proportional to the humidity difference, this being the apparently analogous driving force for moisture transfer. One may write

$$N_V = \beta \phi_M (Y_S - Y_G), \tag{3.23}$$

where the coefficient ϕ_M may be regarded as a "correction factor" to allow for the deviation from direct proportionality. This correction, which has been called the *humidity-potential coefficient* (Keey 1992), is given by

$$\phi_M = \left[\frac{D}{D + Y_G}\right] \frac{\ln(1 + B)}{B}, \tag{3.24}$$

where $B = (Y_S - Y_G)/(D + Y_G)$ and is equivalent to the universal dimensionless driving force used by Spalding (1963) to describe mass-transfer processes. The first term, $D/(D + Y_G)$, allows for the departure of linearity of the humidity potential, and the term involving B for the influence of high mass-transfer rate effects. The latter are negligible under kiln-seasoning conditions; so the humidity-potential coefficient at any place in the kiln is given by

$$\phi_M \cong \frac{D}{D + Y_M}, \qquad (3.25)$$

where Y_M is the arithmetic-mean humidity in the boundary layer and equal to $(Y_S + Y_G)/2$ (Keey and Ma 1986).

In a kiln, where normally there is a high recirculation ratio of the air ($0.95 < r < 1$), the variation in the bulk-air humidity will be small relative to the humidity level itself, and thus the humidity-potential coefficient will be essentially the same everywhere for adiabatic working of a kiln when the surface humidity also remains constant. Under these conditions then, the evaporation rate is indeed directly proportional to the humidity potential or difference in humidities between that next to the wood surface and that in the bulk of the air. The apparent mass-transfer coefficient is thus $\beta\phi_M$. This leads to a considerable simplification in evaluating evaporation rates in the kiln. The coefficient β, however, may still vary throughout the kiln due to fluid-dynamic factors, and this point will now be examined.

A kiln stack may be regarded as a sandwich of thick plates, built up of individual board elements and separated by stickers to form a parallel set of rectangular ducts, as shown in Fig. 3.8. Because the horizontal spacing between stickers is much greater than the thickness of these wooden slats, the airflow over each row of boards might be expected to be similar to that between two infinitely wide plates. (It will be seen later that this assumption is not entirely valid, but it forms a starting point for the analysis of the convection).

The flow over a sharp-edged, flat plate is characterised by the development of a boundary layer of retarded flow which is zero at the leading edge, gradually becoming thicker with distance as the flow is hindered by skin friction. The mass-transfer coefficient is inversely proportional to the thickness of the boundary layer through which the moisture vapour must diffuse to escape into the bulk of the air and be conveyed away. Thus the mass-transfer coefficient as well as the humidity potential dwindles in the direction of the airflow. By solving the equations of motion and fitting a polynomial to the velocity distribution in the boundary layer, Pohlhausen (1921) obtained an expression for the local mass-transfer coefficient β_x as a function of distance x from the leading edge. In non-dimensional form, the local Sherwood number $Sh_x = \beta_x x/cD_V$ a function of the local Reynolds number, $Re_x = ux/v$, and the Schmidt number, $Sc = v/D_V$:

$$Sh_x = 0.332 Re_x^{1/2} Sc^{1/3}, \qquad (3.26)$$

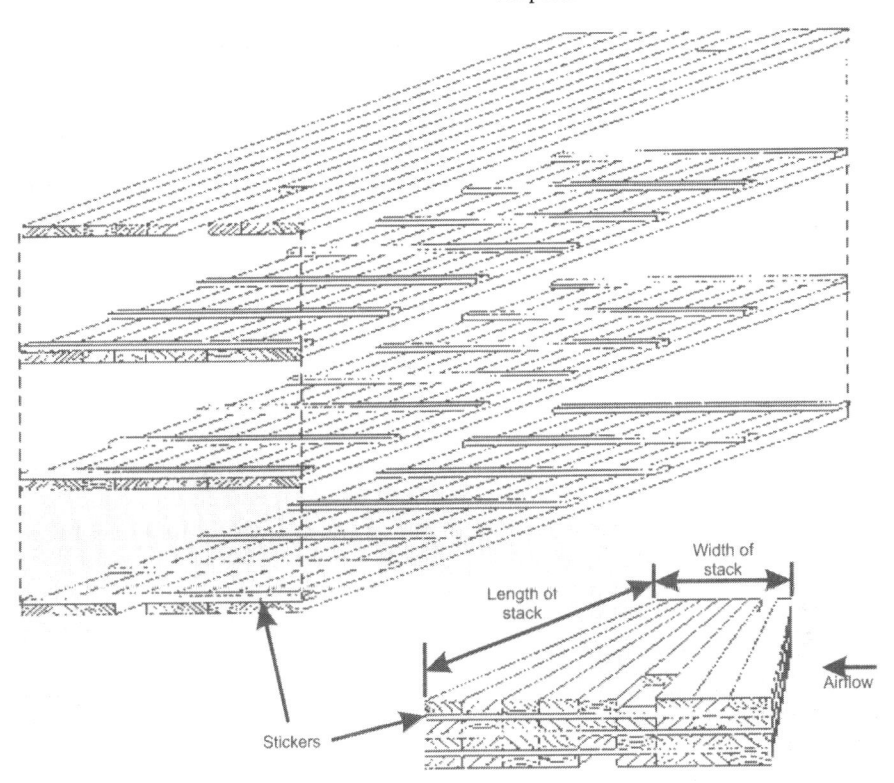

Fig. 3.8. Arrangement of boards in a kiln stack

in which c is the total molal concentration or density, D_V is the moisture-vapour diffusion coefficient, u is the air velocity through the stack and v is the kinematic viscosity of the air. The mean value over a swept length L is twice the local value at $x = L$.

Some authorities, such as Brauer (1971), define (β/c) as a mass-transfer coefficient which represents a mass-transfer velocity (k) having units of m s^{-1}. The conversion of this coefficient to that (β) defined in terms of humidity differences is given by:

$$k = \tilde{\beta}/c = \beta/cM_V. \tag{3.27}$$

When the Reynolds number Re_x reaches a critical value of about 500 000, the boundary layer ceases to be fully laminar and becomes turbulent in its outer regions. For a stack width of 2.4 m, and air velocity of 5 m s^{-1} and a kinematic viscosity of 20×10^{-6} m^2 s^{-1} (corresponding to an air temperature of about 75 °C), the maximum Reynolds number is 700 000, so transition to such turbulence is likely within a stack of that width. In newer kiln arrangements, with even wider stacks, more extensive turbulence would occur with a greater pressure loss. There would also be a greater fall in the mass-transfer coefficients from the air-inlet to the air-outlet face of the stack.

Kestin and Richardson (1963), in investigating heat transfer across turbulent, incompressible boundary layers, have obtained the expression

$$Sh_x = 0.0296 Re_x^{4/5} Sc^{1/3} \qquad (3.28)$$

for the local Sherwood number at a distance x from the leading edge, with

$$Sh_L = 0.037 Re_x^{4/5} Sc^{1/3} \qquad (3.29)$$

for the mean value over a length L.

It thus follows that the mass-transfer coefficient (and thus evaporation rate) will correlate with kiln-air velocity according to the dependence

$$N_v \approx u^n \qquad (3.30)$$

with the exponent n varying between 0.5 or 0.8 since the boundary-layer flow will be transitional from laminar to turbulent. Ogura and Ohnuma (1955), for example, obtain values of n lying between 0.64 and 0.69 in their moisture-evaporation tests with wood. Kollmann and Schneider (1961) find that, in the superheated-steam drying of single boards in the velocity range of 3 to 5 m s^{-1}, the laminar boundary-layer dependence holds, but in the higher velocity range of 9 to 13 m s^{-1} the exponent n is 0.6, which suggests a transition to turbulence along the board.

The view that a stack of lumber boards is a set of sharp-edged parallel plates oversimplifies the geometry somewhat. The boards at the air-inlet face of the stack offer a set of blunt obstacles to the entering airflow. Smoke tests about a single slat 75 mm long and 6.3 mm thick show that a fast-moving airstream will scour the upstream surface of the slat and curl around its leading edge (Dankwearts and Anolick 1962). Such eddying begins when the Reynolds number based on the body thickness is greater than 245 (Sørensen 1969). At a kiln velocity of 5 m s^{-1}, the Reynolds number defined on this basis is over 10 000, and thus strong eddying between the boards at the inlet face is expected. This eddying will lead to greatly enhanced convective transfer coefficients from the surface where the boundary layer is thinned, as the results from experiments of sublimation from naphthalene-coated slabs demonstrate and illustrated in Fig. 3.9.

It might be expected that the effect of the leading-edge eddy would diminish away from the inlet. However, tests by Sparrow et al. (1982) on heat transfer from a string of modules 27 mm square, arranged in rows, 6.3 mm between modules, indicated that the heat-transfer coefficients reached an asymptotic value after several rows, but this limiting value was considerably higher than that predicted from the equation for transfer from a sharp-edged plate.

Even if the boards in a stack are butted together sideways on stacking, inevitably very small board side-gaps will develop during the kiln-seasoning process due to the shrinkage of the wood. While the gaps may be small (of order 1 mm), they appear to be highly significant, as shown by Lee (1990) in following smoke-trace patterns about in-line slabs. Lee found low-frequency oscillations, with a period of 1 to 7 s, at normal kiln-air flow rates. A

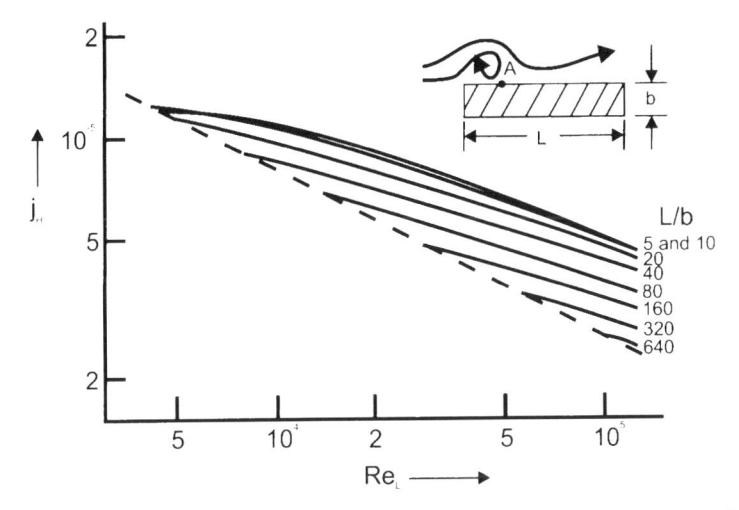

Fig. 3.9. Convective mass-transfer coefficients for sublimation from a naphthalene-coated slab for varying thickness b. Slat length $L = 0.3$ m; the ordinate j_H is equal to $\alpha/C_p u \rho_G$. The Reynolds number Re_L is based on the slat length L. (Reprinted from Sørensen 1969 Chem Eng Sci 24: 1445–1460, with permission from Elsevier Science)

circulating flow in the gap between two boards was seen to spill over periodically into the boundary layer of the upstream board. The consequential oscillatory changes in the thickness of the boundary layer in the zone close to the leading edge of each board would lead to enhanced transfer coefficients there.

The interaction between the circulation within a gap and the boundary layer over the downstream board means that each board effectively behaves as a separate blunt slab swept by an incident airflow. This behaviour is confirmed in the tests reported by Kho et al. (1989, 1990) using a naphthalene-coated test board in a industrial-type pilot kiln, and in the numerical simulations of Langrish et al. (1993) in which the evaporation was estimated through computational fluid dynamics by means of the analogy between fluid friction at a surface and convective transfer of heat or mass. The measured enhancement of the limiting mass-transfer coefficients far from the inlet is considerable, as noted in Table 3.2.

Table 3.2. Mass-transfer coefficients in a pilot-plant kiln. Naphthalene sublimation from a test board 640 × 100 × 20 mm at 41 °C (Reprinted from Langrish et al. 1993, Drying Technol 10: 771, Courtesy of Marcel Dekker Inc)

Velocity between boards (m s^{-1})	Turbulence intensity before stack (%)	Measured mass-transfer coefficient (m s^{-1})	Enhancement over Flat-plate value (%)
3	16	0.0078	66
5	19	0.0115	62
7	23	0.0167	83

These enhancements may be overestimated slightly since the concentration boundary layer only commenced at the front edge of the test board whereas the hydrodynamic boundary layer was more developed from the flow over the upstream boards from the test one. Nevertheless, the numerical simulations follow the trends of the experimental data, with the limiting mass-transfer coefficients much greater than the values suggested from correlations of transfer in turbulent flow over sharp-edged plates.

Salin (1996), in reviewing heat and mass-transfer coefficients for individual boards and board surfaces, recommends that Sørensen's correlation (1969) be adapted to estimate behaviour for multiple slabs, as in a kiln stack. For heat transfer, the expression becomes

$$Nu_L = \frac{0.182 \, Re_L \, Pr^{1/3}}{[Re_l - Re_\Delta]^{0.345}} \tag{3.31}$$

for $Re_L / Re_\Delta > 1.5$ and $Re_\Delta = 29.7 \, Re_b^{0.56}$, and in which Nu_L is the Nusselt number ($\alpha L/\lambda$) based on the streamed length, as is the Reynolds number Re_L, while Re_b is that based on the board's thickness b. This equation may be used for the evaporative mass transfer by replacing Nu_L with Sh_L and Pr and Sc.

Further, Salin (1996) observes that the influence of the gaps between adjacent boards effectively reduces the distance from the leading edge, on average, by a factor of 0.845. The method leads to the prediction of profiles, in accord with experimental sublimation studies of Kho et al. (1990), that show a asymptotic set of values over a board away from the inlet (Fig. 3.10).

On this basis, Salin (1996) calculates the mean board heat-transfer coefficients in the centre of a 1.5-m-wide stack, with 25-mm sticker widths for a kiln temperature of 60 °C, as set out in Table 3.3. The kiln-wide values imply that the mean heat-transfer coefficients vary with kiln-air velocity as $u^{0.65}$ in agreement with the results of Ogura and Ohnuma (1955).

3.7 Subsurface Evaporation

Wood may be regarded as a cellular-porous body. Above fibre saturation, when the cell walls are fully saturated, the removal of free moisture from vessels or tracheids may resemble the way in which a capillary-porous body dries out, particularly in the possibility that moisture might evaporate below the exposed surface. Keey (1978) has reviewed experiments with beds of glass spherules which show that the evaporation falls suddenly as the first amounts of moisture are driven off, and then slowly dwindles until the bed is about 20% saturated. This behaviour is explained by a sudden retreat of the moisture level to just below the exposed surface to the narrowest openings in the bed where the evaporative level dwells for a period, being fed by capillary movement of moisture from the deeper portions of the bed. With the sapwood of softwood, there

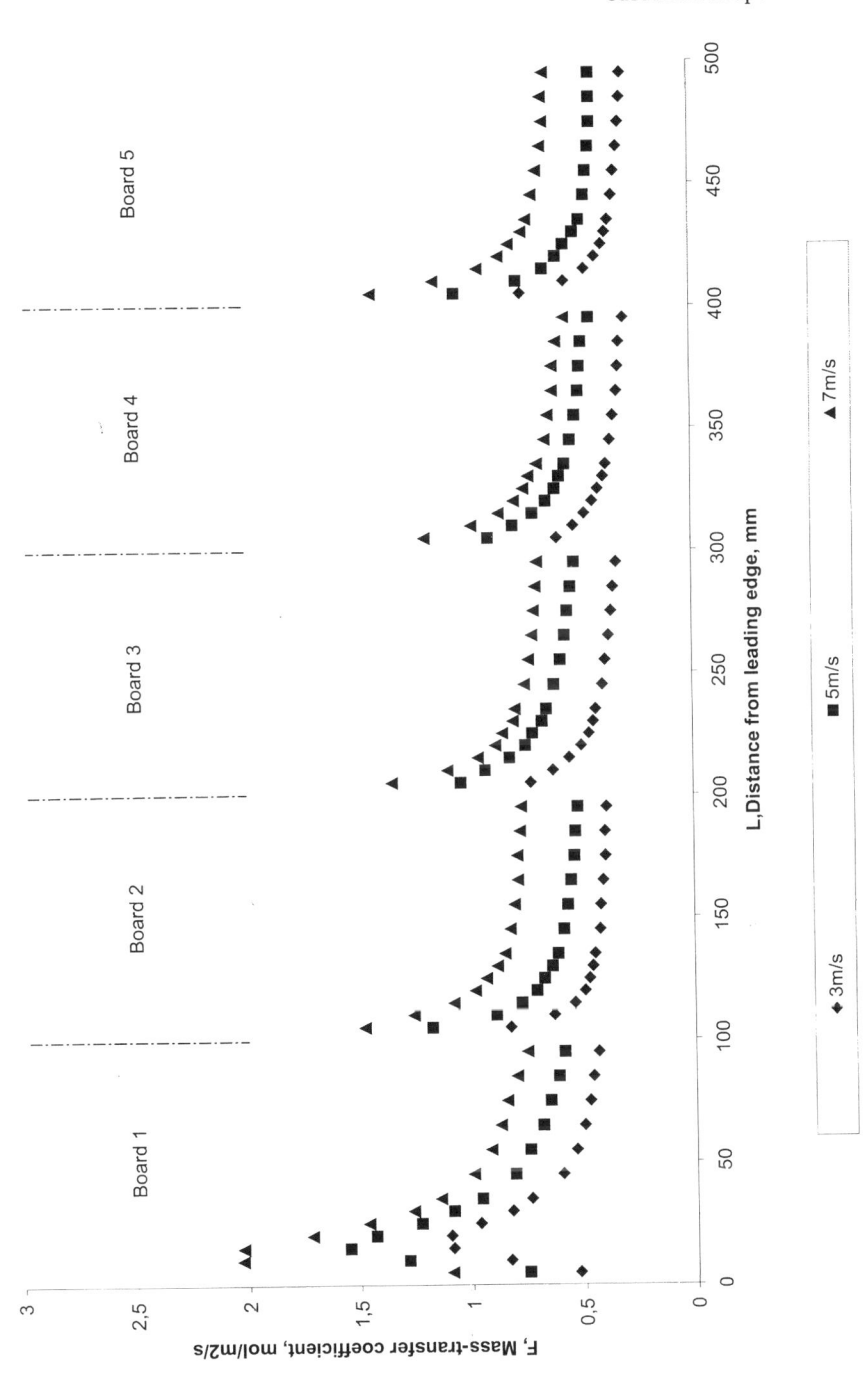

Fig. 3.10. The local mass-transfer coefficient β as a function of distance from the leading edge of a stack made up from 25-mm-thick, 100-mm-wide boards, separated by 25-mm-thick stickers with adjacent board side-gaps of 0.5 to 1 mm. + 3 m s^{-1} □ 5 m s^{-1} ▲ 7 m s^{-1}. (Kho et al. 1990)

Table 3.3. Estimated heat-transfer coefficients ($W\,m^{-2}\,K^{-1}$) based on Sørensen's correlation (1969) corrected for the influence of board side-gaps (Salin 1996)

Board dimension (mm × mm)	25 × 100	75 × 100	75 × 100
Kiln-air velocity ($m\,s^{-1}$)	3	3	3
Board number			
1	41.1	33.2	39.4
2	25.0	22.0	26.4
3	21.0	18.3	22.0
4	19.0	16.6	19.9
5	17.8	15.5	18.7
6	16.9	14.8	17.8
7	16.4	14.2	17.2
8	15.9	13.9	16.7
9	15.6	13.6	16.4
10	15.3	13.3	16.1
11	15.1		
12	14.9		
13	14.8		
14	14.7		
15	14.6		
Kiln-wide mean	18.5	17.5	21.1

is evidence of such a "wet line" through the appearance of kiln brown stain, particularly at higher kiln temperatures (Haslett 1998).

Suppose the evaporation is controlled by vapour diffusion through a surface layer of thickness ζ and a boundary layer of thickness δ. The effective mass-transfer coefficient k_e (in units of $m\,s^{-1}$) is given by

$$k_e = \frac{D_V}{\delta + \xi\zeta} = \frac{k}{1 + Bi_M}, \qquad (3.32)$$

where ξ is a tortuosity coefficient that allows for the irregularities in the flowpath of the vapour from where it evaporates to the exposed surface. The dimensionless ratio, $\xi\zeta/\delta = Bi_M$ (the mass-transfer Biot number), represents the relative resistance to moisture-vapour transport of the dried-out surface layer in the wood to that in the external boundary layer.

The evaporation rate is thus given by

$$N_V = \frac{cM_V k}{[1 + Bi_M]}\phi_M[Y_E - Y_G] = \frac{\beta\phi_M}{[1 + Bi_M]}[Y_E - Y_G], \qquad (3.33)$$

where Y_E is the saturation humidity at the wet-line boundary and Y_G is the bulk-air value.

If the slow heating of the material is ignored, then the wet-line temperature will remain almost constant for a period: the heat absorbed supplies the evaporative load. Whence the difference in temperature between that (T_E) at the wet line and that (T_G) in the bulk of the air becomes

$$(T_G - T_E) = N_V \frac{\Delta H_V}{\alpha_e} = N_V \frac{\Delta H_V}{\alpha}[1 + Bi_H], \qquad (3.34)$$

where α_e is the effective overall heat-transfer coefficient and Bi_H is the heat-transfer Biot number. This equation differs from that for a wet-bulb temperature only in the appearance of the terms involving the Biot number for heat and mass transfer:

$$(T_G - T_E) = \gamma \frac{[1 + Bi_H]}{[1 + Bi_M]} \frac{\Delta H_V}{C_{PY}}[Y_E - Y_G], \qquad (3.35)$$

where $\gamma = C_{PY} \beta/\alpha$ and is almost unity. Since the Biot number function $[1 + Bi_M] / [1 + Bi_H]$ is much greater than 1 (the mass-transfer resistance being greater than the thermal one in response to an impressed potential), the temperature drop between the bulk air and the wet line is much less than wet-bulb depression. In other words, the wet-line temperature T_E is intermediate between the dry- and wet-bulb temperatures. The less permeable the wood, the closer the wet-line temperature will be to that of the bulk of the air.

Under high-temperature drying conditions, with a dry-bulb temperature above 100°C, the wet-line temperature can reach the boiling point at a pressure close to atmospheric. With superheated steam as the drying medium, the temperature will be slightly above 100°C, as demonstrated in the test reported by Perré (1994) for the drying of spruce (*Picea* sp.) at 160°C.

3.8 Mass-Transfer Coefficient Measurements

The vapour-side mass-transfer coefficients for the evaporation of moisture from a wood surface have generally been obtained in two ways. Kho et al. (1989, 1990) have used the rate of sublimation from naphthalene-coated boards as a measure of the vapour-transfer rate through the boundary layer. In the second technique, Choong and Skaar (1969, 1972a,b), for example, use a sorption method, measuring the weight change of a sample following a step change in relative humidity.

If a board is carefully coated with naphthalene, the diminution in thickness of the material over a given period of time under steady-state conditions yields the external mass-transfer coefficient directly, on the assumption that the air is saturated with naphthalene vapour at the surface and there is a negligible amount in the bulk of the air:

$$k = \frac{\rho_N}{M_N} \frac{RT}{p_N^o} \left[\frac{\Delta b}{\Delta t} \right], \tag{3.36}$$

where ρ_N, M_N, and $p^o{}_N$ are respectively the density, molar mass and saturated vapour pressure of naphthalene at the surface temperature and $\Delta b/\Delta t$ is the rate at which the coating dwindles in thickness. The method enables variations in the mass-transfer coefficient over the board's surface to be detected, but requires considerable skill in the preparation of the coating.

With the sorption method, the weight change of a small sample is measured following a step change in relative humidity. The mass-transfer coefficient is deduced from evaluating the time to half of the sorption change, t_h. With two samples, with two different half-thicknesses L, it is possible to calculate both the diffusion coefficient D_C for transfer within the sample and the external transfer coefficient k by using Newman's (1931) solution to diffusion in a body in the presence of an external resistance:

The principal disadvantage of the method lies in the use of small samples (typically 50 × 50 mm) which will have a very different mass-transfer coefficient to those in a kiln at the same cross-flowing velocity over the surface, and the long times (more than 100 h) needed for sorptive equilibrium (Wadsö 1993). Further, the results are also likely to be influenced by possible molecular re-arrangements of the sorbed vapour, which may not be instantaneous, and by swelling of the wood-cell material as moisture is taken up. Christensen and Kelsey (1959a,b), for example, have proposed from considerations of the swelling energy that the time for half-sorption depends upon the magnitude of the initial and final values of the vapour pressure in the step change.

Sometimes moisture-transfer rates to and from a wood surface are reported in terms of *surface-emission coefficients*, based on a concentration driving force (Söderström and Salin 1993). The basis for defining the moisture-content levels is uncertain, and there appears to be no advantage in choosing this description over transfer coefficients that are commonly employed in many fields besides wood science.

4 Wood-Drying Kinetics

In order to predict the rate at which moisture leaves a rough-sawn board, the drying kinetics of the lumber need to be considered. The drying of wood involves the movement of water molecules, either as vapour or liquid (fluids), through a complex biological structure, with changes in phase such as thawing and evaporation. This chapter will contain a qualitative review of moisture movement in lumber, pathways for this movement, and some general considerations in selecting various models for describing this movement. Quantitative features of the different models will be discussed in the following two chapters.

Wood is a highly anisotropic material, permitting the passage of moisture in certain directions more readily than others. Movement along the grain is much faster than that across it. Permeable pines have longitudinal diffusion coefficients approaching that of vapour diffusion in air, while dense hardwoods like teak have coefficients which are ten times smaller. For drying, the most important dimension is across the grain, because the cross-sectional area for water movement is much greater than along the grain, even though longitudinal transport is much faster. The influence of longitudinal movement appears to be confined only to the ends that dry out more rapidly than the majority of the board. For this reason, lumber, especially valuable slow-drying hardwoods, is sometimes end-coated to reduce longitudinal moisture movement (Rasmussen 1961; Wengert and Lamb 1994), and this will further decrease its significance.

4.1 Empirical Observations

The rate of drying for lumber depends on the relative humidity of the air, the air temperature, and the air velocity across the lumber surfaces. The temperature and relative humidity are maintained at higher levels in kiln drying than in the open air, while the air velocity is controlled in kilns to a much greater extent than in air-drying. The *relative humidity* has been discussed in Chapter 3. This affects both the moisture-carrying capacity of the air and its equilibrium moisture content. For example, when considering the drying capacity, 1 kg of dry air with 40 g of moisture would be able to absorb only a further 9.5 g at 40 °C, but at 80 °C, its capacity would be more than an order of magnitude greater (a further 519 g).

The rate of drying is influenced by *temperature* in a number of ways. The rate of moisture transfer to the wood surface increases at higher temperatures and is the principal reason for high-temperature drying (at dry-bulb temperatures above 100 °C). The moisture-carrying capacity of the air also increases at high temperatures, reducing the amount of humid air that needs to be vented. The *airflow* around kilns acts as a heat carrier and as a medium to absorb the evaporated moisture.

Both external evaporation from lumber surfaces and internal moisture movement occur during drying, with external evaporation having been considered in Chapter 3. Internal moisture-content profiles depend on the permeability of the lumber, with both diffusion and mass flow being important. For example, with the sapwood of a species which has a high liquid-phase permeability when the wood is green, moisture movement is dominated by mass flow at first, whereas diffusion alone dominates the drying of impermeable lumbers. Hart (1975) and Walker (1993) have discussed some empirical observations that arise from these considerations, and these will now be summarised.

4.1.1 Permeable Wood

Initially, the moisture content at the surface can be maintained above the fibre-saturation point for some time by liquid flow. During this period, the rate of drying is controlled by the rate of heat transfer from the air to the lumber. An increase in both the air velocity and the wet-bulb depression enhances the heat-transfer and consequential evaporation rate:

- air velocity   (heat transfer and evaporation rate increase with increasing air velocity);
- wet-bulb depression   (heat transfer and evaporation rate are proportional to wet-bulb depression).

In this period, the drying rate is not greatly affected by the wood density (permeability is determined by the pit structure, not by the thickness of the cell walls), nor is it affected by the thickness of the lumber. However, the amount of water is related to the density, with a denser lumber having smaller lumens or vessels and thus a lower initial moisture content. The drying time is proportional to the quantity of moisture to be removed and inversely proportional to the evaporation rate, and so is proportional to:

- wood density, if the quantity of moisture to be removed is defined in terms of the initial moisture content;
- specimen thickness;
- (air velocity)$^{-1}$; and
- (wet-bulb depression)$^{-1}$.

In the later stages of drying, and particularly as the lumber approaches the fibre-saturation point, diffusion becomes more important as a transport mechanism even in the sapwood of permeable species. This is the dominant mechanism for impermeable wood, and the implications will now be discussed.

4.1.2 Impermeable Wood

With many hardwoods and those softwoods for which mass flow is blocked by aspirated pits, diffusion is slow, and the surface moisture content quickly falls below fibre saturation. The drying rate is proportional to:

- (density)$^{-1}$ (more cell-wall material is traversed per unit distance; this offers resistance to diffusion);
- (thickness)$^{-1}$ (during the later stages of drying, the gradient in moisture content is shallower for thicker lumber);
- saturation vapour pressure (this corresponds to a higher equilibrium moisture content).

The amount of moisture to be removed is proportional to the wood density and to the thickness. Hence, the drying time is proportional to both the square of the density and the square of the thickness, as well as the inverse of the saturation vapour pressure of water which rises rapidly with temperature.

However, the drying of any wood is rarely controlled by a single process. Some limiting cases are the sapwood of *Pinus radiata*, for which drying times are proportional to the thickness, on one hand, and the heartwood of *Nothofagus fusca* on the other, for which drying times are proportional to the square of the thickness. Hildebrand (1989) has suggested an empirical approach, in which the drying time is proportional to (density)^n and (thickness)^n, with the exponent n being about 1.5 for many species.

4.1.3 Empirical Models

Given the complexity of wood structure and the natural variability of the raw material, it is not surprising that a number of purely empirical models have been used. Bramhall (1976a,b) has presented an empirical model of the following form:

$$\frac{d\overline{X}}{dt} = \frac{p_{vd} - p_{vw}}{R}, \tag{4.1}$$

where the driving force is the difference between the saturation pressure of water vapour at the dry-bulb temperature (p_{vd}) and that at the wet-bulb temperature (p_{vw}). The resistance to moisture movement is R, found experimentally to be a function of the average moisture content and correlated by Bramhall in one of two forms:

$$R = c_1 \exp(-c_2 \overline{X}) \quad \text{or} \quad R = c_3 \overline{X}^{-c_4}, \tag{4.2}$$

where c_1 to c_4 are fitted constants.

Experiments were carried out on alpine fir (*Abies lasiocarpa*) (1.75" thick, 45 mm), lodgepole pine (*Pinus contorta*) (1.75" thick, 45 mm) and western white spruce (*Picea glauca*) (1.75" thick, 45 mm). The drying conditions and typical values of the constants are given in Table 4.1.

Tschernitz and Simpson (1979a,b) have presented another empirical model, with the characteristic moisture content being correlated by the following equation: –

$$\overline{\Phi} = \exp\left(\frac{-bt}{l^n}\right), \tag{4.3}$$

where t is the time (in days), l is the board thickness (in inches), and b and n are empirical parameters. They performed experiments on 1-inch (25 mm) thick red oak (*Quercus rubra*) boards with an initial moisture content of 85%, considering nine combinations of dry-bulb temperature (120, 150 and 180 °F; 49, 66, 82 °C respectively) and relative humidity (20, 50 and 80%). The air velocity was fixed at 500 ft min^{-1} (2.5 m s^{-1}). They correlated their data with values of b which were a function of the dry-bulb temperature [varying from 0.190 at 120 °F (49 °C) to 0.60 at 180 °F (82 °C)] and a value for n of 1.52.

An empirical model similar to that of Bramhall (1976a,b) was used later by Helmer et al. (1980) as part of a model for a solar kiln:

Table 4.1. Drying conditions and fitted constants for the empirical model of Bramhall (1976a,b)

Species	Range of temperatures (dry bulb/wet bulb) (°F) [°C]	Air velocity (ft min^{-1}) [m s^{-1}]	Expression for R (% mbar^{-1} h^{-1}) [kg kg^{-1} Pa^{-1} s^{-1}]
Alpine fir (*Abies lasiocarpa*)	(140/130) to (185/130) [60/54] to [85/54]	500 [2.5]	4767 exp(−5.9X) [1.324 × 10^{-4} exp(−5.9X)]
Lodgepole pine (*Pinus contorta*)	(155/140) to (200/165) [68/60] to [93/74]	450 [2.3]	1.593 × 10^6 (100X)$^{-2.476}$ [4.942 × 10^{-7} X$^{-2.476}$]
Western white spruce (*Picea glauca*)	(170/160) to (232/205) [77/71] to [111/96]	250 [1.3]	5724 exp(−8.9X) [1.59 × 10^{-4} exp(−8.9X)]

$$\frac{d\overline{X}}{dt} = \frac{Y_w - Y}{R}, \qquad (4.4)$$

where the driving force is the difference between the saturation humidity at the dry-bulb temperature (Y_w) and the bulk-air humidity (Y). The resistance to moisture movement (R) was stated to be a function of the average moisture content and the dry-bulb temperature, as follows:

$$R = \overline{X}^{-2.38}(0.298T_d - 4.18). \qquad (4.5)$$

The overall predictions of the solar-kiln model were reasonable, but the predictions of the drying model were not tested explicitly.

Another empirical model for the drying of boards of angelica (*Dicorynia paraensis*) within a solar kiln has been used by Pallet (1988):

$$\frac{d\overline{X}}{dt} = -(a + b\overline{X} + c\overline{X}^2)(d + el)\exp(f + g/\overline{T})(\overline{X} + X_e) \qquad (4.6)$$

where l is the board thickness (m), X is the average moisture content of the board (kg kg^{-1}), T is the average temperature of the board (°C), and the subscript e refers to the equilibrium moisture content. The seven fitted coefficients were given as

$$a = 5.24; \quad b = 22; \quad c = 31; \quad d = 3; \quad e = -83; \quad f = 0.84; \quad g = -47.45$$

and were fitted to data obtained for temperatures between 20 and 65 °C, board thicknesses between 0.012 and 0.032 m, and moisture contents between 0.2 and 0.6 kg kg^{-1}. Again, the overall predictions of the solar-kiln model were reasonable, but the use of the drying model for extrapolation to other drying conditions is questionable, and the model has no capability for predicting the stress levels in the lumber for the intermittent drying process involved in solar kilns.

Milota and Tschernitz (1990) have developed an empirical model for the drying of loblolly pine (*Pinus taeda*) at high temperatures. Their experiments covered a range of dry-bulb temperatures from 180 to 270 °F (82 to 132 °C), wet-bulb temperatures from 140 to 200 °F (60 to 93 °C), air velocities from 700 to 1900 ft min^{-1} (3.6 to 9.7 m s^{-1}) and board thicknesses from 1 to 2" (25 to 50 mm). The observed drying behaviour was not significantly affected by the presence of juvenile wood or knots. The drying behaviour was divided into two periods: an initial constant-rate drying period and a falling-rate period. During the constant-rate period, they proposed a fitted equation to predict the drying rate (F_{cr}, lb h^{-1} ft^{-2}) as a function of the wet-bulb depression ($T_d - T_w$, °F) and the air velocity (v, ft min^{-1}):

$$F_{cr} = \left[0.01208 + 0.006797(T_d - T_w) - 2.482 \times 10^{-5}(T_d - T_w)^2\right] f_v \qquad (4.7)$$

The parameter f_v reflects the influence of the air velocity (v, ft min^{-1}), as in the following equations:

$$f_v = \left(\frac{v}{1329}\right)^{0.3} \quad \text{for} \quad v < 1329 \text{ ft min}^{-1}(6.8 \text{ m s}^{-1}); \tag{4.8}$$

$$f_v = \left(\frac{v}{1329}\right)^{0.5} \quad \text{for} \quad v > 1329 \text{ ft min}^{-1}(6.8 \text{ m s}^{-1}). \tag{4.9}$$

For the falling-rate period, the drying rate (F_m, lb h^{-1} ft^{-2}) has been fitted to curves of drying rate against moisture content:

$$F_m = \left\{[S_T(M - M_e)]^{-n} + (F_{cr})^{-n}\right\}^{-1/n}, \tag{4.10}$$

where S_T is the slope of the asymptote to the drying rate against moisture content curve through the point at equilibrium moisture content (M_e), and the shape of the drying curve at different external conditions is determined by the parameter n. The air temperature and humidity were found to influence both n and S_T, as follows:

$$n = 0.75 + 0.093(T_d - T_w) \quad \text{for} \quad (T_d - T_w) < 46\,°\text{F}(26\,\text{K}) \tag{4.11}$$

$$n = 5 \quad \text{for} \quad (T_d - T_w) < 46\,°\text{F}(26\,\text{K}) \tag{4.12}$$

where T_d and T_w are the dry and wet-bulb temperatures, respectively;

$$S_T = 150.4 \exp\left(-\frac{7157}{T}\right), \tag{4.13}$$

where T is the dry-bulb temperature, in °R, or $S_T = 150.4 \exp(-3676/T)$ if T is in K. This empirical model is comparatively sophisticated relative to other empirical models and comparable to some mechanistic models, suggesting that the extension of this approach to other species would require extensive experimental work.

Olek et al. (1994) used another empirical model to correlate the drying of 5-mm-thick samples of European beech (*Fagus sylvatica*) when dried at temperatures of 45 and 60°C, with an air velocity of 2.8 m s^{-1}, and various relative humidities:

$$-\frac{d\overline{X}}{dt} = a_1 + \frac{a_2}{(100\overline{X})^n + a_3}, \tag{4.14}$$

where a_1, a_2, a_3 and n are parameters which were fitted to the experimental data. These parameters varied considerably with drying conditions, making the approach of doubtful value for extrapolating observed drying behaviour to other drying conditions. For example, at drying conditions giving an equilibrium moisture content of 12%, the parameters a_1, a_2, a_3 and n were −0.1313, 0.0089, 0.0099 and −1.1265, respectively, while for drying conditions giving an equilibrium moisture content of 4%, these parameters were 0.0379, 0.0137, 0.0072 and −1.3716.

Rosen (1987) quotes approximate kiln-drying times for 25-mm-thick boards of lumber of various species from Rasmussen (1961), who considered

Table 4.2. Typical times to dry 25-mm-thick boards of lumber for different drying strategies involving air-drying and kiln-drying at 65 °C (*After* Rasmussen 1961 and Rosen 1995)

Species	Air-dried to 0.2 kg kg^{-1} & kiln-dried to 0.06 kg kg^{-1} (days)	Kiln-dried to 0.06 kg kg^{-1} (days)
Hardwoods		
Alder, red (*Alnus rubra*)	3–5	6–10
Birch, yellow (*Betula alleghanienis*)	5–8	11–15
Cottonwood (*Populus* sp.)	4–8	8–12
Maple, silver (*Acer saccharinum*)	4–6	7–13
Oak, red (*Quercus rubra*)	5–10	16–28
Oak, white (*Quercus alba*)	6–12	20–30
Sycamore, American (*Acer pseudoplatinus*)	4–7	6–12
Walnut, black (*Juglans nigra*)	5–8	10–16
Softwoods		
Bald cypress (*Cupressus sempervirens*)	4–8	10–20
Redwood, light (*Sequoia sempervirens*)	3–5	10–14
Redwood, heavy (*Sequoia sempervirens*)	5–7	20–24
Western red ceder (*Thuja plicata*)	2–3	6–8
White pine, eastern (*Pinus strobus*)	2–3	4–6

both kiln-drying from the green condition and after an initial air-drying to 20% moisture content. These values are reproduced in Table 4.2.

For drying temperatures under 100 °C, Rosen (1987) also suggested an equation to estimate the drying time t (days) for different board thicknesses l (mm) and temperatures T (°C) based on the kiln-drying times at 65 °C: –

$$t = \frac{1}{\alpha} \ln\left(\frac{X_i}{X_f}\right)\left(\frac{l}{25}\right)^{1.25} \frac{65}{T}, \quad (4.15)$$

where X_i and X_f are the initial and final moisture contents respectively. Rosen notes that the coefficient α depends on the wood species, initial moisture content, humidity, kiln design and other factors. As a guide, he quotes values for α of 0.048 for softwoods and 0.027 for hardwoods, although he notes that the values can vary significantly, for example, with a value of 0.012 being suggested for oak (*Quercus* spp.).

4.1.4 Graphical-Analytical Methods

From 1971 onwards standards for the drying of lumber in kilns have been approved in Russia based on the methods developed at the Moscow State

Forestry University (and described in Chap. 6). Two versions are recommended: one is based on graphical-analytical solutions of the theoretical equations; the second is less universal and is based on the use of tables and nomographs for particular drying conditions specified in the Standard (Shubin 1990).

For example, in low-temperature drying in batch kilns, where the transfer at the wood surface is given by Newton's law, the time of drying becomes:

$$t = \frac{0.233}{D_m}\left[1 + \frac{\pi^2}{4Bi^2}\right]\ln\left[B\frac{(X_S - X_e) + 0.81.(X_0 - X_S)}{\overline{X} - X_e}\right], \tag{4.16}$$

in which D_m is the mean moisture-diffusion coefficient, Bi is the mass-transfer Biot number ($\beta b/D_e$), B is a coefficient, X_S is the surface moisture content when the mean value is $\overline{X}$ at time t, X_e is the corresponding equilibrium value and X_0 is the value in the core at the start of drying. In the case of a three-step drying schedule for beech (*Fagus sylvatica*), the "practical" form of equation (4.16) becomes

$$t = C\left[\ln\frac{0.81(X_0 - X_{e1})}{X_1 - X_{e1}} + \ln\frac{X_{02} - X_{e2}}{X_2 - X_{e2}} + \ln\frac{X_{03} - X_{e3}}{X_3 - X_{e3}}\right], \tag{4.17}$$

where X_i is the mean moisture content at the *i*th stage in the drying schedule for which X_{0i} is the core value at the start of that step and X_{ei} is the corresponding equilibrium moisture content for the specified relative humidity. The coefficient C depends upon the board thickness, the width of the stack and the kiln-air speed, with nomographs being given to determine the influence of the latter two parameters.

Although the use of Eq. (4.17) is claimed to be "relatively accurate", as a further simplification, Shubin (1990) derives a simple logarithmic expression based on the initial and final moisture-content values, with a lumped empirical coefficient determined from the board dimensions, a diffusion coefficient (estimated at the wet-bulb temperature) and a factor to account for the possible use of reversals in direction of the circulation in the kiln.

These procedures represent an enormous investment in the correlation of practical kiln data. However, their application to other species and kiln designs is uncertain.

The empirical observations surveyed in this section can be used to develop drying models for different species, as outlined in Chapters 5 and 6, which are devoted to diffusion and multiple-mechanism models, respectively.

4.2 Normalisation of Drying-Rate Curves

In analysing the behaviour of convective batch drying, van Meel (1958) has suggested that a single characteristic drying curve can be drawn for the material being dried. At each volume-averaged, free moisture content, it is assumed that there is a corresponding specific drying rate relative to the unhindered drying rate in the first drying period that is independent of the external drying conditions. Such a characteristic curve is illustrated in Fig. 4.1. The relative drying rate is defined as

$$f = \frac{N_v}{\hat{N}_v}, \qquad (4.18)$$

where N_v is the drying rate and $\hat{N}_v$ is the rate in the first period of drying limited by the external convection. The characteristic moisture content becomes

$$\Phi = \frac{\overline{X} - X_e}{X_{cr} - X_e}, \qquad (4.19)$$

where $\overline{X}$ is the volume-averaged moisture content, X_{cr} is the moisture content at the critical point, and X_e that at equilibrium. Thus, the drying curve is normalised to pass through the point (1,1) at the critical point of transition in drying behaviour and the point (0,0) at equilibrium.

This representation is attractive because it leads to a simple lumped-parameter expression for the drying rate, namely,

$$N_v = f(\hat{N}_v) = f[\beta \phi_m (Y_w - Y_G)]. \qquad (4.20)$$

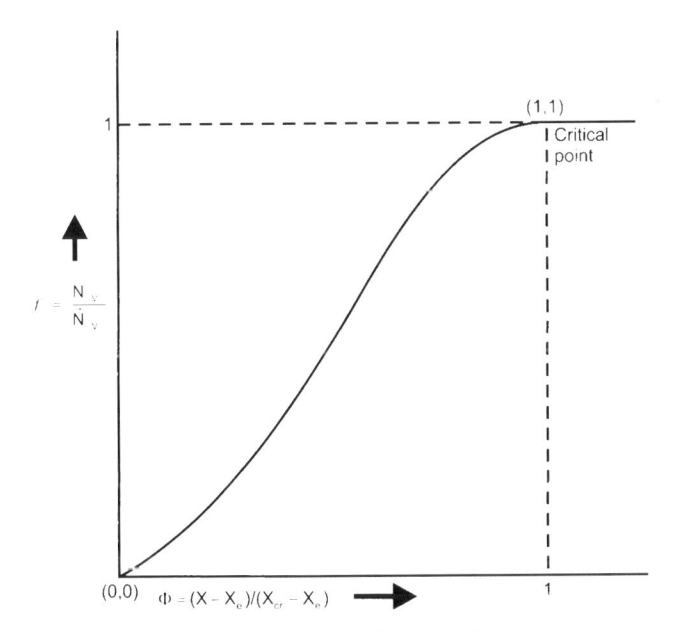

Fig. 4.1. Characteristic drying curve. (Keey 1992)

Here β is the external mass-transfer coefficient, ϕ_m is the humidity-potential coefficient, Y_w is the humidity above a fully-wetted surface and Y_G is the bulk-gas humidity. Equation (4.18) has been used extensively as the basis for understanding the behaviour of industrial drying plants owing to its simplicity and the separation of the parameters that influence the drying process: the material itself f, the design of the dryer β, and the process conditions $\phi_m (Y_w - Y_G)$.

The characteristic drying curve, however, is clearly a gross approximation. A common drying curve will be found only if the volume-averaged moisture content reflects the moistness of the surface in some fixed way. For the drying of impermeable species, for which the surface moisture content reaches equilibrium quickly, there is unlikely to be any significant connection between the volume-averaged and the surface moisture contents, so the concept is unlikely to apply under these conditions. Even in the drying of sapwood of permeable species, the surface moisture content does not relate to the average moisture content in any simple way, as noted in Chapter 6. However, despite these doubts, the concept has proved useful in exploring the complex influence of process variables on kiln-drying.

Keey and Suzuki (1974) have explored the conditions for which a characteristic curve might apply, using a simplified analysis based on an evaporative front receding through a porous mass of non-hygroscopic material such as a bed of sand. Their analysis shows that a unique curve pertains only when the material is spread into a thin layer and the permeability of the bed to moisture movement is large. One might expect, then, to find characteristic drying curves for small, microporous particles dried individually, or a thin layer composed of such particles, and there is a sufficient body of data to suggest that a characteristic drying curve may be found to describe the drying of discrete particles below 20 mm in diameter over a range of conditions that normally exist within a commercial dryer (Keey 1992). Nevertheless, Ashworth (1977) has shown that a characteristic drying curve could be used to describe kiln-drying behaviour of softwood boards provided the curve was based on small-scale drying tests with a sample of the same thickness. More recently, Nijdam (1998) has come to essentially the same conclusion from a detailed analysis of the mass-transfer processes in and about a softwood board.

The shape of the characteristic drying curve reflects the permeability of the wood. The curve is concave-downwards when the wood is very permeable, as exhibited in the drying of the sapwood of a softwood species: whereas it is concave-upwards when moisture movement through the wood is slow, as in the drying of hardwoods with a high extractives content. This difference in behaviour is shown in Fig. 4.2.

The concept of a characteristic drying curve has been applied by Rosen (1978, 1980) for the drying of lumber boards, by Kayihan (1993) for the analysis of variability in kiln drying, by Nijdam and Keey (1996) for investigating the impact of airflow maldistribution on end-moisture variability, and by Keey and Pang (1994) and Pang (1996a) for the kiln-wide analysis of flow reversals in drying softwood.

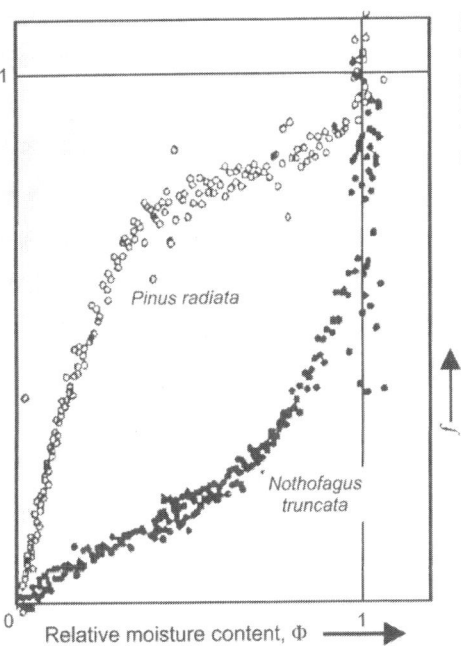

Fig. 4.2. Comparative drying rates for samples of *Pinus radiata* and *Nothofagus truncata*, 220 × 100 × 5 mm thick, at a dry-bulb temperature of 45 °C and a wet-bulb temperature of 39 °C, with air velocities in the range 2.2 to 3.0 m s^{-1}. (Keey and Walker 1988)

Rosen (1978) has developed an empirical model to characterise drying under constant external conditions, expressed as an equation for the characteristic moisture content:

$$\Phi = 1 - j_{wo} \int_0^\tau (-a\tau^{1/b}) d\tau, \qquad (4.21)$$

in which a and b are coefficients to be determined experimentally, and j_{wo} is the initial drying rate, which is related to the coefficients a and b by the equation:

$$j_{wo} = \frac{a^b}{b\Gamma(b)}, \qquad (4.22)$$

where Γ is the delta function. The model was fitted to experimental data on the impingement drying of 27-mm and 54-mm-thick boards of silver maple (*Acer saccharinum*). Dry-bulb temperatures between 107 and 204 °C and air velocities (through the nozzles) between 5 and 45 m s^{-1} were used, with a constant wet-bulb temperature of 82 °C. The values of a and b depended on dry-bulb temperature, air velocity and board thickness, with the values of a ranging from 0.11 to 1.03 and the values of b varying from 0.43 to 1.61.

This model was also fitted to experimental data by Rosen (1980) for the kiln-drying of 50-mm and 100-mm-thick southern pine at a dry-bulb temperature of 116 °C, a wet-bulb temperature of 71 °C and an air velocity of 5 m s^{-1}; the coefficients a and b were 3.79 and 1.03 respectively for 50-mm-thick lumber; and 2.63 and 1.49 respectively for 100-mm-thick lumber. The difference between these coefficients at different board thicknesses suggests that different characteristic curves arise for different thicknesses of lumber, in agreement with Ashworth's earlier finding.

Rosen (1987) notes that his empirical model may be plotted as a curve of the normalised drying rate (f) against the characteristic moisture content, ($\overline{\Phi}$), as follows: –

$$\overline{\Phi} = 1 - \frac{(-\ln f)}{b\Gamma(b)} \sum_{n=0}^{\infty} \frac{(-1)^n (-\ln f)^n}{(n/b+1)n!}. \tag{4.23}$$

Here f is the ratio of the instantaneous drying rate to the initial value. Simplified forms of this equation for limiting cases are

$$\overline{\Phi} = 1 - \frac{(-\ln f)^b}{b\Gamma(b)} \left(1 - \frac{b(-\ln f)}{b+1}\right) \quad \text{for} \quad -\ln f < 0.5 \text{ (short times);} \tag{4.24}$$

$$\overline{\Phi} = \frac{(-\ln f)^{b-2} f(-\ln f + b - 1)}{\Gamma(b)} \quad \text{for} \quad -\ln f > 2 \text{ (long times).} \tag{4.25}$$

The shapes of the fitted curves of drying rate against moisture content are dependent on the parameter b. Drying-rate curves from experiments on silver maple (*Acer saccharinum*) lumber dried over a range of conditions (Rosen 1978) show a wide range of shapes (Fig. 4.3), a feature which indicates that the concept of a characteristic drying curve is not appropriate for this species, since the normalised drying rate curves should have the same shape.

Kayihan (1993) describes an empirical model which incorporates the influences of the external conditions (through a constant-rate drying parameter a) and the internal moisture-transport process (through a diffusion parameter b) to predict the rate of change of the average moisture content X as follows: –

$$-\frac{d\overline{X}}{dt} = \frac{b\overline{X}}{\left[1 + (b\overline{X}/a)^3\right]^{1/3}} \tag{4.26}$$

This expression has a form similar to that suggested by Filoneko and Lebedev (1960) for a number of hygroscopic materials and that derived by Hallström and Wimmerstadt (1983) for the drying of fertiliser granules, who obtained

$$f = \frac{j_v}{j_{vmax}} = \frac{(\overline{X} - X_e)^m}{A + B(\overline{X} - X_e)^m}, \tag{4.27}$$

where A, B and m are constants.

In Eq. (4.26), the activation energy dependence of the drying behaviour on temperature was incorporated in the parameter b, which was correlated by the equation:

$$b = b_o \exp\left(-\frac{2700}{T}\right) \tag{4.28}$$

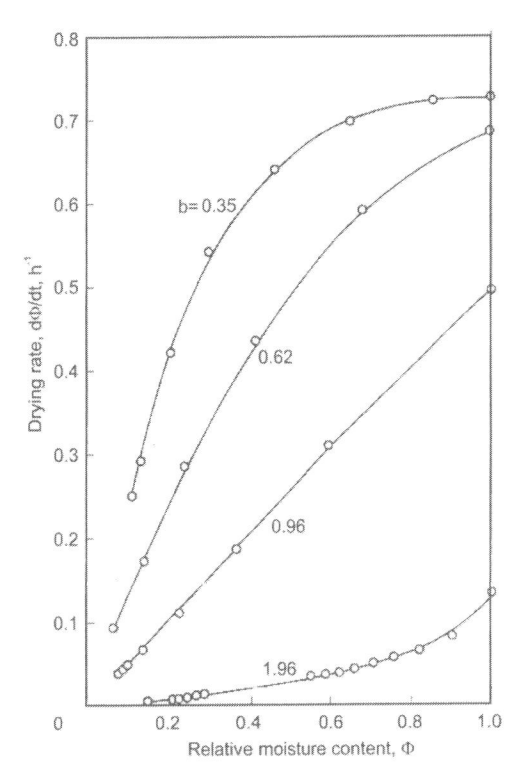

Fig. 4.3. Drying-rate curves for the impingement drying of silver maple (*Acer saccharinum*). (After Rosen 1978)

with T in K.

Keey and Pang (1994) described the drying kinetics of a softwood, radiata pine (*Pinus radiata*), using a characteristic drying curve. They started with a rigorous, physiologically based model which had been developed for the high-temperature drying of a single softwood board. A characteristic drying curve was fitted for drying at temperatures ranging from 70 to 200 °C and air velocities between 5 and 9 m s^{-1}. Analysis of the mechanisms of moisture movement identified three stages in the drying of a sapwood board and two for a heartwood board. The drying kinetics were represented by a dual characteristic curve covering the two falling-rate periods, to yield simplified expressions for the drying kinetics. The constant-rate period, for sapwood boards only, applied from the initial moisture content (X_o) down to the first critical point (X_1), for which the drying rate over unit exposed surface was by:

$$j_{w1} = \varphi K_o (Y_S - Y_g^o). \tag{4.29}$$

Here $(Y_S - Y_g^o)$ is the humidity potential, K_o is a mass-transfer coefficient and φ is the humidity-potential coefficient which is a function of the wet-bulb temperature T_w, given in Table 4.3.

Table 4.3. Characteristic functions for the drying of *Pinus radiata* (Keey and Pang 1994)

Heartwood	Sapwood
$f_1 = \dfrac{4.53(1-\Phi_1)^{0.34}+1.12}{1+45.3\left[1-0.22(1-\Phi_1)^{0.33}\right](1-\Phi_1)}-0.12$	$f_1 = \dfrac{2.11(1-\Phi_1)^{0.84}+1.22}{1+3.80(1-\Phi_1)\exp[0.98(1-\Phi_1)]}-0.22$
$f_2 + -\Phi_2^2 + 2.0\Phi_2$	f_2 as for heartwood
X_1 = initial moisture content	$X_1 = 0.81 - 1.02\,\text{kg kg}^{-1}$ depending on drying schedule
$X_2 = 0.07 - 0.12\,\text{kg kg}^{-1}$ depending on drying schedule	X_2 as for heartwood
$\varphi = 0.413 + 0.0014(T_w - 343)$ $- 0.00047(T_w - 343)^2$	$\varphi = 0.581 + 0.0056(T_w - 343)$ $- 0.00043(T_w - 343)^2$
$j_{w2} = 0.11\,j_{w1}$	$j_{w2} = 0.181\,j_{w1}$

The first falling-rate period applies from moisture contents and drying rates of X_1 and j_{w1} respectively to X_2 (the second critical point) and j_{w2} respectively. The normalised moisture content (Φ_1) and relative drying rate (f_1) are defined in terms of the absolute moisture content X and the actual drying rate j_w by the expressions:

$$\Phi_1 = \frac{X - X_2}{X_1 - X_2}, \quad X_2 < X < X_1; \qquad (4.30)$$

$$f_1 = \frac{j_w - j_{w2}}{j_{w1} - j_{w2}}, \quad j_{w2} < j_w < j_{w1}. \qquad (4.31)$$

The second falling-rate period applies down to the stage when the lumber has reached the equilibrium moisture content and drying has ceased. The normalised moisture content (Φ_2) and relative drying rate (f_2) are defined as:

$$\Phi_2 = \frac{X - X_e}{X_2 - X_e}, \quad X_2 > X > X_e, \qquad (4.32)$$

$$f = \frac{j_w}{j_{w2}}, \quad j_{w2} > j_w > 0. \qquad (4.33)$$

The parameters X_1, X_2 and j_{w2} are given in Table 4.3.

The direct experimental evidence, in terms of normalised drying rate curves, to support the general use of a characteristic drying curve for lumber drying is limited. The data of Rosen (1978) for the impingement drying of silver maple (*Acer saccharinum*) show a wide range of shapes for the normalised drying curves, suggesting that a single set of functions to describe these curves is not appropriate. For the normal cross-flow drying of 27-mm-thick boards of white fir (*Abies alba*) at temperatures of 152–184 °C and air velocities of 6.2–18.6 m s^{-1}, Basilico et al. (1982) show drying curves which are monotonically concave downwards at high temperatures and air velocities, and increasingly sigmoidal at lower air temperatures and velocities. No explanation for these changes was given, but the results may reflect the changing shape of the moisture-content profiles with temperature (and wet-bulb depression).

For *Pinus radiata*, Pang (1996a) has suggested that a characteristic drying curve describes the drying kinetics well for a range of dry-bulb temperatures from 70 °C to 120 °C and air velocities from 6.9 to 9.6 m s^{-1}. The range of wet-bulb depressions studied was not reported, and this point is important because the air which flows through a stack of lumber becomes increasingly humid in the streamwise direction, so that the wet-bulb depression decreases. When the wet-bulb depression is reduced, the moisture-content gradients inside the lumber are likely to be different in shape to those at higher wet-bulb depressions, because the evaporation rate from the surface will be less and it will be easier for capillary action to keep the surface supplied with moisture. The different moisture-content gradients would lead to different shapes of the normalised drying-rate curves.

In spite of this uncertainty, Pang (1996a) showed that the drying-rate predictions of a model based on the use of this concept to describe the behaviour of an entire stack of lumber (through which the air becomes increasingly humid) agreed well with measurements for average moisture contents over a stack. Hence the use of a characteristic curve in this context appears to be reasonably appropriate. The spread of relative drying rates found by Basilico et al. (1982) also appears to be limited to a range of 30% of the mean relative drying rate at any particular characteristic moisture content, so the spread is not as large as seen in the data of Rosen (1978). The results of Ashworth (1977) also show that the extensiveness of the stack (of order one transfer unit wide in a lumber kiln) reduces the predicted sensitivity of the overall drying behaviour to the use of such different characteristic drying curves as $f = \Phi$ and $f = \sqrt{\Phi}$. This insensitivity suggests that variations in the shapes of the normalised drying-rate curves may not have too much impact on the overall drying behaviour of the kiln.

4.3 Pathways for Moisture Movement in Wood

In principle, the same pathways are available for the diffusion of water molecules as are available for the flow of water down a pressure gradient, but their relative contributions are very different. The reason is that the rate of diffusion along a capillary of radius r is proportional to its cross-sectional area, πr^2, whereas the rate of flow is proportional to πr^4 from the Hagen-Poiseuille equation for laminar flow in tubes, so the small capillaries in a distribution of capillaries will make a proportionally greater contribution to diffusional processes than to permeability. The rate of diffusion along 100 tubes of radius r

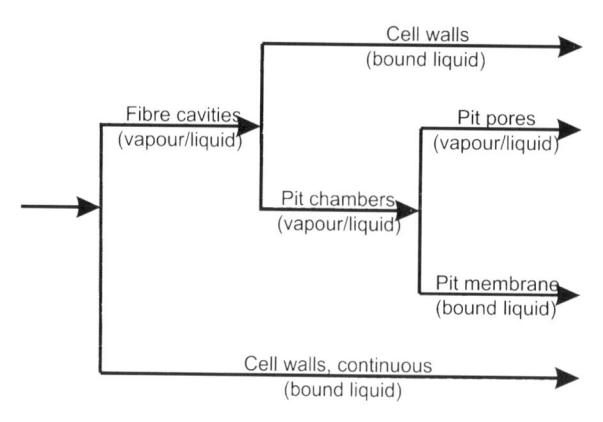

Fig. 4.4. Various flow paths through softwoods. (After Stamm 1967a)

($A = 100\,\pi r^2$) will be identical to that along a single tube of radius $10r$. However, the permeation rate of the array of small capillaries will be proportional to $100r^4$, which is 10^{-2} of the permeability of the large capillary, since the permeability of the latter will be proportional to $(10r)^4$ or $10\,000r^4$. Consequently, in diffusion, the movement of water through the swollen cell wall is of far greater significance than the diffusion of water across the pits, because the total area occupied by the openings in the pit membranes is small compared with the area available for diffusion through the cell wall itself. This feature applies even more to the permeability of the lumber to the flow of liquid moisture.

Possible pathways for the movement of fluid through lumber are shown in Fig. 4.4 (Stamm 1967a). One is the cavity-wall pathway, which is a series combination of bound-water diffusion through the cell wall and vapour diffusion across the cell cavity. A second possibility is the "cavity-pits" pathway, in which the diffusion occurs through the pit system in series with the fibre cavities. The third is the continuous-wall pathway where the diffusion takes place through the continuous portion of the cell wall around the fibre cavity. The movement of free moisture by capillary action effectively ceases once the pits become aspirated. The actual movement of fluid involves all three paths to varying degrees. For example, in high-density species the third pathway becomes relatively more significant. Similarly, the first pathway becomes less significant as pits aspirate.

4.3.1 Longitudinal Movement

Since in softwoods the ratio of tracheid length to tracheid diameter is relatively high, typically about 140 (Stamm 1967a), there are comparatively few cross-walls to offer resistance to longitudinal movement.

Whenever the cells' walls contain enough moisture, at moisture contents more than $0.2\,kg\,kg^{-1}$, the cross-walls offer little resistance to diffusion since the water molecules can diffuse through them without too much difficulty ($D_{lumen}/D_{cell\ wall} \rightarrow 10$). The thickness of cell wall traversed per unit distance is small: the cell wall thickness is about $3.5\,\mu m$, compared with a tracheid length of about $3.5\,mm$. Diffusion is determined primarily by the diffusion along the lumen. Only at low moisture contents (below $0.2\,kg\,kg^{-1}$) does the resistance of the cross walls become significant.

In hardwoods, longitudinal movement of fluids occurs through the vessels, as described in Chapter 1. Their diameter is generally between 20 and $300\,\mu m$ (Walker 1993). The large perforations at each end of the vessels offer comparatively little resistance to longitudinal diffusion relative to the resistance to transverse diffusion through the walls between vessels.

4.3.2 Transverse Movement

When the wood is above the fibre-saturation point, the diffusion coefficient of water vapour in the lumens is about ten times greater than that of adsorbed water in the cell wall (Walker 1993). However, when the moisture content is below the fibre-saturation point, the difference is least when the moisture content of the cell wall is close to the fibre-saturation point and greatest when the moisture content becomes very small as the wood approaches the oven-dry condition. Thus, with dry wood (5% moisture content), the diffusion coefficient of water vapour in the lumen is about 1000 times greater than that of water in the cell wall.

Since the cell walls offer considerably more resistance to diffusion than the lumens do, the transverse diffusion is essentially determined by the diffusion coefficient of moisture through the cell walls and by the thickness of cell wall traversed per unit distance, which is reflected in the basic density of the wood. The presence of pits and the condition of their pit membranes do not influence diffusion very much. Pits only become important at very low moisture contents and for species having a high basic density, for which the diffusion coefficient through the cell wall is very small and where the cell walls are thick.

Stamm (1967b) has calculated the theoretical contribution to diffusion by the pathways shown in Fig. 4.4 for wood of different densities and at different temperatures (Table 4.4).

The pathway involving water vapour diffusion through the fibre cavities in series with bound water diffusion through the cell wall (path 1) accounts for most of the diffusion, although this pathway becomes less important as the basic density increases. Pits (path 2) play a comparatively insignificant role: if no diffusion occurred through the pits, the transverse diffusion coefficient would be reduced by only about 10%, and if the pit membranes were removed

Table 4.4. The proportion of the diffusion that occurs through each of the possible pathways illustrated in Fig. 4.4 for fluid movement in the transverse direction at different temperatures (After Stamm 1967b)

Temperature (°C)	Basic density (kg m^{-3})	Proportion of diffusion through		
		Path 1 Fibre cavity and cell wall in combination (%)	Path 2 Fibre cavity and pits in combination (%)	Path 3 Cell walls only (%)
50	200	95	4.2	0.8
	400	86	10.1	3.9
	800	55	22.2	22.8
120	200	94	5.3	0.7
	400	85	11.7	3.3
	800	57	24.8	18.2

entirely, the diffusion coefficient would increase only threefold (Stamm 1967b). The state of the pits is a minor factor in diffusion, and impermeable heartwood will dry from the fibre-saturation point to the equilibrium moisture content in virtually the same time as highly-permeable sapwood, providing that they are of the same basic density.

The situation is different when drying lumber from above the fibre-saturation point, since mass flow of moisture through the pits is possible in a permeable wood whereas an impermeable one must still rely entirely on diffusion. Under these conditions, a permeable wood loses moisture much faster early in the schedule than an impermeable species.

4.4 Selection of Drying Models

Rosen (1987) has presented a very useful state-of-the art review of drying models for wood, dividing the models into approaches based on diffusion, empiricism and transport mechanisms. In a sense, the diffusion approach is based on assuming that a single transport mechanism, diffusion, is acting during drying, but this view is a simple one and deserves treatment separate from the approaches incorporating multiple transport mechanisms. More accurately, this approach may be described as a Fickian one, since it assumes that moisture migrates as described by Fick's Second Law, regardless of the actual moisture transport mechanism:

$$\frac{\partial X}{\partial t} = \frac{\partial}{\partial z}\left(D\frac{\partial(\text{driving force})}{\partial z}\right). \tag{4.34}$$

Nevertheless, most authors refer to this approach as a diffusion model, and this terminology will be continued here. The driving forces used by various authors have included moisture content, Kirchhoff's flow potential, vapour pressure and chemical potential, and the basis for the use of these driving forces for diffusion will be explored in the discussion of the diffusion model in Chapter 5. Transport mechanisms which have been included in the more sophisticated models include the capillary-driven flow of free liquid, bound-water diffusion, and the permeation of water vapour. These are considered in Chapter 6.

The question as to which approach to wood-drying kinetics is appropriate depends on both the desired outcome and the required degree of fundamental physical understanding. If a correlation of experimental data for drying time is all that is necessary, then an empirical approach may be satisfactory. Prediction of stress levels due to differential shrinkage, however, require moisture-content gradient information that is rarely obtained from empirical approaches, and this step requires the use of either a diffusion model or a "multiple-mechanism" model. The choice then depends on how well the model fits moisture-gradient data. The need to fit temperature-gradient data is less important, since the temperature gradients in lumber dried at less than 100°C are usually very small compared with the moisture-content gradients, so that the lumber is effectively isothermal at any time in the drying process. Even at higher drying temperatures, when a receding evaporative plane may occur, the stresses induced by thermal expansion are usually less than 10% of those induced by shrinkage due to moisture-content changes. The multiple-mechanism approaches also contribute useful information to the physical appreciation of the problem, and hence offer the most promise for analysing more radical approaches to changing the way in which lumber is dried. For example, it is difficult to see how the analysis of microwave and radio-frequency drying of lumber could be done using an empirical model.

Most empirical models also give no information on moisture-content gradients, information which is required for stress analysis. The highly complex mathematics of the transport-property or multiple-mechanism approaches can make the interpretation of some of the results difficult, but can also provide useful information for simplified approaches while still retaining a basis in physical principles of drying theory. Despite the advances in computer power over recent years, the conclusion regarding the need for relatively expensive computing power when using multiple-mechanism models is still of some validity today. Kayihan (1993) remarked that:

"Fundamental drying models developed specifically for wood describing it as a hygroscopic porous media (e.g. Stanish et al. 1986) may be too complicated and computationally too slow for the purposes of process analysis".

Kamke and Vanek (1994) have compared the performance of a number of lumber-drying models, representing mainly diffusion-like and multiple-transport mechanism approaches, for predicting average moisture contents and moisture-content profiles. Four data sets were used, with three sets representing idealised problems. The fourth data set was the experimental results of drying 40-mm-thick boards of Norway spruce (*Picea abies*) from initial moisture contents of 29–66% at a dry-bulb temperature of 60 °C, wet-bulb depressions of 8–25 °C, and an air velocity of 6 m s^{-1}. The required inputs for the models, including physical properties of the lumber and initial and boundary conditions, were supplied to the authors of the models, most of whom performed the simulations. Requests for computer simulations were submitted to 31 authors in 16 countries, and simulation results were obtained for 12 models.

There was significant variability between the predictions of most models. Many models did not predict the data well, and there was no single "best" model. Uncertain coefficients for the models, different degrees of simplifications and different methods for solving the heat and mass-transfer problems were given as possible reasons for the variations between the predictions of the models and for the discrepancies compared with the measurements. Kamke and Vanek (1994) concluded that:

"*Prediction results from drying models are only as good as the input data supplied. The more 'sophisticated' models do not perform any better than the simple models if the physical property data is inadequate. . . . A simple model will work quite well if good physical property data is available for the species and the drying conditions are within the range for which the model was developed. For research purposes where detailed heat and mass transfer information is required (such as predicting stress and strain behaviour) the models that separate the transport mechanisms may prove more useful.*"

Nevertheless, simple diffusion models have been found to predict moisture-content gradients adequately in some cases, as Wu (1989) reports for Tasmanian eucalypts for example, and the successful prediction of these gradients is the main requirement for stress modelling. Implicit in the comments of Kamke and Vanek (1994) is the need to balance the time and expense of obtaining physical property data for complex multiple-mechanism models with the required outcomes from the simulation. They also felt that computing power was not a significant restriction on simulations.

To a significant extent, the question of whether computing power is a limitation depends on the application, and there is some truth in the statements of both Kamke and Vanek (1994) and those of Kayihan (1993). Computing power does not significantly restrict simulations for estimating the drying time (Kamke and Vanek 1994), but the use of simulations within an optimisation scheme, where the simulation may be called upon several times, still restricts the use of the sophisticated models (Kayihan 1993).

5 Moisture Diffusion

Diffusion is a molecular process, brought about by random wanderings of individual molecules. If all the water molecules in a material are free to migrate, they tend to diffuse from a region of high moisture concentration to one of lower moisture concentration, thereby reducing the moisture gradient and equalising the concentration of moisture. In wood, the binding of moisture to the cell-wall matrix depends on moisture content, becoming stronger as the moisture content falls below fibre saturation, so all molecules are not equally free to migrate.

It is useful to discuss a diffusion model for moisture movement in wood first for a number of reasons. This was the first approach taken to modelling the process of wood drying using a transport mechanism (Sherwood 1929), and it is still in widespread use today (Simpson 1993; Bramhall 1995). One of its attractions is its relative simplicity compared with more complex models for moisture movement in lumber. This renders the approach tractable for relatively simple calculations, which are often appropriate given the wide variability in diffusion coefficients and permeabilities both within and between species. The simplicity of this approach is also useful when optimising kiln-drying schedules, where the number and length of calculations, even with modern workstations, are considerable. Moreover, this diffusion approach works well for predicting both average moisture contents and moisture-content profiles for some hardwood species, a feature which is essential when predicting stress levels in lumber. A number of authors (including Stamm 1964 and Pang 1996b) have suggested that the success of a simple diffusion model for describing the drying of hardwoods stems from the impermeability of such species to liquid moisture movement, implying that the moisture-transport process in these lumbers is comparatively simple relative to that in softwoods.

The three main driving forces that have been used within diffusion models to describe the drying of wood (moisture content, partial pressure of water vapour and chemical potential) will now be discussed. Attempts to predict diffusion coefficients theoretically will also be reviewed, together with experimental data for fitted diffusion coefficients and their dependence on temperature and moisture content.

5.1 Driving Forces for Diffusion

Waananen et al. (1993), in their review of drying models, note that most models in their final form express the driving force for moisture movement in terms of a moisture-concentration gradient. However, the true potential for transfer may be different.

For clarity in the subsequent discussion, moisture movement in only one direction is considered, but this geometrical simplification has no bearing on the arguments regarding the transfer potentials.

At equilibrium, with layerwise adsorption of moisture (and low moisture contents compared with green lumber), the chemical potential of the adsorbed moisture is the same as that of the host material on which it is attached. Any departure from equilibrium will thus result in a difference in chemical potential between the interface and the "bulk" of the adsorbate. The flux of moisture attempting to restore the equilibrium state is assumed to be proportional to this difference in potential, and inversely proportional to the path length over which the potential difference acts. Thus, one may write for the moisture-transfer flux j

$$j = -Bc\frac{\partial \mu}{\partial z}, \tag{5.1}$$

where B is a coefficient, c is a concentration, μ is the chemical potential and z is a dimension in the direction of transfer. On introducing the activity, a, one may get an alternative expression involving the moisture-concentration gradient

$$j = -\left(BRT\frac{\partial \ln a}{\partial \ln c}\right)\frac{\partial c}{\partial z}. \tag{5.2}$$

Here, the term in the brackets is often called the *moisture diffusivity*. Equation (5.2) may be put into an alternative form involving the moisture content X, since $c = \rho_w X$:

$$j = -(D)\frac{\partial(\rho_w X)}{\partial z}, \tag{5.3}$$

where ρ_w is the oven-dry density of the wood. The moisture diffusivity in Eq. (5.3) should only be independent of moisture concentration if the moisture is unbound, when $\partial \ln a/\partial \ln c$ is unity. However, concentration-independent diffusion coefficients have been used successfully in some cases over a wide range of moisture contents, including moisture contents below the fibre-saturation point when moisture binding to the cell-wall structure might be expected to be significant. These cases will be reviewed in a subsequent section on the effect of moisture content on diffusion coefficients.

There is another feature of hygroscopic mass transfer. Since the true driving force is the chemical potential difference, transfer will occur between two moist bodies in the direction of falling chemical potential rather than decreasing moisture content. Moisture may flow from a drier body to a wetter.

It is difficult to see how diffusion is likely to occur at high moisture contents in wood, even though experimental evidence will be reviewed later in this chapter which indicates that it is a reasonable empirical description of moisture-content profiles in some impermeable hardwood species. For permeable species, free-water movement, driven by gradients in capillary pressure, is likely to be more important.

Whitaker (1977, 1980, 1984) has derived volume-averaged relationships for transport in a three-phase medium, such as a porous body, from the laws of continuum physics. Local thermodynamic equilibrium is usually assumed. This analysis has been the most recent major attempt to devise a sound theoretical foundation for drying theory. This approach has been applied to many materials, including wood. For liquid permeation when capillary forces are dominant in a three-phase, isothermal system of inert solids, moisture liquid and moisture vapour, the constitutive equation of Chen and Whitaker (1986) becomes:

$$\frac{\partial s}{\partial t} = \nabla(K_s \nabla s), \tag{5.4}$$

where s is the fractional saturation of the pores, and K_s is the moisture permeability, given by

$$K_s = D_m \frac{\partial X}{\partial s}. \tag{5.5}$$

Here, D_m is a conductivity. The moisture permeability will not be invariant with moisture content in the hygroscopic-moisture regime when $\partial X/\partial s$ is not equal to unity even if the conductivity is (which it is not). Nevertheless, diffusion-like expressions arise when liquid diffusion is the dominant moisture-transport mechanism and capillary forces are dominant. Wiley and Choong (1975) successfully used this approach when considering the drying of softwood lumber at high moisture contents (greater than 0.6 kg kg^{-1}). A diffusion-like expression also arises when considering gradients in partial pressure:

$$j_v = (D) \frac{\rho_w}{\rho_o} \frac{\partial X}{\partial \varphi} \frac{\partial p}{\partial z}, \tag{5.6}$$

where the relative humidity φ is the ratio of the vapour pressure p to the total pressure p_o, and $\partial \varphi/\partial X$ is the slope of the moisture isotherm. As with the flow of free water, the apparent diffusion coefficient should not be constant in the hygroscopic regime where the slope of the moisture isotherm will vary significantly with moisture content.

Another interesting application of volume-averaging techniques was undertaken by Whitaker and coworkers (Crapiste et al. 1988), who investigated the drying of cellular plant material which shrinks on drying. Although they were primarily interested in the drying of fruits and vegetables, similar considerations should apply to the cellular structure of wood which also undergoes shrinkage. The final form of their expression for moisture movement is diffusion-like (with a moisture-content driving force), suggesting that the rigorous model might be used as the basis of a simplified theory which could then be used for further process analysis. In this respect, a diffusion-like model is simpler and easier to apply than the full rigorous model, which incorporates different transport mechanisms for the movement of liquid water, water vapour and bound water.

Skaar (1988) has discussed a number of driving forces for moisture movement in wood, including osmotic pressure and "spreading pressure" as well as chemical potential, moisture content and vapour pressure. The chemical potential has also been used as a driving force within a diffusion-like model by Cloutier and Fortin (1993), who point out that this potential can incorporate other effects such as differences in vapour pressure and osmotic pressure. However, at low moisture contents, Perré and Turner (1996) suggest that there seems to be little difference between the predictions of drying models, whether driving forces are based on gradients in chemical potential, moisture content or partial pressure of water vapour, indicating that the simplest approach (a moisture-content driving force) might be most practical.

The majority of work involving the use of diffusion models has used moisture-content driving forces. Some successful applications (mainly impermeable hardwoods) have covered a wide range of moisture contents, for which diffusion (in the strict sense) is unlikely to occur at all moisture contents. Hence, there is some empirical support for the use of moisture-content driving forces. At low moisture contents, true diffusion is probably best described by chemical potential driving forces.

We will distinguish between "theoretical" and "measured" diffusion coefficients, on the basis that some workers have suggested that the diffusion coefficients may be predicted from first principles.

5.1.1 The Kirchhoff Transformation

Fick's first law (1855) of diffusion yields

$$j = -D(c)\nabla c, \tag{5.7}$$

where c is the moisture concentration (the product of the oven-dry density of the wood ρ_w and the dry-basis moisture content X). A new variable, ψ, may be defined by the expression

$$\psi(c) = \psi_o + \int_{c_o}^{c} D(c)dc, \tag{5.8}$$

which is the so-called *Kirchhoff transformation* (Carslaw and Jaegar 1986). It follows that

$$\frac{d\psi}{dc} = D(c) \tag{5.9}$$

and

$$j = -\left[\frac{d\psi}{dc}\right]\cdot \nabla c = -\nabla\psi \tag{5.10}$$

and thus the moisture-transfer flux j is given directly by the gradient of the potential ψ.

This formulation leads to some mathematical simplifications. For example, for one-dimensional, stready-state flow through a slab of thickness L, one has

$$\psi_L - \psi_0 = -j \cdot L. \tag{5.11}$$

The difference in potential across the slab is simply the product of the flux and the thickness.

However, not all difficulties are resolved by this transformation. Fick's second law becomes

$$C_\psi(\psi)\frac{\partial \psi}{\partial t} = \nabla^2 \psi, \tag{5.12}$$

where C_ψ is the moisture capacity given by

$$C_\psi = \frac{1}{D(c)}. \tag{5.13}$$

Problems with any variability of the diffusion coefficient with moisture content and temperature are just transferred to another parameter, the moisture capacity C_ψ.

Nevertheless, Arfvidsson (1998) shows how Kirchhoff potentials can be used to describe anisotropic moisture flow in mixed heart/sapwood boards. Earlier, Rosenkilde and Arfvidsson (1997) reported the use of Kirchhoff potentials in abstracting diffusion coefficents from the measurement of moisture-content profiles over the range from fibre saturation to the end-use moistness in the drying of 29- and 30-mm cubes of Scots pine (*Pinus sylvestris*). Measurements were done on heartwood and sapwood specimens separately, as well as separate measurements in all three directions to the grain, axial, radial and tangential. From the variation of moisture content with time and with distance, the dependence of the Kirchhoff potential on moisture content could be determined, and the diffusion coefficent estimated directly from the gradient $d\psi/dc$ though Eq. (5.9). However, as Pel (1995) shows, it is not necessary to invoke Kirchhoff potentials to obtain diffusion coefficients in this way. In practice, then, the use of the Kirchhoff potential offers no outstanding advantages and it is a more obscure parameter to employ than a quantity such as moisture content, which can be measured directly.

5.1.2 Moisture-Content Driving Force

An underlying assumption in this model is that moisture migrates by diffusion, driven by a gradient in moisture content, as described by Fick's Second Law, which for one-dimensional movement becomes:

$$\frac{\partial X}{\partial t} = \frac{\partial}{\partial z}\left(D\frac{\partial X}{\partial z}\right), \qquad (5.14)$$

where X is the free moisture content above the equilibrium moisture content, t is time, z is the distance coordinate perpendicular to the airstream from the core, and D is the diffusion coefficient. Figure 5.1 shows the coordinate directions and a typical moisture-content profile. With this choice of coordinate directions, the moisture fluxes have negative numerical values when the lumber is drying, since the moisture content falls from a maximum value at the core to a minimum at the surface.

Sherwood (1929) was the first to use this approach, and he made the following additional assumptions:

- the diffusion coefficient D is constant;
- the initial moisture content in the lumber is uniform; and
- surface fibres come into equilibrium with the surrounding air instantaneously, so that the resistance of the boundary layer outside the board of lumber is negligible.

Translated into mathematical terms, the last two of these assumptions are

$$t = 0; \quad X = X_i \qquad (5.15)$$

$$z = \delta; \quad X = X_e \text{ (at the surface)}, \qquad (5.16)$$

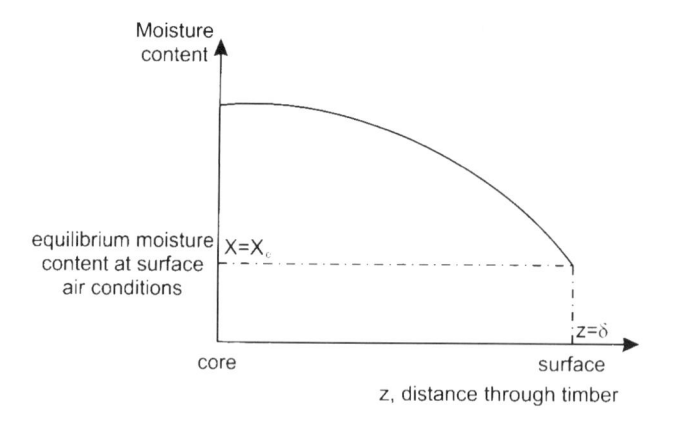

Fig. 5.1. Coordinate directions and a typical moisture-content profile for diffusion

where δ is the half-thickness of the board. This approach allowed Eq. (5.14) to be integrated to yield a predicted moisture-content profile. This profile may be integrated to give an average moisture content, with a characteristic moisture content (Φ) being defined as follows:

$$\Phi = \frac{\overline{X} - X_e}{X_i - X_e}, \tag{5.17}$$

where $\overline{X}$ is the volume-averaged moisture content and X_i and X_e are the initial and equilibrium moisture contents respectively. The equation for the characteristic moisture content is:

$$\Phi = \frac{8}{\pi^2} \sum_{n=0}^{\infty} \frac{1}{(2n+1)^2} \exp\left[-\left(\frac{2n+1}{2}\right)^2 \pi^2 t\right]. \tag{5.18}$$

With this model, a characteristic parameter which governs the extent of drying is the mass-transfer Fourier number (Fo_M), defined as follows:

$$Fo_M = \frac{Dt}{\delta^2}. \tag{5.19}$$

If drying is controlled by diffusion, then for the same drying conditions, doubling the thickness of the material should increase the drying time to the same final moisture content fourfold.

If the diffusion coefficient were constant, the moisture-content profile would be linear over a material for the steady-state movement of moisture through it. However, the drying of a lumber board is not a steady-state process, and the moisture-content profile is non-linear. When the moisture-content change occurs over the entire half-thickness of the board, in other words when there is no longer a fully-wet region in the core, the moisture-content profiles can be shown to be parabolic in shape during drying if the diffusion coefficient is constant.

From Eq. (5.14), at each point in time t,

$$\frac{\partial}{\partial z}\left(D\frac{\partial X}{\partial z}\right) = \text{constant}, \ a, \ \text{say}. \tag{5.20}$$

Hence, integrating once gives

$$\left. D\frac{\partial X}{\partial z}\right|_t = az. \tag{5.21}$$

At this time, the flux of moisture within the material, j_w, is given by

$$j_w = \rho_w D \frac{dX}{dz}, \tag{5.22}$$

where ρ_w is the oven-dry wood density. So for constant D and ρ_w it follows that:

$$j_w = j_{wS}\frac{z}{\delta} = \rho_w D\frac{dX}{dz}, \qquad (5.23)$$

in which j_{wS} is the evaporation rate per unit surface area, at the surface. The free moisture content is zero at the surface, $z = \delta$, so the moisture content X at any position a distance z from the core is given by

$$\rho_w D \int_0^{X-X_e} dX = \frac{j_{wS}}{\delta}\int_\delta^z dz, \qquad (5.24)$$

whence

$$\rho_w D \frac{(X-X_e)^2}{2} = \frac{j_{wS}}{\delta}(z-\delta). \qquad (5.25)$$

This equation yields a parabolic moisture-content profile with distance through the material.

The surface of the lumber does not necessarily come instantly to equilibrium. In order to account for the non-constancy of the surface moisture content, Choong and Skaar (1969) separated the surface resistance to moisture movement from the internal resistance through a "*surface-emission coefficient*" (S), which is defined by the following mass-transfer boundary condition:

$$D\frac{\partial X}{\partial z} = -S(X_s - X_e). \qquad (5.26)$$

The solution to the diffusion equation then becomes:

$$\Phi = 2L^2 \sum_{n=1}^{\infty} \frac{\exp(-\beta_n^2 \tau)}{\beta_n^2(\beta_n^2 + L^2 + L)}. \qquad (5.27)$$

Here the coefficients β_n are the positive roots of the equation

$$\beta_n \tan\beta_n = L \qquad (5.28)$$

where $L = (S\delta)/D$ and δ is the half-thickness of the lumber board. They have shown that it is possible to estimate D and S from experiments on boards of different thicknesses by estimating an apparent diffusion coefficient D' on the assumption that there is no external resistance ($L = \infty$) and plotting δ/D' against δ. The slope of the plot yields D, and the intercept is $3.5/S$ (Choong and Skaar 1972a,b).

The definition of a surface-emission coefficient allows the difference between the surface moisture content and the equilibrium moisture content to be allowed for explicitly. This difference arises because of the external resistance to mass transfer in the boundary layer above the surfaces of the lumber boards, in which the moisture-vapour content adjacent to a surface is taken to be in equilibrium with the surface moisture content. The mass-transfer boundary condition at the surface of the boards would normally have a form in which an internal moisture flux ($D\,\partial X/\partial z$) would be equated with an external one,

$\beta(p_{vS} - p_{vG})$, where β is the external mass-transfer coefficient, p_{vS} is the vapour pressure above the surface of the board and p_{vG} is the partial pressure of vapour in the bulk gas. The use of a mass-transfer coefficient is a more conventional approach for analysing the boundary-layer resistance to mass transfer than a surface-emission coefficient (Treybal 1968; Brauer 1971). Moreover, there are well-established procedures for relating mass-transfer coefficients to friction factors (for fluid flow) and heat-transfer coefficients, which are often better known than their mass-transfer equivalents. This relationship becomes a useful way of estimating the convective behaviour about lumber boards from a computation of the airflow.

Peck and Kauh (1969) have modified this mass-transfer boundary condition as follows:

$$D\frac{\partial X}{\partial z} = -\beta\gamma(p_{vS} - p_{vG}), \qquad (5.29)$$

where γ is a variable parameter, the use of which assumes that drying rates lessen due to the smaller fraction of the surface remaining wet as drying proceeds. As a first estimate, the area available for evaporation varies as the square of some equivalent length of the object being dried, whereas the moisture concentration should vary as this length cubed. Whence,

$$\gamma = \left(\frac{X_S}{X_{cr}}\right)^{2/3}. \qquad (5.30)$$

Here, X_{cr} is the critical moisture content, often taken as the initial moisture content for materials such as wood, which shows no appreciable constant-rate drying period. The evaporative interface may recede into the material, so the exponent in the equation above may be somewhat less than 2/3, and the "thickness function" becomes:

$$\gamma = \left(\frac{X_S}{X_{cr}}\right)^{n}, \qquad (5.31)$$

where the exponent n is determined experimentally. A value of $n = 0.6$ has been found experimentally in the drying of balsa wood slats up to 9.5 mm in thickness by Peck and Kauh (1969). With this mass-transfer boundary condition, the coupled diffusion treatment for internal mass transfer must be solved numerically.

However, this "thickness function" reduces the method to an empirical one because of the fitting parameter n. Moreover, the treatment is based on a physically unrealistic picture of the way in which porous solids dry, except when they are very thin. Thicknesses of lumber boards in the range under 9.5 mm tested by Peck and Kauh (1969) are unusual in kiln-drying practice. Although dry patches have been seen and photographed on the surface of moist granular beds as they dry out (Oliver and Clarke 1973), fine porous material can have a significant fraction of its exposed surface dry before the evaporation from

5.1.3 Vapour-Pressure Driving Force

Bramhall (1976a, 1979a, b, 1995) has suggested that the driving force for moisture migration is the partial pressure of water vapour and that Fick's Second Law should be modified to read:

$$\frac{\partial X}{\partial t} = \frac{\partial}{\partial z}\left(D\frac{\partial p_v}{\partial z}\right), \tag{5.32}$$

where p_v is the partial pressure of water vapour at the local temperature and moisture content. He claims that the diffusion coefficient obtained in this way should be more nearly constant than that obtained using a moisture-content driving force. By dividing the diffusion coefficients obtained by Biggerstaff (1965) and Choong (1965) by the vapour pressure of water at the given temperatures, he found that the diffusion coefficients of both workers were independent of temperature over the range of experimental conditions used (Biggerstaff 1965, 42 to 98 °C; Choong 1965, 40 to 60 °C). The diffusion coefficients of Biggerstaff (1965), when converted for the vapour-pressure driving force, gave values between 3×10^{-8} cm^{2}s^{-1} (mm Hg)$^{-1}$ and 8.2×10^{-8} cm^{2}s^{-1} (mm Hg)$^{-1}$ (2.3–6.2 $\times$ 10^{-14} m^{2}s^{-1}Pa^{-1}). While the data of Choong (1965), when converted in this way, give diffusion coefficients between 1.40×10^{-8} cm^{2}s^{-1} (mm Hg)$^{-1}$ (1.05×10^{-14} m^{2}s^{-1}Pa^{-1}) at a moisture content of 0.05 kg kg^{-1} and 6.22×10^{-8} cm^{2}s^{-1} (mm Hg)$^{-1}$ (4.67×10^{-14} m^{2}s^{-1}Pa^{-1}) at a moisture content of 0.15 kg kg^{-1}.

Hunter (1993) supported this work by analysing the data of Martley (1926) and Voigt et al. (1940) for steady-state diffusion of moisture through Scots pine (*P. sylvestris*), presenting a diffusion coefficient which depends on the relative humidity inside the lumber in the following way:

$$10^{12}D = 1.66 + \frac{64.4526}{1 + 42.473(\ln\psi)^2}, \tag{5.33}$$

where ψ is the relative humidity. However, this diffusion coefficient is not independent of temperature, since the relative humidity inside lumber is dependent on both moisture content and temperature (Simpson and Rosen 1981).

It is true that the vapour pressure is proportional to the moisture content for the range of moisture contents chosen by Biggerstaff (1965) and Choong (1965). However, the vapour pressure is dependent on the temperature alone above the fibre-saturation point, which occurs at a moisture content of around 0.25 kg kg^{-1} at normal kiln temperatures. The practical success of the diffusion model with a moisture-content driving force in the work of Wu (1989) and Doe et al. (1994), together with their observed moisture-content profiles that are consistent with a transport process which is driven by the total concentration of water, suggests that the conclusion may not be so clear as stated by Bramhall (1995). With a number of moisture-transport mechanisms operating in wood, the use of a vapour-pressure driving force is only likely to be an appropriate driving force for the diffusion of water vapour and bound water, and unlikely to be related to the flow of liquid water in any simple way. The use of a vapour-pressure driving force explains the observed temperature dependence of the diffusion coefficient in the work of Biggerstaff (1965) and Choong (1965) for moisture contents below the fibre-saturation point, for which the vapour pressure is proportional to the moisture content. However, a vapour-pressure driving force does not explain the temperature dependence of the diffusion coefficient in the work of Wu (1989) and Doe et al. (1994), for which the moisture contents were above fibre saturation, nor the diffusion-like moisture-content profiles over the entire range of moisture contents.

Inspection of Fig. 5.2 from Wu (1989) illustrates these points. The agreement between the moisture-content profiles predicted by the diffusion model based on moisture-content gradients and the observed profiles is no worse at high moisture contents than at low ones, so that even if there are different moisture-transport mechanisms operating above the fibre-saturation point to those

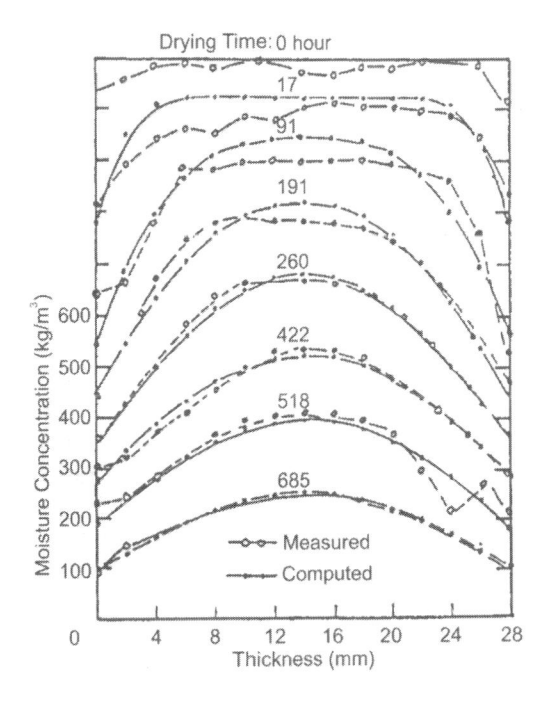

Fig. 5.2. Moisture-content profiles through Tasmanian eucalypts during drying. (After Wu 1989)

below it, these are not evident from this comparison. Also, at high moisture contents, the vapour pressure in the lumber is likely to be within a few percent of the vapour pressure of pure water at the same temperature from the sorption isotherm of Simpson and Rosen (1981). In addition, the lumber was virtually isothermal for these drying conditions, with no significant profile of temperature with distance into the board. Nevertheless, at these high moisture contents, there is a noticeable gradient of moisture content, and it is difficult to see how this could arise from a vapour-pressure driving force alone.

Langrish and Bohm (1997) have analysed drying data for three species of Australian eucalypts, yellow stringybark (*E. muelleriana*), spotted gum (*E. maculata*) and ironbark (*E. paniculata*), in terms of both moisture-content and vapour-pressure gradients. Moisture-content gradients were found to fit the drying data for all species and drying schedules better than vapour-pressure gradients, and the diffusion coefficients based on moisture-content gradients also extrapolated one set of drying data (moisture contents against time) better than coefficients based on vapour-pressure gradients.

An overall drying model has been developed using vapour-pressure driving forces (Bramhall 1979b). This model incorporates mechanisms for the internal movement of moisture by both diffusion (driven by a gradient in vapour pressure) and liquid transport (driven by a gradient in free moisture content, above the fibre-saturation point). The model was applied to the drying of 50-mm-thick boards of alpine fir (*Abies lasiocarpa*) for a conventional stepped drying schedule (with dry-bulb temperatures from 53 to 80°C), a constantly-increasing temperature schedule (dry-bulb temperatures from 49 to 100°C) and two constant-temperature schedules with dry- and wet-bulb temperatures of 93°C/71°C and 71°C/53°C, respectively. The diffusion coefficient in Eq. (5.25) was fitted by the following correlations:

$$D = \exp\left[\frac{X}{(0.1792 - 2.553X)} - 9.2\right]\left(\frac{0.248}{X_{FSP}}\right) \quad \text{for } X > 0.08, \tag{5.34}$$

$$D = 6.64 \times 10^{-4} X^2 \left(\frac{0.248}{X_{FSP}}\right) \quad \text{for } X < 0.08, \tag{5.35}$$

where the moisture content (X) and the moisture content at fibre saturation (X_{FSP}) are expressed as the mass of moisture per unit dry mass of the lumber. As pointed out by Bramhall, the term involving the fibre-saturation point incorporates a weak temperature dependence through the relationship between the local lumber temperature (T, °C) and the fibre-saturation point, correlated by Bramhall with the expression:

$$X_{FSP} = 0.341 - 0.00133T, \tag{5.36}$$

where T is the temperature in degrees Celsius. The movement of liquid water was assumed to be proportional to the local gradient in the free water content and to the surface tension, with the flux expression for free water movement being

$$j_{wf} = D_c \gamma \frac{\partial X_f}{\partial z}. \tag{5.37}$$

Here X_f is the free water content, which is the difference between the local moisture content and that at fibre saturation, calculated on a dry basis. This quantity must always be positive or zero. The effective conductivity coefficient D_c was fitted by the relationship

$$D_c = 2.6 \times 10^{-3}(0.145 + 0.0335T)X_f. \tag{5.38}$$

The dependence on free-water content was assumed because the total conductance should be proportional to the number of flow paths available, and this was assumed to be proportional to the free-water content. The surface tension (γ, mN m^{-1}) in equation (5.37) is given by the equation

$$\gamma = 75.6 - 0.1625T \tag{5.39}$$

with T again in degrees Celsius.

Typical moisture-content profiles predicted by this model are shown in Fig. 5.3, with an inflection point in the moisture-content profiles occurring at moisture contents at which liquid flow ceases ($X \approx 0.6$). At lower moisture contents, the moisture profiles reflect diffusional movement.

The overall drying model fits the drying rate curves well for the constant-temperature schedules with batches of 50 boards (50-mm thick) which were carefully end-coated before drying at an air velocity of 1.8 m s^{-1}. For the constantly increasing temperature schedule and the stepped schedule, the experi-

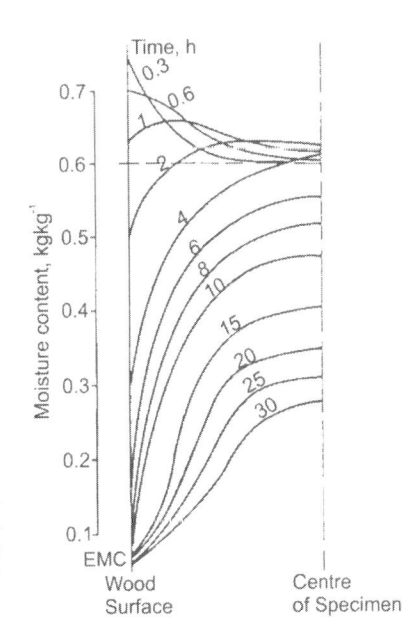

Fig. 5.3. Predicted moisture-content profiles from a drying model including liquid movement and vapour-pressure-driven diffusion for 50-mm-thick boards of alpine fir (*Abies lasiocarpa*). (After Bramhall 1979b)

mental data were used to test the model independently. Other required inputs, such as the external transfer coefficients, were required for the model for air velocities other than $1.8\,\mathrm{m\,s^{-1}}$. While this approach has a basis of fundamental principles, the correlations for the diffusion coefficient are still complex relative to an Arrhenius-type relationship for the relationship between the diffusion coefficient and the temperature. The driving force for liquid movement is likely to be the capillary-pressure gradient (Spolek and Plumb 1981), which is related to, but not directly proportional to, the free-moisture content. Also, only average drying rate curves were shown, and no moisture-content gradients were measured, although predicted profiles were given. No temperature profiles or changes with time were shown, and it is uncertain how well this model would work for the high-temperature drying of softwoods, where substantial temperature gradients within such lumber have been measured (Northway 1989).

At low moisture contents, the use of vapour-pressure driving forces within a diffusion model allows diffusion coefficients to be used which are not strongly temperature-dependent. However, at higher moisture contents, particularly for relatively impermeable species, the use of moisture-content driving forces within diffusion models is simpler in practice and has some experimental support for predicting moisture-content gradients.

5.1.4 Chemical-Potential Driving Force

Within a comprehensive drying model, Stanish et al. (1986) have presented a fundamental picture of the drying process in lumber which treated water vapour, free water and bound water separately. They considered bound-water migration as a molecular diffusion process. The flux of bound water, j_{wb}, was assumed to be driven by the chemical potential ($\partial\mu/\partial z$) of adsorbed water molecules:

$$j_{wb} = D_b(1-\varepsilon_d)\frac{\partial\mu}{\partial z}, \qquad (5.40)$$

where the proportionality constant is a diffusion coefficient for bound moisture, D_b, and $(1-\varepsilon_d)$ is the volume fraction taken up by the solid matrix.

In addition, they assumed that there is local thermodynamic equilibrium between the vapour and the components of water. For example, if free water were present, the vapour should remain saturated, and the bound-water content should remain at the fibre-saturation point at the local temperature. If free water were absent, the bound-water and vapour-phase compositions should obey the sorption equilibrium. The implication of local thermodynamic equilibrium is that the chemical potential of bound water equals the chemical potential of the water vapour, so that the bound water flux may be expressed in terms of the chemical potential of water vapour. Water vapour is a gas, so the thermodynamic relation for the chemical potential of gases applies:

$$M\,d\mu = -S\,dT + V\,dp_v, \qquad (5.41)$$

where M is the molar mass of the water vapour, S is the water vapour entropy, T is the local vapour temperature, V is the molar volume of the vapour (ε_d/ρ_v, and ρ_v is the vapour density) and p_v is the vapour pressure. Combination of Eq. (5.40) and (5.41) yields:

$$j_{wb} = D_b(1-\varepsilon_d)\left[-\left(\frac{S}{M}\right)\frac{\partial T}{\partial z} + \left(\frac{\varepsilon_d}{\rho_v}\right)\frac{\partial p_v}{\partial z}\right]. \qquad (5.42)$$

The entropy of water vapour may be estimated by treating it as an ideal gas, since the pressure and temperature in lumber-processing operations are usually substantially below those where non-ideality is significant, as noted in Chapter 2. This assumption yields the expression:

$$S = 187 + 35.1\ln\left(\frac{T}{298.15}\right) - 8.314\ln\left(\frac{p_v}{101325}\right), \qquad (5.43)$$

where the temperature T is in Kelvin and the vapour pressure p_v in Pa. Equation (5.43) implies that, if the temperature and moisture-content profiles through a piece of lumber (and therefore the vapour-pressure profile) are known, then the water-vapour entropy and the bound-water flux (n_b) may be estimated from the equations above. No assumptions regarding temperature-driven diffusion are necessary, because a contribution from the temperature gradient arises automatically in these equations, suggesting the use of the chemical potential as a driving force for diffusion rather than vapour pressure alone.

Cloutier and Fortin (1991) extended this concept to include other contributions to the chemical potential (called the "*water potential*" by them), such as capillary, sorptive, osmotic and gravity forces. They measured the relationship between the chemical potential and the moisture content for desorption in aspen (*Populus tremuloides*) sapwood over a range of moisture contents from 0.05 to 1.8 kg kg^{-1} and temperatures from 20 to 45 °C. The relationship between the water potential and the moisture content was found to be very non-linear, with values of the water potential ranging from -10^{-6} J kg^{-1} for utterly dry wood to -10^{-1} J kg^{-1} at a moisture content of 1.8 kg kg^{-1}. They have applied the concept of the chemical potential as the driving force for moisture movement over the entire moisture-content range within various drying models (Cloutier et al. 1992; Cloutier and Fortin 1993,1994).

The "*effective water conductivities*", which are essentially the diffusion coefficients D_b shown above, were reported by Cloutier and Fortin in 1993. Their measurements were carried out over a range of dry-bulb temperatures from 20–50 °C, a range of relative humidities from 29–69%, and air velocities from 0.3–0.5 m s^{-1} on 45-mm cubes of aspen (*Populus* sp.). The conductivities varied by over six orders of magnitude over a range of moisture contents from 0.1 to 1.4 kg kg^{-1} (with a small temperature dependence), suggesting from a practical viewpoint that considerable experimental work would be required to measure the conductivities for other species, since these conductivities would be expected to be species-dependent. From a fundamental perspective, this variation suggests that the chemical potential may have limited applicability as a driving force over the entire moisture-content range from green to dry. However, it does appear to be an appropriate driving force for the movement of the bound-water component (Stanish et al. 1986).

5.2 Penetration Periods and Regular-Regime Drying

A number of workers (Liou and Bruin 1982a,b; Thijssen and Coumans 1984) have developed analytical solutions to the diffusion equation with moisture-content dependent diffusion coefficients. The following summary of the method is taken from Keey (1992).

For concentration-dependent diffusion through material in any form (slab, cylinder or sphere), Liou and Bruin (1982a) noted that a general formulation can be expressed by the equation

$$\frac{\partial m}{\partial \tau} = \frac{\partial}{\partial \phi}\left(D_r X^2 \frac{\partial m}{\partial \phi}\right). \tag{5.44}$$

For a non-shrinking particle, the variables in this equation have the following meaning:

- m is the ratio of the moisture concentration c compared with its initial value c_o.
- τ is the Biot number ($D_o t/R^2$) based on a reference value (D_o) for the diffusion coefficient and the half-thickness or radius R of the body.
- ϕ is a dimensionless space coordinate, defined by

$$\phi = \left(\frac{r}{R}\right)^{\upsilon+1}, \tag{5.45}$$

where υ is a parameter that takes the value 0, 1, or 2 for a slab, cylinder, or sphere, respectively.
- X is a dimensionless geometric variable, which is the following function of ϕ:

$$X = (\upsilon+1)\phi^{\upsilon/(\upsilon+1)}. \tag{5.46}$$

- D_r is the relative diffusion coefficient D/D_o.

A similar set of definitions apply for shrinking systems.

The non-dimensional drying rate is given by

$$\frac{dm}{d\tau} = -FX_i, \tag{5.47}$$

in which F is a dimensionless flux parameter

$$F = -D_r X_i \left(\frac{\partial m}{\partial \phi}\right)_i, \tag{5.48}$$

with X_i being the value of the geometric parameter at the moisture-gas interface (where $\phi = 1$) and thus,

$$X_i = v + 1. \tag{5.49}$$

The relative diffusion coefficient D_r is assumed to be a simple algebraic function of the moisture content, namely,

$$D_r = m^a \tag{5.50}$$

for which a is a constant.

The actual drying flux is found from F:

$$N_v = \left(\frac{\rho_L D_o}{R_o}\right) F. \tag{5.51}$$

The generalised diffusion Eq. (5.44) is solved by analogy with the short-time approximation for a slab. To this end, two further auxiliary variables are introduced: the extent of moisture loss being

$$E = 1 - m \tag{5.52}$$

and a flux function G defined by

$$G = \frac{E.F}{v+1}. \tag{5.53}$$

Thijssen and Coumans (1984) show how an experimentally obtained drying curve, transformed with the coordinates E and G, can be used as a generalised kinetic curve for the material being dried, irrespective of its shape. Two isothermal drying experiments are needed, normally on a small slab of material. A typical transformed drying curve is shown in Fig. 5.4.

At first, when the drying rate is invariant with falling moisture content, the flux function G increases linearly with E until the critical point is reached (at E_1 in Fig. 5.4). Thereafter, a penetration period exists in which a moisture-denuded zone ($0 < m < 1$) penetrates through the material towards the core from a surface, with the flux function G being independent of the extent of moisture loss, until this extent has reached E_2. At that point, moisture has just begun to be lost from the whole material. In the final period, or regular regime, the value of G falls until it is zero at equilibrium when $E = 1$.

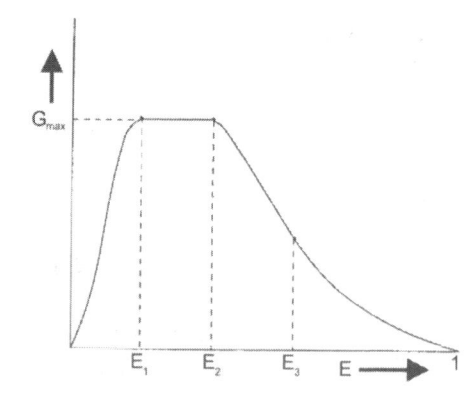

Fig. 5.4. Transformed drying curve. The flux function G is defined by Eq. (5.48) and (5.53). The abscissa E is the extent of moisture loss, Eq. (5.52). (Adapted from Thijssen and Coumans 1984)

For the penetration period, the solution of Eq. (5.44) yields

$$\left(\frac{G}{G_o}\right)_{max} = 1 - \frac{E}{0.82^a[1+1/(v+1)]}, \quad (5.54)$$

where $[G_o]_{max}$ is the maximum value of the flux function for a slab. In the regular regime,

$$G = \frac{Sh\, E(1-E)^{a+1}}{2(a+1)(v+1)}, \quad (5.55)$$

where the limiting values of the Sherwood number Sh are given in Table 5.1. Liou and Bruin (1982a) suggest that intermediate values can be interpolated as a function of $a/(a+2)$. The transformed flux function G has an inflexion point E_3 given by

$$E_3 = \frac{2}{2+a}. \quad (5.56)$$

Thus, an inflexion point does not exist for a material with a constant moisture-diffusion coefficient ($a = 0$).

Liou and Bruin (1982b) applied these transformations to determine moisture-content profiles in both shrinking and non-shrinking particles. They

Table 5.1. Limiting values of the Sherwood number in Eq. (5.55) calculated by Thijssen and Coumans (1984)

Geometry	v	$a = 0$ (constant diffusivity)	$a = \infty$ (infinitely variable diffusivity)
Slab	0	$\pi^2/2$	e^2
Cylinder	1	5.783	$4e$
Sphere	2	$2\pi^2/3$	$e^{8/3}$

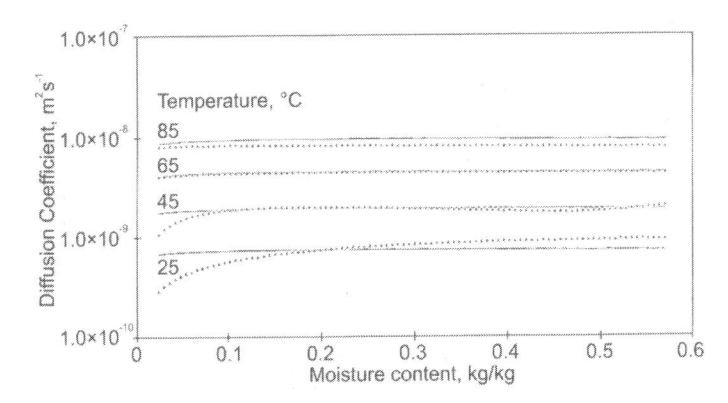

Fig. 5.5. Temperature and moisture-content dependence of diffusion coefficient in *Pinus radiata*. (Kroes and Kerkhof 1996, pers. comm.)

noted that the profiles at a given value of E do not depend greatly on whether the material shrinks or not, and this difference becomes even less as the value of the exponent a increases, that is, as the diffusion coefficient becomes more concentration-dependent. Indeed, failure to account for shrinkage in defining a coordinate system shows up as such a dependence, as Viollez and Suarez (1985) point out.

This method is powerful and relatively rigorous, enabling moisture-content profiles as well as drying rates to be extrapolated to other drying conditions from two slab-drying experiments in the laboratory. However, there are few reports of the technique being applied for the analysis of lumber drying, apart from the work of Kroes and Kerkhof (1996, pers.comm.), who used this technique to extract effective diffusion coefficients for *Pinus radiata* (softwood) and *Nothofagus fusca* (red beech, a hardwood) at moisture contents below complete liquid saturation. Their work demonstrates that the diffusion coefficient is virtually independent of moisture content and is dependent on temperature only, as shown in Fig. 5.5. The main dependence of the measured diffusion coefficients on moisture contents is at low moisture contents (below about 8%) and at relatively low temperatures (below 40°C). This result may be due to multilayer sorption of water molecules on cell walls breaking down at low moisture contents, leaving a monolayer. When the temperature is raised, less dependence on moisture content may be observed, consistent with a loss of sorptivity.

5.3 Theoretical Modelling of Diffusion Coefficients

Stamm (1967b) has presented the results of an extensive effort (Stamm 1946, 1956a,b, 1959, 1960a,b, 1962, 1964; Stamm and Nelson 1961; Tarkow and Stamm

1961; Choong 1965) to predict diffusion coefficients in wood at low moisture contents (below the fibre-saturation point) in terms of different moisture-transport pathways through a "typical" cell as discussed in Section 4.4.

Stamm and Nelson (1961) carried out drying experiments on loblolly pine (*Pinus taeda*) specimens in a drying oven to test this model. The samples were 6" (150mm) long, 2" (50mm) wide and 0.25" (6.1mm) thick, and were equilibrated in an atmosphere of 50% relative humidity before being soaked to give an initial moisture content of around 30%. At temperatures from 50 to 120°C, the experimentally fitted tangential diffusion coefficients ranged from 0.49×10^{-9} to $4.47 \times 10^{-9} \, m^2 s^{-1}$, while the radial coefficients ranged from 0.60×10^{-9} to $6.55 \times 10^{-9} \, m^2 s^{-1}$, respectively. The experimental values were between 45% and 82% of those predicted by the theoretical model, with the relative constancy of the ratio over the tested temperature range, indicating that the model incorporates the main physical processes involved. However, the authors concluded that "more experimental data and further refinements in the theory are needed before predictions in drying rates can be made truly quantitative". The theoretical prediction relies on experimental measurements of bound-water diffusion coefficients as a function of moisture content, and this is a significant limitation to the wider use of the model. Nevertheless, it does provide a useful guide to the relative amount of moisture movement through different moisture-transport pathways. The theoretical model also allows the temperature and moisture-content dependency of the diffusion coefficient to be predicted, as will be discussed in subsequent sections.

5.4 Experimental Measurements of Diffusion Coefficients

A number of researchers have reported experiments that were designed to yield estimates of diffusion coefficients, mainly for wood at or below the fibre-saturation point. Since these data have often been used by subsequent workers in their own work, some description of the corresponding experimental conditions is useful before discussing the effects of temperature and moisture content on moisture diffusion. The range of diffusion coefficients also gives some feeling for the orders of magnitude involved.

Sherwood (1929) discussed a very limited amount of drying data for lumber. One set of data (Tuttle 1925) for sitka spruce (*Picea sitchensis*, a softwood) gave a diffusion coefficient of $1.12 \times 10^{-9} \, m^2 s^{-1}$ for 50.8-mm-thick boards dried at 71.1°C under a relative humidity of 75% from an initial moisture content of $0.51 \, kg \, kg^{-1}$ (dry basis). This diffusion coefficient was derived by curve-fitting measurements of moisture-content gradients, which were parabolic in form. Sherwood's own drying-rate data for 12.7- and 19.0-mm-thick boards of poplar (*Populus* sp., a hardwood), dried at 31–35°C with an air velocity between boards of $1 \, m \, s^{-1}$ from initial moisture contents of 0.434 and

0.397 kg kg^{-1}, respectively, were fitted well with a similar diffusion coefficient of 9.4 × 10^{-10} m^2 s^{-1} despite the twofold difference in mass-transfer Fourier number.

Another study in which diffusion coefficients were derived by curve-fitting measurements of moisture-content gradients has been described by Skaar (1958), who performed measurements with the heartwood of American beech (*Fagus grandifolia*) at moisture contents below the fibre-saturation point. He acknowledged the role of both water-vapour and bound-water diffusion under these conditions.

Choong (1963) carried out absorption experiments on pieces of western fir (*Abies grandis*) which were 15 × 15 mm in lengths of 40, 50, 67, 100 and 200 mm in order to assess diffusion coefficients due to the combination of bound-water and water-vapour movement below the fibre-saturation point. Unsteady-state measurements were carried out at temperatures of between 30 and 60 °C to determine diffusion coefficients in both radial and longitudinal directions. The fitted radial diffusion coefficients varied from 1.05 × 10^{-11} m^2 s^{-1} at 40 °C and 5% moisture content to 3.45 × 10^{-11} m^2 s^{-1} at 60 °C and 15% moisture content, while the tangential coefficients varied from 2.9 × 10^{-9} m^2 s^{-1} at 40 °C and 5% moisture content to 0.3 × 10^{-9} m^2 s^{-1} at 60 °C and 15% moisture content.

Stamm (1964, 1967b) reported the application of the diffusion model below the fibre-saturation point to a wide range of softwoods and hardwoods including Scots (*Pinus sylvestris*) and loblolly pine (*P. taeda*), birch (*Betula* sp.), spruce (*Picea* sp.), Douglas-fir (*Pseudotseuga menziesii*) and oak (*Quercus* sp.), giving some data for both radial and tangential directions. For the 39 data points including these species, the scatter at any given temperature and moisture content in the diffusion coefficients was up to half an order of magnitude, which might be considered to be a relatively small amount of variation given the amount of biological variability between lumber taken from a given species. Other data mentioned in Stamm (1964) show that the variation between species may be even greater. He suggests that diffusion of both water vapour and bound water occurs below the fibre-saturation point for most wood and at all moisture contents for species which are impermeable to the flow of liquids, and that the fitted diffusion coefficients are lumped parameters which include both bound-water and water-vapour movement. For permeable species above the fibre-saturation point, capillary-controlled movement of liquid water is likely to occur.

In 1965, Choong carried out experiments using 68-mm-diameter discs, 5 mm thick, of the same species (western fir, *Abies grandis*) at moisture contents below the fibre-saturation point using a steady-state method. He placed a saturated salt solution (with a known vapour pressure) in a container, which was then placed in a saturated atmosphere of water vapour at a constant temperature. A disc of wood was fitted over the top of the container, so that vapour passed through the disc and condensed in the saturated salt solution. At regular time intervals, the apparatus was weighed, and the experiment was repeated at various temperatures. This technique has some elegance in comparison with unsteady-state drying experiments.

Biggerstaff (1965) performed drying experiments on flat-sawn eastern hemlock (*Tsuga canadensis*) sapwood at temperatures from 50 to 120 °C. The samples were 6" (150 mm) long, 2" (50 mm) wide and 0.25" (6.1 mm) thick and were predried to near the fibre-saturation point. His fitted coefficients ranged from $0.53 \times 10^{-9} \, m^2 \, s^{-1}$ at 50 °C to $3.33 \times 10^{-9} \, m^2 \, s^{-1}$ at 120 °C.

Koponen (1987) carried out drying experiments on birch and spruce sapwood and heartwood at temperatures between 50 °C and 120 °C and moisture contents below 30% (close to fibre saturation), with diffusion coefficients being given in Table 5.1. Further results (Koponen 1988) for the drying of birch (*Betula* sp.), pine (*Pinus* sp.) and spruce (*Picea* sp.) at 20 °C and a relative humidity of 67% showed that the transverse diffusion coefficients were similar for these species ($3.3 \times 10^{-10} \, m^2 \, s^{-1}$ for birch, $4 \times 10^{-10} \, m^2 \, s^{-1}$ for both pine and spruce), while the longitudinal coefficients were much higher ($7 \times 10^{-10} \, m^2 \, s^{-1}$ for birch, $12 \times 10^{-10} \, m^2 \, s^{-1}$ for pine and $15 \times 10^{-10} \, m^2 \, s^{-1}$ for spruce).

Both desorption and absorption experiments on northern red oak (*Quercus rubra*) have been carried out by Simpson (1993). All the experiments were performed at 43 °C, with the desorption experiments beginning from an initial (equilibrated) percentage moisture content of 23.3% and moving down to final moisture contents of between 5.5% and 15.9%. The absorption experiments were carried out from an equilibrated moisture content of 4.5% to final moisture contents from 7.5% to 16.8%. Slicing experiments, that were carried out to assess the extent to which the diffusion model based on moisture-content driving forces can predict the moisture-content gradients, found excellent agreement at all the sampling times (at 32, 73, 122, 168 and 265 h in a 320-h experiment).

Choong et al. (1994) performed experiments on $38.1 \times 38.1 \times 25.4$ mm samples of both sapwood and heartwood from six species of hardwood (American elm, *Ulmus americana*; hackberry, *Celtis occidentalis*; red oak, *Quercus rubra*; white ash, *Fraxinus americana*; sweetgum, *Liquidamber styraciflua*; sycamore, *Platanus occidentalis*). Samples were orientated so that longitudinal and transverse diffusion coefficients could be fitted, with different coefficients being obtained above and below the fibre-saturation point. While there were large differences between longitudinal and transverse coefficients above and below the fibre-saturation point, the variation between species was surprisingly small. Above the fibre-saturation point, the longitudinal coefficients varied from $2.37 \times 10^{-8} \, m^2 \, s^{-1}$ to $3.43 \times 10^{-8} \, m^2 \, s^{-1}$, while the tangential coefficients ranged from $1.14 \times 10^{-8} \, m^2 \, s^{-1}$ to $2.16 \times 10^{-8} \, m^2 \, s^{-1}$. The corresponding ranges below the fibre-saturation point were $1.27-2.38 \times 10^{-8} \, m^2 \, s^{-1}$ (longitudinal) and $3.29-5.74 \times 10^{-8} \, m^2 \, s^{-1}$ (tangential).

Langrish et al. (1997) have reported drying experiments on a number of Australian hardwoods, with lumber samples which were 25 mm thick. The drying schedules used involved temperatures from 30–80 °C and air velocities around $0.3-0.5 \, m \, s^{-1}$. These data have been further analysed by Langrish and Bohm (1997), with the diffusion coefficients lying in the range 10^{-9} to $10^{-10}\,m^2\,s^{-1}$, depending on temperature.

Grace (1996) has determined diffusion coefficients for presteamed and untreated samples of a New Zealand southern beech (*Nothofagus truncata*) at 20°C. The diffusion coefficients for this impermeable lumber varied between 10^{-10} and $10^{-11}\,m^2\,s^{-1}$, with the maximum values appearing at about 40% moisture content (dry basis). Presteaming enhanced the permeability, more particularly at the lower moisture contents. Diffusion coefficents obtained from wood which was presteamed at 70°C are greater than the values for untreated wood by a factor of about 1.5.

In summary, most of these measured transverse diffusion coefficients lie between 10^{-8} and $10^{-10}\,m^2\,s^{-1}$. In softwoods, the radial coefficients are higher than the tangential ones. Stamm and Nelson (1961) suggest a ratio of radial:tangential coefficients of about 1.55:1. This difference is thought to be due to the combination of a number of features, including the contribution of ray cells to radial diffusion and the concentration of pits on the radial faces relative to those on the tangential faces of the fibres. The longitudinal coefficients are larger than the transverse ones, with Stamm (1960a) indicating that the ratio of longitudinal:transverse coefficients is around 2.5:1.

5.4.1 The Effect of Temperature on Diffusion Coefficients

The temperature dependence of the diffusion coefficient has been assessed by several researchers, including Skaar (1958), Stamm (1964), Choong (1965), Keyihan (1989), Morén (1989), Wu (1989), Simpson (1993) and Kroes and Kerkhof (1996, pers.comm.). They all used a relationship of the form:

$$D = D_o \exp\left(-\frac{B}{T}\right), \tag{5.57}$$

where T is the absolute temperature of the wood and D_o ($m^2\,s^{-1}$) and B (K) are constants. The parameter D_o is sometimes described as a *preactivation factor*, while B has the significance of the activation energy dependence of moisture diffusion, since this equation could be written as an Arrhenius relationship as follows:

$$D = D_o \exp\left(-\frac{E}{RT}\right) \tag{5.58}$$

with E as the activation energy and R as the gas constant (8.314 J mol^{-1} K^{-1}). Many workers report values of B rather than activation energies, as such, so both values will be given here. This temperature behaviour is consistent with Eyring's theory of liquids (Glasstone et al. 1941) if the diffusivity is considered to be proportional to the frequency that molecules can jump into vacant holes in a molecular lattice.

The temperature dependence of the diffusion coefficient from the above researchers and other workers is summarised in Tables 5.2 and 5.3, with some data below (Table 5.2) and other data above (Table 5.3) the fibre-saturation point.

Some of the variation in activation energies within these tables may arise because of differences between steady-state (such as Martley 1926; Choong 1965) and unsteady-state (from drying experiments) measurements, since Evans et al. (1994), in permeability measurements on *Pinus radiata*, have found that the unsteady-state permeability was much greater than the steady-state permeability.

Table 5.2. Activation energies for diffusion below the fibre-saturation point reported by various workers, together with experimental conditions

Author	Activation energy, kJ mol^{-1} (B, K)	Temperature range (°C)	Comments
Biggerstaff (1965)	27.9 (3350)	50–120	Eastern hemlock (*Tsuga canadensis*)
Choong (1965)	35.55 (4280)	40–60	Western fir (*Abies grandis*), transverse
Cloutier and Fortin (1993)	33.4 (4020)	20–50	Aspen (*Populus tremuloides*), tangential, $X = 0.2\,\mathrm{kg\,kg^{-1}}$
	48.5 (5830)	20–50	Ditto, tangential, $X = 1.2\,\mathrm{kg\,kg^{-1}}$
	49.3 (5930)	20–50	Ditto, radial, $X = 0.2\,\mathrm{kg\,kg^{-1}}$
	69.4 (8350)	20–50	Ditto, radial, $X = 1.2\,\mathrm{kg\,kg^{-1}}$
Kayihan (1989)	22.5 (2700)	50–120	Western hemlock (*Tsuga heterophylla*)
Koponen (1987)	17.6 (2120)	50–165	Spruce (*Picea* sp.) heartwood, longitudinal, $X = 0.3\,\mathrm{kg\,kg^{-1}}$
	42.6 (5120)	50–165	Birch (*Betula* sp.), radial, dry
Michel et al. (1987)	35.8 (4300)	10–90	Scots pine (*P. sylvestris*)
Skaar (1958)	23.9 (2880)	10–60	American beech (*Fagus grandifolia*), tangential
	39.45 (4740)	10–60	Ditto, radial
Stamm (1956)	50.2 (6040)	4.5–50.5	Cellulose
Stamm (1959)	44.2 (5320)	3.5–56.5	Sitka spruce (*Picea sitchensis*), longitudinal direction
Stamm and Nelson (1961)	32.8 (3940)	50–120	Loblolly pine (*Pinus taeda*), tangential
	36.1 (4240)	50–120	Ditto, radial
Stamm (1967b)	35.15 (4230)	0–120	Many species, transverse

Table 5.3. Activation energies for diffusion above the fibre-saturation point reported by various workers, together with experimental conditions

Author	Activation energy, kJ mol^{-1} (B, K)	Temperature range (°C)	Comments
Langrish and Bohm (1997)	31.4–35.8 (3780–4310)	30–80	Yellow stringybark (*Eucalyptus muelleriana*)
	31.3–42.9 (3770–5160)	30–80	Spotted gum (*E. maculata*)
	31.5–38.5 (3790–4630)	30–80	Ironbark (*E. paniculata*)
Morén (1989)	36.2–3.3X (4354–3969X)	40–55	Scots pine (*P. sylvestris*), drying from above the fibre-saturation point
Wu (1989)	31.6 (3800)	15–30	Tasmanian eucalypts, drying from above fibre-saturation point

Note: Tasmanian eucalypts include alpine ash (*Eucalyptus delegatensis*), messmate (*E. obliqua*) and mountain ash (*E. regnans*).

Evans and coworkers explained this difference in terms of a structural model for this species, in which it was assumed that there are both

- long-length paths of low resistance along radial and axial resin canals that carry the fluid into the boards; and
- short-length paths of high resistance that carry the fluid from the radial resin canals to the tracheids.

Only the mechanism involving the longer-length paths of low resistance is active during flow under a constant pressure gradient which occurs in a steady-state experiment. On the other hand, in the unsteady-state case, when there will be accumulation of vapour within the wood structure, both mechanisms are active.

Bramhall (1976a) has criticised some of the data sources in Table 5.2. Choong's data (1965) was criticised on the basis that

- the upper and lower surfaces of the sample may not have been in equilibrium with their respective atmospheres because of boundary-layer resistance;
- heat of condensation may have raised the temperature of the salt solution, opposing the inward diffusion of vapour; and
- the surface of the salt solution may have been significantly diluted by condensation.

Since the last two effects would be more significant at higher temperatures because of the higher condensation rates in these experiments, Bramhall (1976a, 1979a) considered that the activation energy reported by Stamm (1956) of 50.2 kJ mol^{-1} ($B = 6040$ K) was more reliable than the value of 35.55 kJ mol^{-1} ($B = 4280$ K) found by Choong (1965).

The data of Stamm and Nelson (1961) were also re-examined, Bramhall (1976a, 1979a) feeling that the values of 32.8–36.1 kJ mol^{-1} ($B = 3950$–4235 K) were lower than the true values because the oven temperature was significantly above that of the wood samples. This feature would have the effect of shifting more the curve of the diffusion rate against temperature at higher temperatures, raising the true value of the activation energy (Fig. 5.6).

Bramhall concluded that the value of the activation energy was at least as high as the value measured by Stamm (1956) of 50.2 kJ mol^{-1} ($B = 6040$ K), arguing that the activation energy should be the sum of the latent heat of vaporisation (45 kJ mol^{-1}, $B = 5410$ K) and the heat of sorption. However, almost all subsequent measurements of activation energy (except those of Cloutier and Fortin 1993) have significantly lower values than those measured by Stamm (1956), and are indeed lower than the heat of vaporisation. The criticisms made by Bramhall cannot apply to all these measurements. In particular, many of the other measurements were made in situations with significant air velocities where the boundary-layer resistance would be small, and where the true wood temperature was measured.

What is not completely clear is why bound water molecules must vaporise before they diffuse, since it is possible to envisage some kind of surface diffusion without vaporisation of the moisture, that is the breaking of a single hydrogen bond, so allowing the water molecule to "walk" across the surface. This surface diffusion would be expected to be relatively easier at higher moisture contents at which there are several layers of water molecules over the cell walls (multimolecular adsorption) and water molecules may be arranged in clusters. This is consistent with the apparent activation energies

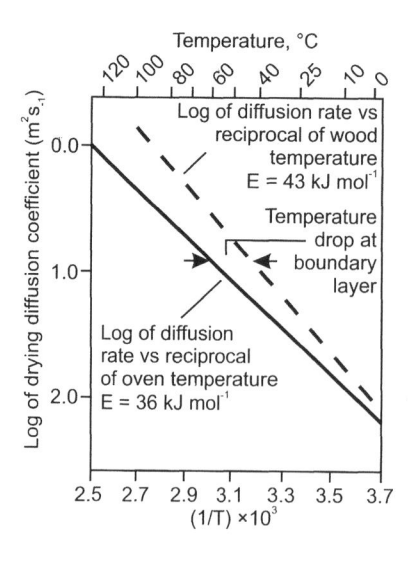

Fig. 5.6. Diffusion rate as a function of temperature, with data normalised to a basic density of 400 kg m^{-3} and an average moisture content of 0.2 kg kg^{-1}. (Bramhall 1979a)

derived from drying experiments which were started above the fibre-saturation point [Morén 1989, 36.2–3.3X kJ mol^{-1} (B = 4354–3969X); Wu, 1989, 31.6 kJ mol^{-1} (B = 3800 K)] being lower than the latent heat of vaporisation for water (45 kJ mol^{-1}).

In summary, the activation energy should be a property of the cell-wall matrix itself, which has a limited range of chemical composition, and so would not be expected to vary greatly. The data in Tables 5.2 and 5.3 generally support this view. The fitted activation energies appear to be a little below that for the evaporation of water, while the theoretical model of Stamm (1967b) and the experimental data which he reviewed (as described in the previous section) suggests an activation energy of around 35 kJ mol^{-1} (B = 4230 K). Most values are in the range 23 to 43 kJ mol^{-1} (B = 2800–5200 K). Hence the increase in the diffusion coefficient with increasing temperature (with activation energy 35 kJ mol^{-1}) appears to be significantly less than that expected for the evaporation of water (45 kJ mol^{-1}).

A slightly different approach to allowing for the temperature variation of the diffusion coefficient was taken by Ashworth (1980), but the same order of temperature dependence was found to that given by the Arrhenius relationship [Eq. (5.58)]. He assumed that the diffusion coefficient was a function of both the moisture content and temperature, and correlated diffusion data by an overall diffusion coefficient (D) which was the product of a contribution due to the moisture content and a contribution due to temperature. The temperature contribution has been correlated in the following way:

$$\frac{D_T}{D_{Ti}} = \left(\frac{T_s}{T_w}\right)^{10} \tag{5.59}$$

with

$$T_s = \omega T_w + (1-\omega)T_d, \tag{5.60}$$

where T_s, T_d and T_w are the surface, dry-bulb and wet-bulb temperatures respectively; in absolute units, D_{ti} is a reference value for the diffusion coefficient, and ω is a parameter which is obtained in the following way:

$$\omega(\varepsilon_s) = \begin{cases} 1 & X_s \geq X_h \\ \omega(\varepsilon_h) & X_e < X_s < X_h \\ 0 & X_s = X_e \end{cases} \tag{5.61}$$

$$\varepsilon_h = \frac{X_s - X_e}{X_i - X_e} \tag{5.62}$$

and X_s is the surface moisture content. The function $\omega(\varepsilon_h)$ was calculated from psychrometric relationships (Ashworth 1980) and desorption isotherms of wood (Kollmann and Cote 1968). This temperature dependence is based on the work of Luikov (1966) and gives a temperature dependence which is consistent with the Arrhenius expression of Eq. (5.58).

5.4.2 The Effect of Moisture Content on Diffusion Coefficients

The theoretical model of Stamm (1967b) (as described in Section 4.4) for the prediction of the diffusion coefficient suggests that the diffusion coefficient should be proportional to the exponential of the moisture content, particularly within the range of moisture contents below the fibre-saturation point (around 0.3 kg kg^{-1}) which are usually encountered in lumber drying (greater than 0.05 kg kg^{-1}). If the relationship between the diffusion coefficient and the moisture content (X, kg kg^{-1}) is taken as $D = \exp(CX)$, where C is a constant, then some of the fitted values for C are given in Table 5.4. Some of these values are close to that of 13 given by the theoretical model (Stamm 1967b) for predicting diffusion coefficients, as analysed by Choong (1965). Kawai et al. (1978) also used an exponential relationship between the moisture content and the diffusion coefficient to predict moisture-content distributions in the longitudinal direction when drying the heartwood of hinoki (*Chamaecyparis obtusa*).

However, some applications of diffusion models do not need the assumption of any relationship between the moisture content and the diffusion coefficient, while some measurements have also failed to find an effect of moisture content on the diffusion coefficient over a range of moisture contents. Koponen (1987) found no significant consistent effect of the moisture content on the diffusion coefficient in his study. Wu (1989), on fitting drying data for Tasmanian eucalypts, in terms of both the change of the average moisture content with time and in terms of the moisture-content profiles, used a

Table 5.4. The effect of moisture content on the diffusion coefficient (the parameter C) found by various workers

Author	Value of C	Species
Choong (1965)	13.5	Western fir (*Abies* sp.)
Furuyama et al. (1994)	19.6	Akamatsu (*Pinus densiflora*); Douglas-fir (*Pseudotsuga menziesii*); Hinoki (*Chamaecyparis obtusa*); Sugi (*Crytomeria japonica*)
Simpson (1993)	1.15	Desorption, red oak (*Quercus rubra*)
	1.45	Absorption, red oak (*Quercus rubra*)
Skaar (1958)	0	Tangential, American beech (*Fagus grandiflora*)
	7.27	Radial, American beech (*Fagus grandiflora*)
Stamm (1959), Skaar (1988)	11	Sitka spruce (*Picea sitchensis*)

diffusion coefficient which did not depend on the moisture content. This approach was employed subsequently by Doe et al. (1994, 1996) and Langrish et al. (1997). They did not specify what (liquid water, water vapour, bound water) was considered to be diffusing, but the success of the model at all moisture contents was very evident. A good fit of the diffusion model to the moisture-content profiles through the lumber was found at different drying times, as shown in Fig. 5.2. There are no discontinuities in the profiles at or near the fibre-saturation point (a moisture content of around $0.3\,kg\,kg^{-1}$) or at moisture contents up to $0.6\,kg\,kg^{-1}$, where continuity of free water would be expected to break down if this had occurred. Similar parabolic moisture-content profiles were found by Harris and Taras (1984) for both conventional and radio-frequency/vacuum drying of 50-mm-thick boards of red oak (*Quercus rubra*). The initial moisture contents were about $0.7\,kg\,kg^{-1}$, and no inflection points were found in the profiles near the fibre-saturation point. The profiles are characteristic of those which would be expected for Fickian diffusion.

Most of the studies cited in Table 5.4 that show an effect of moisture content on the diffusion coefficient have been carried out on wood samples which are below the fibre-saturation point. At these low moisture contents, most values for the parameter C relating the moisture content and the diffusion coefficient have been in the range 11–20, where the moisture content is expressed in kg water per kg dry wood. Above the fibre-saturation point, the evidence suggests that there is no significant effect of moisture content on the apparent diffusion coefficient.

5.5 Conclusions

Wood-drying studies, where a diffusion model has been used to analyse the data, have generally shown that the diffusion coefficients lie in the range from $10^{-8}\,m^2\,s^{-1}$ to $10^{-10}\,m^2\,s^{-1}$. An application of this approach could be the estimation of drying times under constant drying conditions, for which an estimate of the diffusion coefficient (if available) could be used in the following way. For values of the mass-transfer Fourier number $Fo_M\,(=Dt/\delta^2)$ greater than 0.1, corresponding to low final moisture contents, the solution to the diffusion Eq. (5.7) becomes

$$t = \frac{4\delta^2}{\pi^2 D} \ln\left[\frac{8(X_i - X_e)}{\pi^2(X_f - X_e)}\right], \tag{5.63}$$

since only the first term in the infinite series is significant. For example, when drying 25-mm-thick ($\delta = 0.0125\,m$), flat-sawn eastern hemlock (*Tsuga canadensis*) at 50 °C between moisture contents of $0.6\,kg\,kg^{-1}$ to $0.06\,kg\,kg^{-1}$ (with $X_e = 0.03\,kg\,kg^{-1}$), for which the diffusion coefficient is $0.53 \times 10^{-9}\,m^2\,s^{-1}$ (Biggerstaff 1965), the predicted drying time is 4 days. However, Hougen et al.

(1940) point out that the drying of very wet wood is likely, at least for permeable species, to involve capillary movement of moisture as well as diffusion, so the use of the diffusion equation is questionable in these cases.

The use of the diffusion model has been more successful with highly impermeable species (particularly hardwoods), with some work showing good agreement between the model predictions and observations for both moisture-content profiles and average moisture contents. Where the lumber is very permeable, free water movement is probable at much higher moisture contents than the fibre-saturation point, and diffusion is unlikely to be the transport mechanism for this process. This suggests that a diffusion model is unlikely to be appropriate for describing the drying of highly permeable species from green, particularly easily-dried softwoods, and is more likely to be useful for describing the drying of some hardwood and impermeable softwood species.

A characteristic feature of the most successful applications of a diffusion model, and an implicit comment on the semiempirical nature of the applications, has been that it has been possible to fit diffusion coefficients as functions of temperature and moisture content using comparatively simple relationships in these cases. The use of more complex fitting relationships in other cases creates the suspicion that diffusion is not the principal moisture-transport mechanism, and that a diffusion-like model may not be appropriate. For example, Collignan et al. (1993) presented a drying model for maritime pine (*Pinus pinaster*) which consists of dividing the drying time into three periods. A diffusion model was applied throughout the drying time, with a different equation being used to fit the diffusion coefficient for each period. Each of the three equations used for the diffusion coefficient was a complicated function of the temperature, moisture content and density, resulting in an overall model of considerable complexity.

Whenever diffusion data are analysed assuming moisture-content driving forces, both temperature and moisture content appear to be exponentially related to the diffusion coefficient:

$$D = D_o \exp\left(CX + \frac{E}{RT}\right). \tag{5.64}$$

Most fitted values for the activation energies (E) are in the range 23–43 kJ mol^{-1}, which are just below those for the evaporation of water (45 kJ mol^{-1}), while the theoretical model of Stamm (1967b) suggests an activation energy of around 35 kJ mol^{-1} ($B = 4230$ K). As the temperature is raised, the increase in the diffusion coefficient appears to be virtually proportional to the increase in vapour pressure of water, a finding which lends weight to the use of the partial pressure of water vapour as the driving force for diffusion. However, the effect of temperature on the diffusion coefficient persists even at moisture contents for which the vapour pressure is independent of moisture content (above the fibre-saturation point), suggesting that the use of a vapour-pressure driving force should not be an automatic choice. The use of a chemical-potential driving force results in effective water conductivities which remain strong functions of moisture content, so the use of this driving force on its own does not appear to offer significant advantages over the use of a moisture-content driving force.

The effect of moisture content on the diffusion coefficient is less clear than the effect of temperature. Most of the studies showing an effect of moisture content on the diffusion coefficient have been carried out on wood samples which are below the fibre-saturation point, for which the parameter C has been in the range 11–20. Above the fibre-saturation point, the experimental evidence suggests that there is no significant effect of moisture content on the apparent diffusion coefficient.

6 Multiple-Mechanism Models

Although describing moisture movement in terms of diffusion has been satisfactory in many practical cases, more attention has been paid in recent years to models in which several transport mechanisms for the movement of moisture are involved. It has been hoped thereby to gain a better understanding of the lumber-drying process, particularly for permeable species. As a long-term aim, such understanding should enable drying schedules to be improved while still maintaining product quality without the need for extensive pilot-scale kiln trials.

The first attempts to use a transport-based model for the drying of wood were based on the equations developed by Luikov (1966) for heat and mass-transfer processes in capillary-porous bodies. This theoretical basis was employed by Shubin (1990) to produce nomographs in terms of dimensionless parameters for estimating one-dimensional temperature and moisture-content distributions during the drying and preheating of lumber boards. The methods are extended to calculate two-dimensional transfer in anisotropic boards. Shubin (1990) also provides a more elaborate set of equations based on Luikov's ideas (1968) of using irreversible thermodynamics as the starting-point.

Moisture can be present as vapour or liquid and, should the ambient temperature be low enough, it can be frozen as ice. Moisture can move under moisture-content, temperature or pressure gradients. The various transport parameters, the thermal gradient coefficient for moisture transport and the pressure-induced permeation term, are all complex functions of the biophysical system as well as the moisture-content and temperature levels.

Since the derived equations are non-linear and contain coefficients that need to be determined from experiments, such a transport-based model is very complex and is strongly dependent on experimental data. Also, since bound water is not distinguished from liquid water or water vapour, Puiggali and Quintard (1992) suggest that the model should be modified to take into account the effect of sorbed water, particularly when the moisture content is below the fibre-saturation point.

Many recent drying studies (Kayahan 1982; Plumb et al. 1985; Stanish et al. 1986; Perré 1987; Chen and Pei 1989; Turner 1990) have described the drying process in lumber by considering different transport mechanisms for bound water, free liquid water and water vapour. Normally, each lumber board has been treated as a homogeneous, hygroscopic, porous material. However, these models can still become very detailed and unwieldy since physical properties, such as the way in which the aspiration of bordered pits affects the permeability, and differences between heartwood and sapwood, play an important role in the drying of lumber. While this approach is potentially capable of accounting for differences between heartwood and sapwood, as well as between earlywood and latewood, through the use of different transport parameters (Turner and Ferguson 1995a,b; Perré 1996a), some new studies (e.g. Slade et al. 1996) have continued to treat wood as homogeneous. The large number of transport parameters required in a "realistic" biophysical framework for the model may be a reason for the continuing use of a homogenous treatment. In some instances, however, with softwood species having an extensive growing season in temperate climates, the proportion of earlywood in the whole board can be very high (about 85% in the sapwood of *Pinus radiata* grown in favourable sites in New Zealand); under these conditions, a model assuming a homogeneous material may be considered a reasonable approximation.

The detailed model equations of Perré (1987) and Turner (1990), as compared by Turner and Perré (1995), both have the form of Luikov's equations, with their many transport parameters, which may be complex functions of temperature, moisture content, place within the log, site and species. Hence, extensive experimentation is likely to be required for each species and for lumber of the same species taken from different places in order to determine appropriate functional relationships. The resultant complexity of these models and the computing power needed to solve them reduce the utility of these ideas and limit their range of application in analysing lumber-drying practice (Kayihan 1993).

6.1 Fundamental Equations

This description is based on De'ev's translation (1999) of Shubin's monograph (1990) on the drying and heat treatment of wood.

From mass and energy balances over a volume element, a set of equations can be derived for the change of moisture content X, temperature T and pressure P with respect to time t:

$$\frac{\partial X}{\partial t} = D\rho_b \nabla^2 X + D\delta\rho_b \nabla^2 T \tag{6.1}$$

$$\frac{\partial T}{\partial t} = \lambda \nabla^2 T + \left[\frac{\varepsilon \Delta H_{LV}}{C_X}\right]\frac{\partial X}{\partial t} \tag{6.2}$$

$$\frac{\partial P}{\partial t} = \frac{RT}{\psi M_V}\left[\nabla\left(\frac{K}{\mu}\nabla P\right) + \varepsilon\rho_b \frac{\partial X}{\partial t}\right] + \frac{P}{T}\cdot\frac{\partial T}{\partial t}. \tag{6.3}$$

In the foregoing equations, the various parameters have the following meanings: D is a diffusion coefficient; ρ_b is the basic wood density, δ is the *thermal gradient coefficient*, which governs the amount of moisture being transferred under the temperature gradient; λ is the thermal conductivity; ε is the *phase-transformation criterion*, equal to the ratio of the amount of moisture being transferred as vapour to the total amount; ΔH_{LV} is the heat of phase transformation (desorption and evaporation); C_X is a modified heat capacity of unit mass of dry woody matter and its associated moisture; R is the universal gas constant; ψ is the fraction of the tracheids or vessels occupied by moisture vapour, M_V is the molar mass of the vapour; K is a *hydraulic conductivity*; and μ is the vapour viscosity.

These equations are very complicated to solve, even if the various coefficients were constant over the temperature and moisture-content ranges of interest, which they are not. Moreover, there are also a number of other difficulties hard to resolve. It is not entirely clear from Shubin's text what some of these parameters, such as the diffusion coefficient and thermal conductivity, actually refer to. Local thermal equilibrium is assumed, which may not hold under intensive high-temperature conditions. Equation (6.2) for the temperature gradient has been derived on the assumption that the fraction of convective heat transfer through the tracheids or vessels is negligible small. Another problem in solving these equations arises because the phase-transformation criterion ε varies from 0 to 1 and is difficult to determine experimentally with any precision.

To circumvent some of these difficulties, Luikov (1966) has suggested that the three-equation system can be replaced by a two-equation equivalent system:

$$\frac{\partial T}{\partial t} = \lambda_{ef} \nabla^2 T + \kappa \cdot (\nabla T)^2 \tag{6.4}$$

$$\frac{\partial X}{\partial t} = D_V \nabla^2 X + D_V \delta_{ef} \nabla^2 T, \tag{6.5}$$

in which the three new coefficients, λ_{ef}, δ_{ef} and κ, are defined in the following way:

$$\lambda_{ef} = \lambda + \frac{\Delta H_{LV} K}{C_X \rho_b \mu} \cdot \frac{\partial P}{\partial T} \tag{6.6}$$

$$\delta_{ef} = \delta_L + \frac{K}{\rho_b \mu D_L} \cdot \frac{\partial P}{\partial T} \tag{6.7}$$

$$\kappa = \frac{C_V K}{C_X \rho_b \mu} \cdot \frac{\partial P}{\partial T}. \tag{6.8}$$

Shubin (1990) notes, however, that the effective thermal conductivity λ_{ef} is little different from the actual measured value λ.

Rosen (1987) produced a set of expressions similar to Eq. (6.1) to (6.3), except that the moisture-content gradient is also considered to be a function of the pressure difference. A further parameter, a *pressure-gradient coefficient* analogous to δ for thermal effects, is incorporated in the pressure-difference term. Significant filtration effects are only expected, however, under intensive drying conditions with moderately permeable wood when the internal pressures in the drying lumber boards can rise significantly above ambient values. As discussed by Rosen (1987), if the flow of moisture is unidirectional, there is local thermodynamic equilibrium, and water vapour is only a small part of the total moisture content. This leads to a simplification of the starting equations and enables the local moisture content and relative humidity to be coupled through the moisture isotherm.

The ponderous nature of the Luikov-type equations has led workers to find other frameworks for analysis through a mechanistic interpretation of observed biophysical phenomena observed on drying lumber, and we now turn to these approaches.

6.2 Experimental Observations

6.2.1 Temperature Profiles

Shubin (1990) notes that the temperature fields during the drying of lumber boards or veneers may take one of four forms (Fig. 6.1). If the moisture-content is below the fibre-saturation point at the start, the temperature field develops progressively without reaching any intermediate steady-state values, as shown in Fig. 6.1a. The rate of temperature development depends primarily on the board thickness; the thicker the board the earlier the surface temperature reaches the dry-bulb temperature. The profiles shown in Fig. 6.1b represent the limit of a process governed by a so-called *constant drying rate* when there is no temperature gradient within the material. This limit is approached in the drying of veneers. More commonly observed in the drying of board lumber from green are the temperature profiles shown in Fig. 6.1c. The surface temperature rises progressively from the initial value to the dry-bulb temperature, but temperatures below the surface show a steady-state temperature intermediate between the wet- and dry-bulb temperatures for varying periods of time. This intermediate temperature is the *pseudo-wet-bulb temperature* seen in the drying of thick porous materials (Nissan et al. 1959). In high-temperature drying, that temperature can be the boiling point. The curves shown in Fig 6.1d represent an intermediate behaviour between that sketched in Figs. 6.1b and 6.1c, when, in the drying of thin pieces and veneers at high temperatures (>160°C), there is almost no intermediate steady-state temperatures, but the material temperature may hover at the wet-bulb value for a short period at the start of drying.

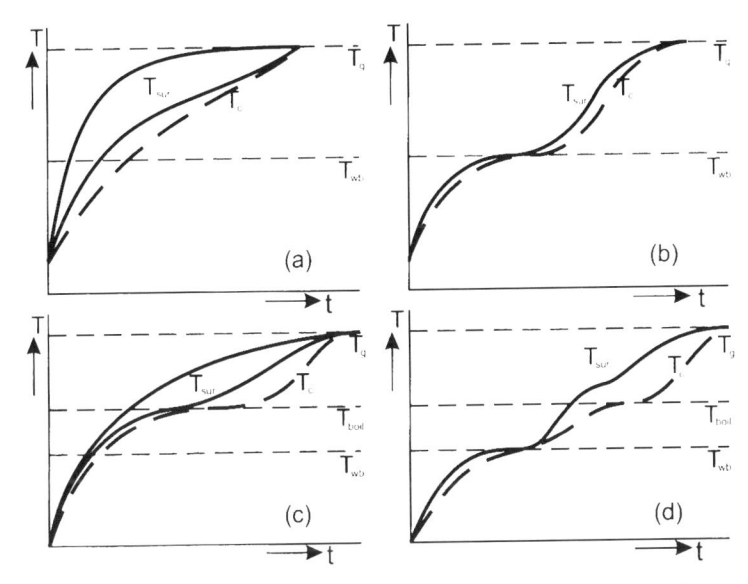

Fig. 6.1a–d. Temperature-rise curves on drying wood. a Drying from below fibre saturation; Biot number for curve *1* is less than that for *2*. b Drying of veneers and thin samples at low temperatures. c Drying of thick board lumber. d Drying of veneers and thin samples at high temperatures. (After Shubin 1990)

Stanish et al. (1986) used their model to predict temperature profiles in boards of southern pine (*P. palustris* etc) and Douglas-fir (*Pseudotsuga menziesii*) during high-temperature drying. The resulting simulated profiles are complex. However, their results indicate that the core temperature approaches a plateau temperature of about 100 °C and remains there for a substantial part of the process in a similar way to the profile sketched in Fig. 6.1c. Similar findings are presented by Northway (1989) and Beard et al. (1982, 1985) who have carried out high-temperature drying tests on *Pinus radiata* and yellow poplar (*Liriodendron tulipifera*). Their results show that the temperature at different depths of the drying board initially reaches a steady value of 100 °C before beginning to rise progressively. This pattern of temperature-time profiles at different depths in the timber suggests the presence of a receding evaporative plane which may remain quasistationary.

6.2.2 Moisture-Content Profiles

Slade et al. (1996) justify a complex mathematical model of moisture movement based on the comparison of the final measured and predicted moisture-content profiles. This is not a severe test. Indeed, it could be argued that almost any model, including a simple diffusional one, is likely to yield satisfactory agreement at the end of drying, when the final average moisture content is under 20% and close to equilibrium values. A critical feature of any comparison between theory and experiment is the extent to which the model can predict the measured moisture-content profile at intermediate times.

Hardwoods and the heartwood of softwoods present very different moisture-content profiles to the sapwood of softwood, which is a reflection of the relative permeability and ultrastructure of the various woods. Shubin (1990) presents data, reproduced in Fig. 6.2, for the moisture-content distribution in drying beech (*Fagus sylvatica*) and the heartwood of pine (*Pinus* sp.) which illustrate diffusion-like parabolic profiles. By contrast, the drying of larch (*Larix* sp.) exhibits steep moisture gradients which are consistent with the presence of an evaporative plane close to the surface.

The parabolic moisture-content profiles associated with diffusional movement are observed in the low-temperature experiments of Wu (1989), who dried Tasmanian hardwoods (*Eucalyptus* spp.) at temperatures of less than 30 °C. However, not all hardwoods show such simple profiles. Grace (1996), who examined under room conditions the drying behaviour of a southern beech, *Nothofagus truncata*, found bell-shaped moisture-content distributions in wood samples 50 mm thick (Fig. 6.3). He attributed the shape of these profiles to the relative speed at which the exposed surfaces reached near-equilibrium conditions with this very impermeable wood (the moisture-diffusion coefficients varying between 10^{-10} and 10^{-11} $m^2 s^{-1}$) compared with the slow diffusion within the core. This diffusion appeared to be a two-stage process, which Grace attributed to a primary diffusion from the vessels to the cell-wall matrix and a secondary migration through or along it. A similar set of features, with both characteristics of diffusional and receding-plane behaviour, was found by Cloutier et al. (1992) when drying aspen (*Populus* sp.) from moisture contents of 160 to 180% at temperatures between 20 and 50 °C. Such receding-plane behaviour can be explained, however, in terms of a penetrating diffusion front rather than an evaporative one characteristic of softwoods.

A receding evaporative plane implies a steep moisture-content gradient over a narrow zone that moves into a board of lumber as drying proceeds, unlike a diffusional process, for which the gradients nearly always become steeper towards the surface. For samples of a softwood, loblolly pine (*Pinus taeda*) dried at temperatures of 105 to 150 °C, the moisture-content profiles measured by Wiley and Choong (1975) show evidence of an evaporative front, as do the experimental data of Plumb et al. (1985) for southern pine and those of Furuyama et al. (1994) for akamatsu (*Pinus densiflora*) sapwood and heartwood, Douglasfir (*Pseudotsuga menzesii*), hinoki (*Chamaecyaris obtusa*) and sugi (*Crytomeria japonica*) dried at 60 °C. The measured moisture-content profile (Stamm 1967a) for a 25-mm-thick board of sitka spruce (*Picea sitkensis*), dried on one side from a water-saturated condition at 40 °C and zero relative humidity shows a receding evaporative front very clearly. The semi-

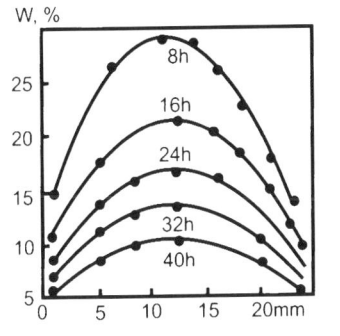

Fig. 6.2a–c. Moisture-content distributions. **a** Drying of 25-mm beech (*Fagus sylvatica*) at 80/33 °C dry/wet-bulb temperatures. **b** Drying of 25-mm pine heartwood at 70/52 °C dry/wet-bulb temperatures. **c** Drying of 54-mm larch (*Larix* sp.) at 108 °C (wet-bulb temperature not given). (After Shubin 1990)

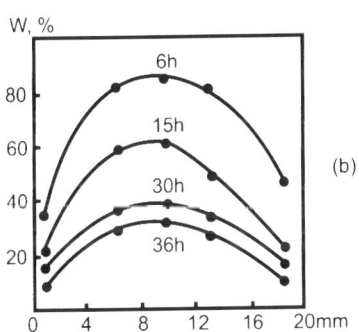

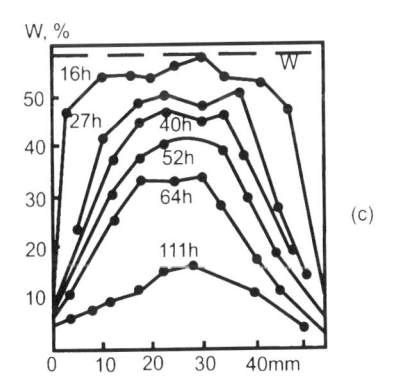

empirical model of Tang et al. (1994), which correlated the press-drying times of loblolly pine (*Pinus taeda*) at temperatures of 350–475 °F (177–246 °C) incorporated the assumption that:

"moisture evaporates from wood only at an interface where the moisture remains at the fibre-saturation point (FSP) and the temperature remains at the boiling point".

Early versions of the softwood model described by Pang (1994) assumed the presence of an evaporative plane explicitly for ease of programming. However, the presence of a receding evaporative plane is a consequence of the assumed

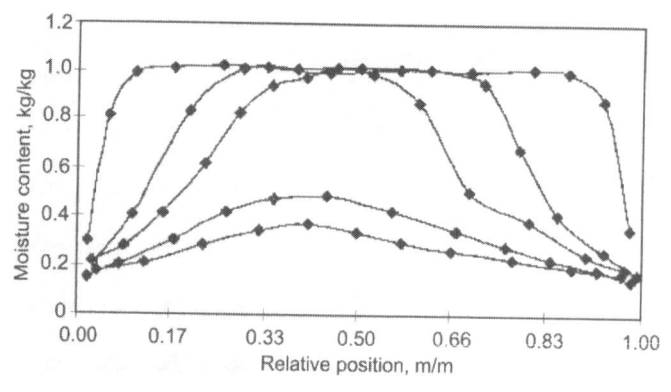

Fig. 6.3. Moisture-content profiles on drying 50-mm-thick samples of heartwood *Nothofagus truncata* at 20 °C. (After Grace 1996)

moisture-content mechanisms, as demonstrated in the later control-volume implementation of this model by Pang (1996b). When a modified version of Pang's early model is used, Chen et al. (1996a) predict moisture-content profiles which resemble the experimentally measured profiles obtained by Plumb et al. (1985), both at intermediate times and towards the end of drying.

Initially, the moisture-content profiles in the core for a permeable species are relatively flat since moisture is assumed to be moving as liquid under a pressure gradient. The gradual loss of permeability, with a slight steepening of the moisture-content gradients in the core, has been modelled for the sapwood of a softwood by assuming that random aspiration of the bordered pits occurs, thus closing off liquid pathways as the pits seal or isolating still-filled, unaspirated tracheids (Chen et al. 1996b). The relative uniformity of the moisture content in the core at moisture contents well above fibre saturation has been found in tests reported by McCurdy and Keey (1998) on the high-temperature drying of a 50-mm board of *Pinus radiata* at 120 °C. These tests with flat-sawn, sapwood boards, having wide growth rings essentially parallel to the drying faces, show another interesting feature: below a moisture content of about $0.6\,kg\,kg^{-1}$ the moisture-loss behaviour suddenly changes; the moisture content falls in stepwise fashion across each growth ring with an apparent resistance across each latewood zone (Fig. 6.4). Interestingly, a transitional moisture content of $0.6\,kg\,kg^{-1}$ is predicted in the model of Chen et al. (1996a) as a result of the breakdown of pressure-driven, free-moisture flow. They term this condition *irreducible saturation*, by analogy with liquid flow in porous media, and which occurs at a moisture content significantly higher than the acknowledged fibre-saturation point.

The actual steepness of the moisture-content profiles around irreducible saturation predicted by Chen et al. (1996a) and Pang (1996b) is much greater than the gradients found in the experimental data of Plumb et al. (1985). This is believed to be an artefact of the models, which assume a sudden transition

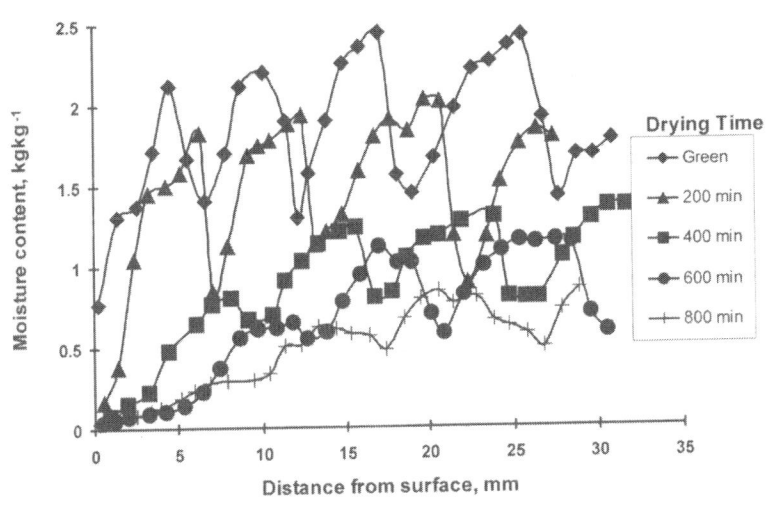

Fig. 6.4. Moisture-content profiles in a flat-sawn 100 × 50 mm board of *Pinus radiata* sapwood on drying in air at 5 m s^{-1} and dry/wet-bulb temperatures of 120/70 °C. (McCurdy and Keey 1998)

in the way that moisture moves at irreducible saturation. Nijdam (1998) shows that by permitting both moisture-vapour flow and bound-moisture movement between irreducible saturation and fibre saturation and to lower moisture contents the discontinuity in the predicted moisture-content profile can be eliminated, with less steep gradients being predicted more like those found experimentally.

In general, then, with impermeable hardwoods, diffusional mechanisms seem to apply. With more permeable species, an evaporative front appears to reside in the material, which may stay close to the surface or withdraw into the board, as seen in the drying of other thick porous materials such as wound textile bobbins (Nissan et al. 1959).

6.2.3 Pressure Profiles

Another implication of the results, both from the simulation by Stanish et al. (1986) and the drying experiments of Beard et al. (1985), as well as those of Northway (1989), is that the core of the board is not significantly above atmospheric pressure since a boiling temperature higher than normal for the water would result otherwise. This feature contradicts the results of the simulation reported by Perré et al. (1988), who predicted centre-line pressures in the drying of a spruce (*Picea rubens*) as high as 0.24 MPa. Figure 6.5 shows the temperatures and pressures recorded by Basilico et al. (1982) in the drying of this wood, when the internal pressure has reached a maximum just above

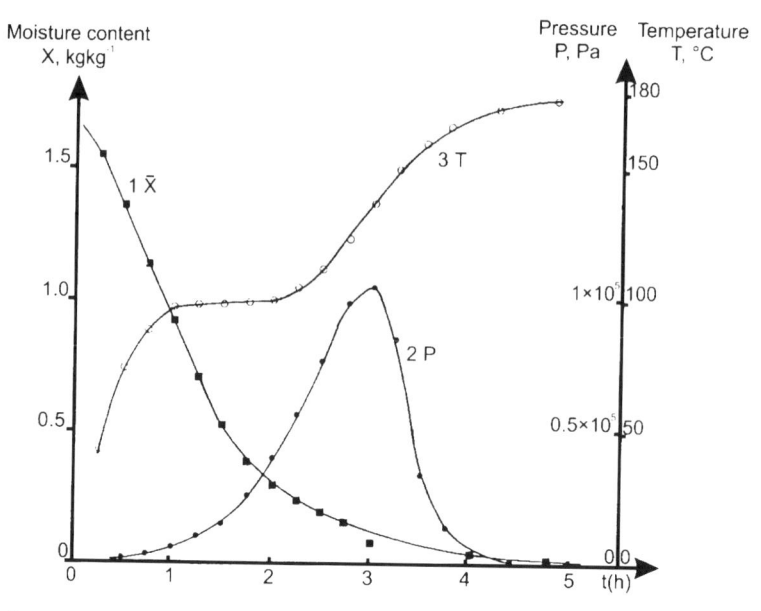

Fig. 6.5. Average moisture contents, wood temperatures and internal pressures in the high-temperature drying of spruce (*Picea rubens*); air velocity 12.4 m s^{-1}; dry-bulb temperature 184 °C (*Picea rubens*). (Basilico et al. 1982)

atmospheric after the temperature has started to rise above the normal boiling point.

Perré (1996a,b) maintains that the internal pressure is an important driving force for the drying of timber at high temperatures, although some theoretical calculations predict higher overpressures than actually observed experimentally. For example, in the drying of fir (*Abies grandis*) in superheated steam at 160 °C, Perré (1996a,b) shows a maximum theoretical overpressure of 0.17 MPa whereas the experimental value is given as 0.09 MPa, but he points out that the predicted overpressure is very sensitive to the value of the transverse permeability. Experimentally-observed overpressures to 0.06 MPa were observed by Antti (1992a,b) for the microwave drying of birch (*Betula* sp.), while overpressures of 0.1 MPa were noted when drying spruce (*Picea* sp.) at 110–138 °C by Lowery (1972). Hence, the existence of these overpressures during drying of some softwoods has experimental support, although much less likely with very permeable species such as *Pinus radiata*.

There is also an apparent contradiction between the appearance of positive pressures and the observed phenomena of collapse and internal checking in the early stages of drying, since this behaviour has been attributed to the action of tensile stresses, or negative pressures, caused by tensile forces in the liquid held in the lumens. Booker (1996) suggested that this apparent contradiction may be overcome by assuming that the liquid moisture in the lumens of greenwood is superstressed to some degree and so can withstand some amount of excess pressure before it vaporises. In such a case, there would be a metastable condition for certain periods in the drying of lumber boards, and that local thermodynamic equilibrium may not hold, particularly at the higher moisture contents in the early stages of drying. In a classical work, water tensions up to 28 MPa were obtained when separating polished anvils (Briggs 1950), while Sedgewick and Trevans (1976) have routinely obtained tensions up to 4 MPa in Berelot tubes for sap and water. Whenever the metastable condition is broken, a spontaneous transition, known as *cavitation*, takes place, creating a vapour bubble inside the lumen, when both the vapour and liquid are at the saturation vaour pressure and local thermodynamic equilibrium is restored.

The process of creating surface area in an expanding bubble requires considerable energy, which can be quantified by the Laplace equation:

$$P_1 - P_2 = 2\gamma/r, \tag{6.9}$$

where P_1 is the pressure inside the bubble and P_2 is the pressure within the surrounding fluid, γ is the surface tension and r is the radius of the bubble assumed spherical. In the context of cavitation, the Laplace equation can be cited to show that the nucleation of a vapour bubble is very difficult (Tabor 1969). To open up a spherical hole of the same radius as a water molecule (0.3 nm), the Laplace equation suggests that a vapour pressure of 480 MPa would be needed. This pressure is so large that cohesive failure between the water and the cell wall is more likely to happen first and initiate cavitation.

6.3 The Physical Process of Drying for a Softwood, *Pinus radiata*

Softwoods such as *Pinus radiata* have a relatively simple system of hollow-celled fibres or tracheids providing both structural support and sap conducting pathways. The bordered pits in sapwood, when open or unaspirated, allow fluid to pass between adjacent tracheids. The outer portion of a tree steam is sapwood, while heartwood begins to be formed within the trunk once a tree has reached a certain age in its development. This stage is reached at different ages for different species, as described in more detail in Chapter 7. The ray parenchyma cells are dead in heartwood, which is characterised by the accumulation of polyphenolic resins and a sharp reduction in moisture content.

The properties of softwood timber vary significantly from sapwood to heartwood. The bordered pits in the cell walls between longitudinal tracheids generally aspirate when conifers such as *Pinus radiata* form heartwood, possibly initiated by the formation of resins (Booker 1996). This process significantly retards liquid flow within the timber. However, the pits in sapwood are usually not aspirated when the sapwood is green, since the pits are required for the transport of liquid material. They aspirate progressively when the sapwood is dried. A dry zone of intermediate wood between the sapwood and the true heartwood has intermediate properties. Aspiration also occurs in sapwood after it is sawn from the log as a physiological response to "heal" the damage. This aspiration process is irreversible.

The degree of pit aspiration governs the permeability of the timber to the flow of liquid. The permeability of the timber to liquid flow becomes very low once the pits are aspirated and the pits are blocked; drying can then proceed only by diffusional mechanisms.

6.3.1 Heartwood

The process of drying for heartwood of a permeable species may be divided into three periods. All the pits are aspirated. During the first and second periods, it is postulated that an evaporative plane (at which all the free water evaporates) exists. This plane will recede into the board of timber as drying proceeds. The evaporative plane divides the material into two parts, a wet zone beneath the plane and a dry zone above it. In the dry zone, moisture is assumed to exist as bound water and water vapour. Bound water is likely to be in local thermodynamic equilibrium with water vapour at the local temperature (Stanish et al. 1986). In the wet zone, the moisture content remains at the initial green value and the partial pressure of water vapour is equal to that at the evaporative plane: however, unlike in sapwood, the green moisture content in the wet zone is usually between 40 and 60%.

For the first period, when the evaporative plane has only just receded into the board, the external transfer process controls drying. However, as the plane withdraws, the moisture-transfer resistance within the wood becomes significant and the drying is now internally controlled in this second period. The third drying period occurs after the evaporative plane has reached the centre-line of the board, when the moisture movement is then maintained by bound-water diffusion and the flow of water vapour.

6.3.2 Sapwood

In the drying of sapwood boards, the flow of moisture just below the surface is insignificant, since board surfaces are damaged during milling, causing the cells in the timber to aspirate close to the surface in response to the damage. Therefore, at the beginning of the first drying period for sapwood, the evaporative plane recedes into the material very quickly. However, when the plane reaches a certain distance (ξ_o) from the surface, most of the pits inside this position are not yet aspirated and the flow of liquid is possible. The liquid flow towards the surface will keep the evaporative plane at the position ξ_o until the moisture content at the plane decreases to the minimum value for liquid continuity. Booker (1989) has postulated that cavitation occurs in the tracheids at random as a result of rising capillary pressures. Because of the very large suction potential across any pits connecting these tracheids to their neighbours, these pits rapidly shut and the pit membranes become dish-shaped. Since initially this cavitation-induced aspiration takes place within only a small fraction of the many tracheids in a given board, there remain enough capillary pathways for liquid to flow towards the exposed surfaces for some time.

After some time, when tracheid cavitation and pit aspiration have become so extensive that there are few contiguous pathways to the surface, the evaporative zone starts to withdraw into the material, and the second period of drying starts. The main difference between the drying behaviour of sapwood and heartwood during the second period (whilst the evaporative plane is receding) is that the liquid flow in the wet zone for sapwood has a significant influence on the velocity of the receding evaporative plane. Thereafter, the drying process for sapwood in the third period is similar to that for heartwood.

There is experimental evidence for the existence of a thin dry layer in sapwood boards. A section through a flat-sawn sapwood board shows a stain line about 1 mm below the surface. This phenomenon is called kiln brownstain, which is a brown discoloration that develops during the kiln-drying of a permeable sapwood but does not affect heartwood (Haslett 1998). The position of this stained zone is consistent with the expected position for the evaporative zone during the first drying period for sapwood when compounds such as cyclitols, sugars, amino acids and other water-soluble materials would migrate to the evaporative zone where they could react and deposit. At higher kiln temperatures, lignin-degradation compounds, carried there by the liquid flow, may add to the colouring.

These three drying periods are illustrated in Figure 6.6.

Although the permeability in the longitudinal direction is many times that across the grain, the geometry of a lumber board is such that the nett moisture movement is essentially one-dimensional, normal to the longest face, except near the ends. The one-dimensional moisture-transport equations have been presented by Chen et al. (1996a) and are summarised in the following section.

6.3.3 The Moisture-Transport Equations

The moisture within wood may exist in three forms: free (liquid) water in the tracheid lumen, bound water in the cell walls and water vapour. In sapwood, the free water will be removed first. It is drawn towards the wood surface by capillary forces and then evaporates at the evaporative front, which initially

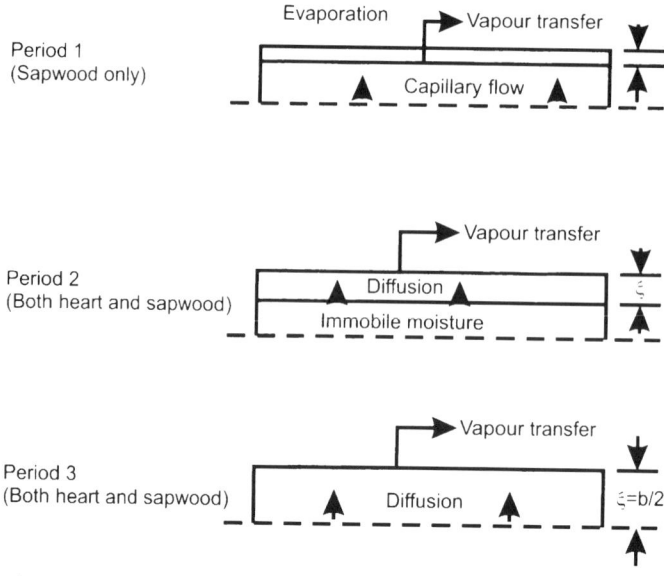

Fig. 6.6. Periods of drying for softwood lumber. (Keey 1994)

will be very close to the surface. This process will continue until liquid continuity ceases. After this stage, the evaporation zone will gradually withdraw into the wood towards the centreline of the board. The evaporation zone divides the board into dry and wet zones (Figure 6.7). The final stage of drying occurs when moisture exists only in an immobile condensed form (bound water) and as water vapour.

6.3.3.1 The Movement of Free Moisture

The flow of mobile, free moisture occurs due to a pressure gradient in the liquid phase, and its flux is governed by the equation:

$$j_{wf} = -E_l \frac{\partial p_l}{\partial z}, \tag{6.10}$$

where p_l is the pressure in the liquid phase and E_l is the effective permeability to liquid flow. This permeability may be related to the commonly-measured liquid permeability k_l by the equation

$$E_l = \frac{k_l \rho_l}{\mu_l}, \tag{6.11}$$

in which ρ_l and μ_l are the density and viscosity of liquid water, respectively.

In any lumen partially filled with water, a meniscus forms between the liquid and gas phases, and a force balance gives

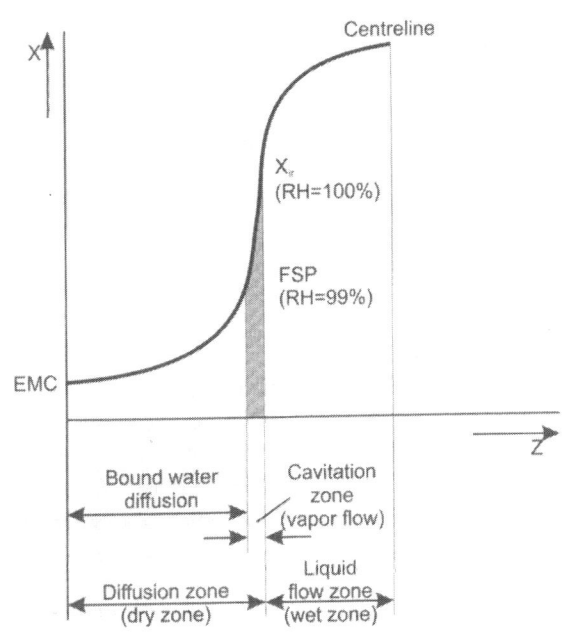

Fig. 6.7. Moisture-flow zones in a softwood board during drying. (Chen et al. 1996a)

$$p_g = p_l + p_c \tag{6.12}$$

or

$$p_l = p_g - p_c, \tag{6.13}$$

where p_c is the capillary pressure, so

$$\frac{\partial p_l}{\partial z} = -\frac{\partial p_c}{\partial z}. \tag{6.14}$$

The capillary pressure p_c is a function of the *saturation s* (Spolek and Plumb 1981), with the saturation defined by

$$s = \frac{\text{Liquid volume}}{\text{Void volume}} = \frac{X - X_{FSP}}{X_{max} - X_{FSP}}, \tag{6.15}$$

where X_{max} is the moisture content of the wood if the entire void structure were filled with water (the maximum moisture content) and X_{FSP} is the moisture content at the fibre-saturation point when the lumens contain no liquid water. For softwoods, Spolek and Plumb (1981) have assumed that the capillary pressure is a simple algebraic function of saturation:

$$p_c = A\, s^B, \tag{6.16}$$

where A and B are constants. For southern pine (*Pinus palustris* etc), Spolek and Plumb (1981) give $A = 12\,400$ Pa and $B = 0.61$, while Chen et al. (1996a) have assumed that these values described the capillary behaviour of *Pinus radiata* as well.

The liquid permeability k_l depends on the degree of pit aspiration. Chen et al. (1996b) assume that the individual tracheids in the wet zone cavitate as the tension build up on free moisture transfer. This reduces the effective volume-averaged permeability as local liquid continuity is disrupted. If the cavitation is presumed to be a random process, then the variation of permeability with moisture saturation is governed by the relation

$$\frac{k_l}{k_l^s} = \exp\left[-\frac{\alpha(X_o - X)}{X_o - X_i}\right], \tag{6.17}$$

where the coefficient α is a constant representing the fractional cavitation rate and k_l^s denotes the liquid permeability at full saturation. Equation (6.17) is valid from the start of cavitation (X_o) down to the so-called *irreducible moisture content* (X_i), when a small amount of remaining moisture may be trapped in the end of the tracheids or filled tracheids have become isolated. Chen et al. (1996a) assumed that cavitation begins once the moisture content fell below 80% of the greenwood value, while the irreducible moisture content was taken to be at 40% of the initial value at $0.6\,\text{kg}\,\text{kg}^{-1}$. A mean cavitation-rate coefficient of 2.5 was also assumed. These considerations lead to the conclusion that about 70–95% of the tracheids will have effectively cavitated at irreducible saturation depending on the cavitation-rate coefficient. These values may be compared with Donaldson's observation (pers. comm.) that about one-third of the pits in kiln-dried *Pinus radiata* are unaspirated.

This approach to estimating the variation in liquid permeability with moisture content is similar to that used by both Stanish et al. (1986) and Pang et al. (1992), who used an expression of the kind

$$\frac{k_l}{k_l^s} = 1 - \cos\left[\frac{\pi}{2}\left(\frac{s - s_{min}}{1 - s_{min}}\right)\right], \tag{6.18}$$

where s_{min} is the minimum saturation for liquid flow within the wood. Minimum values of the saturation in the range of 0.1 to 0.32 were used by Stanish et al. (1986). For *Pinus radiata*, these saturations would correspond to moisture contents between $0.42\,\text{kg}\,\text{kg}^{-1}$ and $0.68\,\text{kg}\,\text{kg}^{-1}$, a range which includes the irreducible moisture content of $0.6\,\text{kg}\,\text{kg}^{-1}$ used by Chen et al. (1996a). The basis for Eq. (6.18) is the experimental finding of Tesoro et al. (1974) that the relative permeability increases with increasing relative saturation, often in a sigmoidal manner.

6.3.3.2 The Movement of Water Vapour

For water-vapour movement in the dry zone, the vapour flux is governed by the equation

$$j_{wv} = -E_b \frac{\partial p_v}{\partial z}, \tag{6.19}$$

in which E_v is the effective permeability to vapour flow and p_v is the vapour partial pressure determined by the local temperature and moisture content. As with Eq. (6.11), the permeability E_v may be related to the commonly-measured vapour permeability k_G by the equation

$$E_v = \frac{k_G \rho_G}{\mu_G} \tag{6.20}$$

in which the vapour density ρ_G and viscosity μ_G depend on the local gas temperature.

Comstock (1968) indicates that the square root of gas permeability decreases linearly as the percentage moisture content increases from 0–20% for eastern white pine (*Pinus stroba*) and western hemlock (*Tsuga heterophylla*). Between percentage moisture contents of 20 and 25%, the permeability drops rapidly, possibly due to condensation of moisture on the pit-membrane openings. This suggests that a functional relationship between gas permeability and moisture content could be used in the dry zone.

6.3.3.3 The Movement of Bound Water

For the diffusion of bound water, the chemical potential, as used by Stanish et al. (1986) and described in Eq. (5.40) to (5.43), has been adopted as the driving force:

$$j_{wb} = -D(1 - \varepsilon_d) \frac{\partial \mu_b}{\partial z}, \tag{6.21}$$

in which μ_b is the chemical potential of bound water, D_b is the bound-water transport coefficient and ε_d is the fractional void space. By using the equation above, the need for assumptions regarding temperature-driven diffusion is eliminated because a contribution from the temperature gradient arises automatically in the calculation of chemical potential. Movement of bound water is assumed to occur below the fibre-saturation point.

6.3.4 The Evaporative Zone

During the drying of sapwood, the water tension inside the water-filled cells will continue to increase until it reaches a maximum threshold value when water cavitates inside a tracheid. Inside the evaporative zone shown in Fig. 6.7, this progressive cavitation will reduce the liquid permeability as the moisture is expelled from the tracheids until the liquid-phase moisture movement to the surface cannot sustain the evaporative loss. Once liquid continuity has broken down, the moisture transfer in this zone will be dominated by the vapour flow instead (Strumillo and Kudra 1986). This transfer will be driven by a small vapour-pressure difference between the breakdown moisture content (X_i) and the fibre-saturation point (X_{FSP}), which can be estimated from the sorption isotherm of Simpson and Rosen (1981):

$$X_{FSP} = \frac{18}{K_0}\left[\frac{K_1 K_2 \psi}{1+K_1 K_2 \psi} + \frac{K_2 \psi}{1-K_2 \psi}\right], \tag{6.22}$$

in which the relative humidity ψ may be taken to be 0.99 (Skaar 1988). The other coefficients in the above equation are temperature-dependent and can be found in the *Dry Kiln Operator's Manual* (Rasmussen 1961). These coefficients are as follows:

$$K_0 = 187.6 + 0.6942T + 0.01853T^2 \tag{6.23}$$

$$K_1 = 9.864 + 4.773 \times 10^{-2}T - 5.012 \times 10^{-4}T^2 \tag{6.24}$$

$$K_2 = 0.7196 + 1.698 \times 10^{-3}T - 5.553 \times 10^{-6}T^2 \tag{6.25}$$

where the temperature is in degrees Celsius.

6.3.5 Mass and Energy Conservation Equations

The conservation equations for mass and energy, as derived by Pang et al. (1994), can be written respectively as:

$$-\rho_w \frac{\partial X}{\partial t} = \frac{\partial j_w}{\partial z} \tag{6.26}$$

$$\frac{\partial[C_p \rho_w (1+X)T]}{\partial t} = \frac{\partial}{\partial z}\left(\lambda \frac{\partial T}{\partial z}\right) + \Phi, \tag{6.27}$$

where ρ_w is the density of dry wood and λ is the thermal conductivity of moist wood. The water flux j_w in Eq. (6.26) is either the vapour and bound water flux ($j_{wv} + j_{wb}$) for the dry zone (including the cavitation zone) or the free moisture flux j_{wf} for the wet zone.

The term Φ in the energy Eq. (6.27) was described by Pang et al. (1992), Pang (1996a) and Chen et al. (1996a) as the source term. Since it is due to bulk movement of all components of moisture together with their associated enthalpy, it is strictly speaking a convective component of energy flow. It is equal to the gradient in the sum over all products of each component of moisture multiplied by its enthalpy:

$$\Phi = \frac{\partial}{\partial z}\left(\sum h_w j_w\right), \quad (6.28)$$

where h_w is the enthalpy of each component of moisture. Equations (6.26) and (6.27) are based on a total energy balance over a control volume. Pang et al. (1992), Pang (1996a) and Chen et al. (1996a) have used an energy balance based on the dry wood, which gives

$$C_p \rho_w \frac{\partial T}{\partial t} = \frac{\partial}{\partial z}\left(\lambda \frac{\partial T}{\partial z}\right) + \Phi, \quad (6.29)$$

with a "source term" which uses the latent heat of vaporisation and the heat of sorption rather than the total vapour enthalpy and bound water enthalpy.

6.3.6 Initial Conditions

The initial temperature and moisture content have usually been assumed to be uniform throughout the material (Chen et al., 1996a; Pang et al., 1994a), but variations through the material could be incorporated if desired.

6.3.7 Boundary Conditions

The boundary conditions are independent of the internal model used, whether this be based on diffusion or a receding evaporative interface.

At the surface, the flux of moisture is the flux of water vapour, and this may be specified as

$$j_{wv} = \beta(p_{vs} - p_{vG}), \quad (6.30)$$

where β is the external mass-transfer coefficient, p_{vs} is the vapour pressure above the surface of the board and p_{vG} is the partial pressure of vapour in the bulk gas. The energy flux at the surface of the board (E_S) consists of a convective energy component and a component due to the flux of vapour away from the surface:

$$E_S = h(T_G - T_S) - h_{wv} j_{wv} \quad (6.31)$$

in which h is the external heat-transfer coefficient, T_G is the temperature of the gas, T_S is the surface temperature and h_{wv} is the vapour enthalpy for the vapour leaving the surface.

At the centre-line of the board, there should be no nett flux and the gradients in both temperature and moisture content should be zero there:

$$\frac{\partial X}{\partial z} = 0 \quad (6.32)$$

$$\frac{\partial T}{\partial z} = 0 \tag{6.33}$$

A control-volume technique, as described in Pang (1996b) and Wan and Langrish (1995), is an appropriate one for solving this set of equations.

6.4 Mixed-Wood Boards

In practice, often a number of boards sawn from a softwood log will contain both heartwood and sapwood which have quite different permeabilities. Between the heartwood and sapwood, there is a transition layer in which the ray parenchyma cells are alive and not impregnated with polyphenols, but it differs from sapwood in that moisture has been lost and the pits are aspirated. For *Pinus radiata*, Booker (1990) finds this transition layer to be only one growth-ring wide, which is too narrow for its permeability to be measured. However, earlier Harris (1954) gave some experimental data on the fraction of aspirated bordered pits in greenwood from the last-formed growth ring to the pith. He observed that the transition zone between the sapwood and heartwood has almost the same percentage of aspirated pits as does the heartwood, while the moisture content in that layer drops abruptly to a level similar to that in heartwood. Effectively, the transition layer is as impermeable to liquid moisture movement as the heartwood zone itself.

Pang et al. (1994a) note that with a mixed-wood board, in the case where the heartwood layer is less than that of the sapwood, the moisture in the sapwood must either flow through a greater distance to the exposed sapwood face or migrate a shorter distance through the more impermeable layers to the heartwood face. As the sapwood is very wet when green, the sapwood adjacent to the transitional layer dries very slowly and is usually the wettest part of the board at the later stages of kiln-drying. When the thickness of the heartwood zone is equal to or greater than that of the sapwood, the drying of such a mixed board is similar to that of a wholly heartwood board since the considerably lower moisture content of the heartwood offsets its lower permeability.

The drying of a board containing mostly sapwood, then, is the one that is most likely to offer more difficulty than a board that does not contain mixed wood. In the first phase of drying of a mixed board, an evaporative front will withdraw into the heartwood zone, whereas the one in the sapwood remains close to the exposed surface, since the majority of the pits are unaspirated and liquid flow is possible. In the second phase, the heartwood's evaporative zone has reached the transition layer. Liquid in the wet sapwood now moves to both this layer and the exposed sapwood surface. In the subsequent drying, once the moisture reaches irreducible saturation at a front, it begins to move towards the other one; so both fronts start to converge, one originally from the transition layer and one from the vicinity of the exposed sapwood surface. In the final phase, the two evaporative fronts have met within the sapwood; thenceforth bound-moisture and water-vapour diffusion to both exposed surfaces govern the drying.

With this picture of moisture movement in mind, Pang et al. (1994a) adapted their earlier softwood-drying model (Pang et al. 1992, 1994b) to describe the drying of a 50-mm-thick mixed board containing mainly sapwood in a band 44-mm thick. Their predictions indicated that the drying of such a board is more difficult than either a pure sap or heartwood board. To reach a final moisture content of $0.06\,kg\,kg^{-1}$ at a dry-bulb temperature of 140 °C and a wet-bulb temperature of 90 °C, the drying time with an air velocity of $5\,m\,s^{-1}$ for an entirely heartwood board is 10 h, for a sapwood one 11 h, compared with that of 14 h for a mixed board.

These predictions were consistent with experiments on drying end-coated, flat-sawn boards, $100 \times 50 \times 350$ mm long, with the transition layer parallel to the two wider drying surfaces. However, the experimental fall in moisture content with time did not track the predicted profile exactly, probably due to deficiencies in the early form of the receding-front model employed. On the other hand, it was noted that the predicted surface temperature on the sapwood face of the board was lower than that on the heartwood face for about half of the drying process, while the centre-line temperature remained at approximately 100 °C for two-thirds of the whole period, in agreement with the asymmetrical temperature profiles measured.

The drying of mixed boards of hardwood species is unlikely to show such marked asymmetries in drying behaviour, as there would appear to be no equivalent period of drying akin to the early period in the drying of the sapwood of a permeable softwood species when moisture is moving as liquid flow.

6.5 Conclusions

A number of mechanisms appears to be involved in the drying of softwoods. In the heartwood of softwood species, only bound-water diffusion and water vapour movement are possible. In the sapwood, however, when the volume-averaged saturation of the lumens is about 40% or higher, liquid flow is possible. Overall, a qualitative description of moisture movement in softwoods predicts the presence of a receding evaporative zone, after dwelling near the surface for some time from the beginning of drying in the case of sapwood, but withdrawing immediately from the start in the case of heartwood. This prediction is consistent with the appearance of a brown stain-line during drying, particularly under high-temperature conditions. It is also consistent with experimental moisture-content and temperature profiles which show large differences over relatively small changes in depth within the board, and the shift of these differences with time.

The drying of hardwoods can generally be described in terms of diffusional mechanisms. A drying front, ostensibly similar to a receding evaporative zone, has been observed in the early stages of drying from greenwood. This behaviour corresponds to the penetration period (as discussed in Chapter 5) when moisture transfer is taking place only through a surface zone which is slowly "penetrating" the wet wood.

7 Lumber Quality

Lumber quality is thought of in terms of gross defects and intrinsic wood properties. Defects include features such as knots, checks, distortion and wane at the edge of boards cut from the outer wood, while the wood properties include anisotropy, density and stiffness. Drying, which causes shrinkage, interacts with particular combinations of these wood features to produce dried lumber which has a specific quality, defined in terms of a grade.

The attitude of the kiln operator to any gross defects and these intrinsic wood properties, in terms of the drying conditions to be imposed, is determined by the final products that the operator is seeking to supply. There is a world of difference between drying decorative woods and structural-grade lumber, as there is between drying hardwoods and softwoods.

Thinking on wood quality, and the influence of drying on it, is coloured by cultural fashions as much as by intellectual arguments. People can only relate to their experience with local species and traditional ways of doing things and, indeed, to their own concept of quality. For example, many people invariably value indigenous timbers more highly than an objective assessment of their intrinsic properties might indicate, and there is still a preference for solid wood over most veneered products. When one thinks of the significance of knots, then those familiar with the arboreal forests of Northern Europe or America will have a totally different perspective to someone from the southern United States or South America.

Moreover, wood from younger stands or wood grown under different environmental conditions is likely to behave differently in the kiln from mature old-growth or even second-growth forests of the same species. As ever younger stands are being exploited, so kiln operators are having to adapt their processes to cope with the new kinds of wood.

7.1 Gross Features of Wood

7.1.1 Greenwood Moisture Content

Mackay and Oliveira (1989) have tabulated average moisture-content values for freshly-felled heartwood, sapwood and mixed wood of commercial species grown in British Colombia, reproduced in Table 7.1. Clearly, these authors are

Table 7.1. Average green moisture contents of some commercial British Columbian softwoods (Adapted from Mackay and Oliveira 1989)

Species	Heartwood %	Moisture content kg kg^{-1}		
		Heartwood	Sapwood	Mixed wood
Cedar, western red *Thuja plicata*	98	0.58	2.49	0.62
Douglas-fir, coastal *Pseudotsuga menziesii*	94	0.39	1.51	0.45
Douglas-fir, interior *Pseudotsuga menziesii*	88	0.31	1.32	0.43
Fir, amabilis *Abies amabilis*	87	0.55	1.64	0.69
Hemlock, western *Tsuga heterophylla*	66	0.55	1.43	0.53
Larch, western *Larix occidentalis*	83	0.35	1.19	0.49
Pine, lodgepole *Pinus contorta*	84	0.38	1.15	0.50
Pine, ponderosa *Pinus pondorosa*	67	0.40	1.48	0.76
Spruce, sitka *Pices sitchensis*	98	0.41	1.42	0.43
Spruce, white *Picea glauca*	93	0.51	1.63	0.93

concerned with the natural forest as the proportion of heartwood is so great, ranging from about 90%, except in the cases of western hemlock (*Tsuga heterophylla*) and *Pinus ponderosa* where the amount is around two-thirds. As expected, the moisture content of heartwood is very much less than that of sapwood.

Most hard pines and Douglas-fir (*Pseudotsuga menziesii*) show large differences in green density and the amount of water within logs, varying with age and height up the stem. Figure 7.1 illustrates these effects for *Pinus radiata*. The enormous quantity of water, especially in young trees and in the top log of older trees, contributes to the cost of transport from the forest to the mill and the subsequent kiln-drying. However, the drying of sapwood lumber is relatively easy, as the permeability of sapwood is often significantly higher than that for heartwood.

Unlike softwoods, the green moisture contents of sapwood and heartwood in hardwoods are roughly comparable; thus the moisture contents are, in general, little influenced by tree age and position of the log within the stem, as shown in Table 7.2.

Gross Features of Wood 141

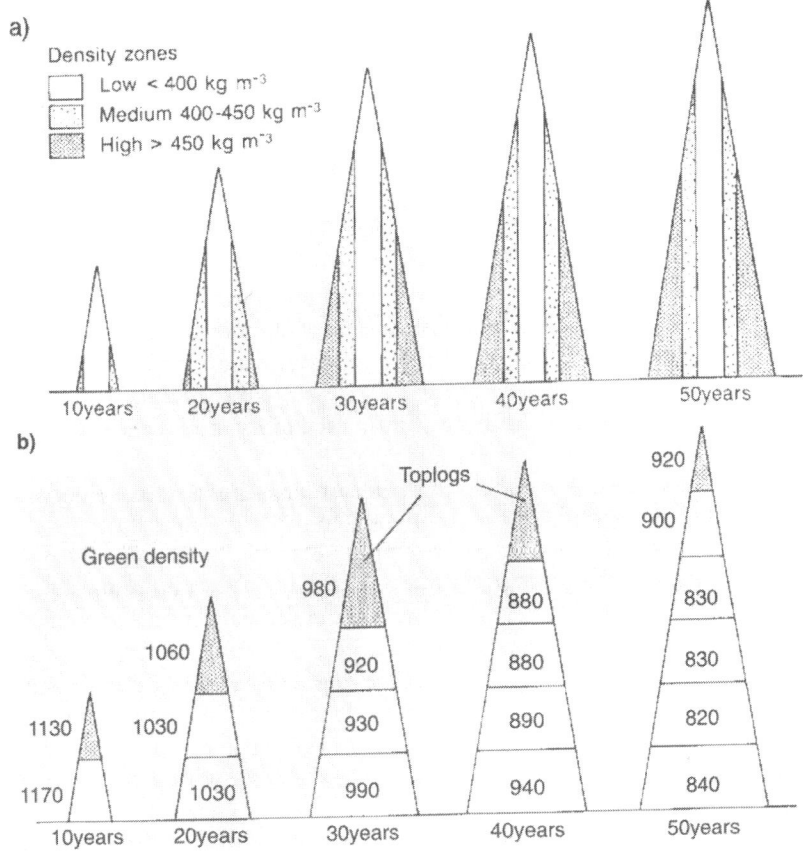

Fig. 7.1. a Density zones. b Within-tree variations in green density for New Zealand-grown *Pinus radiata*. (After Cown 1992b)

Table 7.2. Green moisture content values for certain American hardwoods (USDA 1987)

Species	Moisture Content, $kg\,kg^{-1}$	
	Heartwood	Sapwood
Beech, American (*Fagus grandiflora*)	0.55	0.72
Birch, yellow (*Betula alleghaniensis*)	0.74	0.72
Elm, American (*Ulmus americana*)	0.95	0.92
Oak, northern red (*Quercus rubra*)	0.69	0.80
Sycamore, American (*Platanus occidentalis*)	1.14	1.30

7.1.2 Wetwood

The term *wetwood* was introduced in the 1930s to describe the condition of discoloration and high moisture content found in pines (*Pinus* spp.) and spruces (*Picea* spp.) in Sweden. Other species, such as poplars (*Populus* spp.), willows (*Salix* spp.) and American sycamore (*Platanus occidentalis*) amongst hardwoods and many of the firs (*Abies* spp.) and hemlocks (*Tsuga* spp.) amongst American softwoods, often discolour in this way. Important commercial species worldwide that are susceptible include both red and white oaks (*Quercus rubra* and *Q. alba*) and white pines (*Pinus strobus* and *P. monticola*) of America, Scots pine (*P. sylvestris*) in Scandanavia, hoop pine (*Araucaria cunninghamii*) in Australia and elm (*Ulmus* sp.) and ash (*Fraxinus* sp.) from Japan. In all these instances, both the unpredictability and prevalence of wetwood are features of economic significance.

Wetwood differs from wood held in wet storage, which is described in Chapter 11. Wetwood is confined predominately to the heartwood and the wet tissue is slow-drying: whereas, in ponded wood, bacteria make the sapwood much more permeable due to degradation of unlignified pit membranes.

Wetwood has a unpleasant, acrid odour, is generally darker and may be of higher moisture content than sound heartwood. Facultative bacteria, able to live with very little oxygen, and anaerobic bacteria are associated with wetwood. These bacteria live mainly on the cell contents and only partly on the cell-wall materials, causing no noticeable loss of basic density. However, Verkasalo et al. (1993) report a weakening of the cell wall, with a 40% loss in tensile strength perpendicular to the grain for red oak (*Quercus rubra*). In the standing tree, wetwood is often associated with ring failure and frost cracks. In turn, this incipient weakening of intercellular bonding and cohesion results in shake, collapse and honeycombing during subsequent drying. The latter is not the direct fault of the kiln operator, except in so far as the potential problem has not been anticipated. Countermeasures include the adoption of lower temperatures during the early and middle stages of drying, or even preliminary air-drying of the wetwood. Ward and Pong (1980) state that annual losses owing to wetwood during drying factory-grade oak are at least 3%, but it is not unusual for 10 to 25%, or even half the lumber from individual kiln loads to be lost because of honeycomb and ring failure.

Wetwood is difficult to dry, primarily because of the reduced vapour pressure caused by the secretions of microorganisms making it harder to draw moisture from this tissue, and to a lesser degree by the very low moisture-transfer rates in the drier tissue (Chafe 1996). The weakened cell walls account for the susceptibility of wetwood, even under mild schedules, to collapse, to surface-checking and honeycombing, and ring failure. The practical difficulty with drying wetwood lies in the danger of overdrying the bulk of the lumber in order to dry the wetwood. The compensating strategy calls for a long equalising period at the end of the drying schedule.

7.1.3 Heartwood

In a tree, the wood in the outermost growth rings is light in colour and is involved in certain physiological functions including conduction of sap. In general, at a few growth rings in from the cambium there is a more or less abrupt change in colour in many species which corresponds to the transition to heartwood, but there is a surprising range in the number of sapwood growth rings from 1–2 for northern catalpa (*Catalpa speciosa*) to 80–100 for black tupelo (*Nyssa sylvatica*), and correspondingly in the width of the sapwood zone, from as little as 10 to 150 mm or more (Hoadley 1980). The heartwood region corresponds to a central tapered column or cone in which all cells are dead and physiologically inactive, including the parenchyma cells, as described in Chapter 1. Where wood is enriched by coloured extractives, such as oils, resins and gums, their mass can range as high as 20% of the oven-dry weight. These chemicals permeate both the cell wall and the lumen, but are not chemically attached to the wall. The term extractives includes compounds which are soluble in organic solvents as well as carbohydrates that can be extracted by water. Where pale heartwood is indistinguishable from sapwood, the heartwood, like the sapwood, is generally not decay-resistant. There are exceptions, however; the pale heartwood of white cedar (*Thuja occidentalis*) withstands decay.

The presence of heartwood impacts on drying lumber in four major ways:

- In reduced moisture diffusivity and minimal permeability, necessitating slower drying if checking is to be avoided.
- In reduced shrinkage, due to bulking of cell walls by extractives, in itself a desirable feature.
- In greater volatile emissions, especially in high-temperature drying, challenging environmental and health standards.
- Through corrosion of some kiln parts due to the release of wood acids, for instance in the case of western hemlock (*Tsuga heterophylla*).

7.1.4 Knots

Hoadley (1980) makes the striking comment that commercial hardwood grading presumes every knot to be a defect and proceeds to grade every board on the basis of the size and number of clear areas, which are free of knots and other blemishes, that can be cut from a board (NHLA 1988); yet many highly prized items contain knots and other blemishes.

In many instances, small knots add life and visual character and present no problem in drying, whereas large knots and clusters of small knots create difficulties in both drying and subsequent manufacturing that deservedly lead to knots being described as defects.

A knot is simply that part within the stem which emerges to become a branch. The gradual inclination of the adjacent tissue permits functional continuity between stemwood and branchwood. While the branch is alive, each successive growth ring forms in a continuous sheath at the cambium of both stem and branch. If the branch is alive at the time of felling, the lumber will contain a *live knot*, where there is no discontinuity between the knotwood and the stemwood. However, as the tree continues to grow, the lower branches will gradually be suppressed and die back toward the stem. Over the years, subsequent enlargement of the stem gradually encases the dead branch and its bark. The result is a *dead knot*.

Even live, steeply angled branches are undesirable as they can entrap bark between the upper side of the branch and the stem. Also, they occupy a larger part of the stem compared to a horizontal branch and, when sawn, result in unattractive and weak *"spike"* knots. A horizontally branched tree is a desired breeding strategy.

In a forest stand, the live branches extend initially to the base of the tree. Once there is canopy closure, the lower branches subsequently die. With heavily stocked stands, canopy closure occurs early and the base of the live crown ascends the stem more rapidly than in a more open stand. All branches below the live crown are no longer alive and contribute "dead" knots in outer wood: whereas the knots in the juvenile wood that are "live" reflect an earlier time when live crown extended to the base of the tree.

Some species are self-pruning, shedding their branches early and providing abundant clearwood in the lower part of the stem in old-growth stands. However, in most plantations, the stands never become old enough to yield much clearwood through self-pruning. If forest managers seek clearwood, they are obliged to prune their stands.

Requirements for structural lumber pay little attention to visual characteristics and so are not concerned whether there is a knot or a knothole: a blanked or rough-sawn surface is all that is required. Since lumber with dead knots or knotholes must have been cut from the mature wood in the lower part of the stem, such material is likely to display superior structural performance, as the mature wood of many commercial softwoods is stiffer and stronger, more stable, and has a higher basic density (Senft et al. 1985). Conversely, material with live knots, coming from the juvenile core, is more likely to produce inferior wood for structural purposes.

In millwork, large dead knots are major defects, as the knots are liable to fall out during drying. Even if they remain in place, the occluded bark is too brittle to plane to a smooth finish. Large live knots present less of a problem, but they tend to check during drying and will require filling. Loose knots and checking of live knots are both due to differential shrinkage between the knotwood and surrounding tissue.

This differential shrinkage is simply a consequence of the different orientation of the fibres in the knot and in the surrounding wood. Generally, wood shrinks little along the grain and significantly across it. The knotwood, whose fibres lie roughly perpendicular to the length of the board, will shrink noticeably in diameter and remain circular. This means that when a dead knot dries and shrinks it will be held by the surrounding material only at points on a line perpendicular to the grain, creating two crescent-shaped gaps. A dead knot can easily fall out, leaving a knothole. The same differential shrinkage occurs with live knots, but the continuity between the stem and branch fibres means that there is no obvious line of weakness; instead, live knots check radially across their grain from their pith to the cambium.

Both knot types dry faster than stem tissue, partly because of the very short, high-conductivity path along their grain and partly because they have a lower initial moisture content. Thus, early in any kiln schedule, the knot tissue is drier than the surrounding tissue and wants to shrink first, aggravating the differential shrinkage and making the knots more susceptible to falling out or checking. Both knotholes and knot-checking can be minimised by drying more slowly, with a reduction in kiln capacity, and by avoiding overdrying lumber, which is desirable anyway.

Lumber with a large face knot in the margin of the board is liable to kink if allowed to dry unrestrained or check badly if restrained. Kinking arises because the significant volume of wild grain or cross-grain around a large knot seeks to shrink perpendicular to the local grain orientation. This results in a large local component of shrinkage along the length of the board due to the severe cross-grain which can be inclined at a slope of about 1 in 3. This longitudinal shrinkage component is more significant than the transverse shrinkage of the knot itself because the volume of surrounding cross-grain is much greater than the knot volume.

Differential shrinkage means that a knot which runs from face to face in a board will tend to emerge from the face of the dried lumber. This is because the knot will not shrink in length on drying, whereas the board itself will shrink noticeably in thickness. This is rarely a problem, as lumber is either acceptable in the rough-sawn condition or it is dressed after drying. However, one example of a potential problem is with a glue-laminated member where an individual lamination has been underdried; further drying in service may result in the knot emerging and forcing the laminations to split apart locally. Hence, there is the need to accurately specify and supply lumber having the required moisture content.

Major differences between knotwood and stemwood are the higher resin content of and amount of compression wood in knots, which mean that the unextracted basic density is high. Knots have a high resin content, which increases further as branches die, sometimes to 30% or more by weight (Timell 1986, p. 496). Moreover, a large proportion of the knot consists of heavier compression wood, in softwoods or tension wood in hardwoods. In general, knotwood is between 25 and 100% denser than stemwood, occasionally exceeding 1000 kg m^{-3}, although there are some rare exceptions with pines (*Pinus* spp.).

Knot volume appears to range between 1 and 2% of the stemwood in genera such as *Abies*, *Larix*, *Picea* and *Pinus*, with knot size reaching a maximum at the base of the live crown (Timell 1986, p. 925). The percentage and absolute knot volume increase with reduced forest stocking as the live crown can extend further down the stem under these conditions. While knot volume appears insignificant, it is the volume and properties of the distorted grain around the knot that are so influential in processing and utilisation. The volume of this disturbed grain is roughly three times greater than that of the knot. This cross-grain is the cause of low strength in knotty timber. Lumber fails normally in the vicinity of the knot in this cross-grained wood and less usually at the knot itself. Also, the grain deviation results in distortion and warping during drying. Finally, cross-grain is hard to machine to a fine finish.

Not only does the surrounding tissue display extremes of cross-grain, but the same tissue has abundant compression wood. A detailed study by von Wedel et al. (1968) has found the ratio of the associated compression stem-wood to knot volume has an average value of 7:1 in 11-year-old *Pinus taeda*. Much of this compression wood lies immediately under the branch and extends some distance down the stem: for this reason, basic density measurements of lumber should be taken at least two knot diameters from any knot.

7.1.5 Spiral Grain

Harris (1989) noted that it is difficult to generalise about spiral grain, but indicates that there is a tendency for the pattern in hardwoods to be the reverse of that in softwoods. He states that, typically in softwoods, the wood immediately adjacent to the pith is straight-grained, but a left-hand spiral develops soon afterwards, increasing to reach a maximum value within the first ten growth rings. Thereafter, the grain angle decreases to zero before increasing again for the rest of the tree's life, but with a change in direction to a right-hand spiral. In the case of *Pseudotsuga menziesii* maximum left-hand spirality has been found to occur between ring 20 and 200 (Northcott 1957), with *Larix kaempferi* between ring two and nine (Ozawa 1972) and with *Pinus radiata* the maximum is detected by the end of the third growth ring (Cown et al. 1991). The sense of direction, left-hand or right-hand, corresponds to the direction of the central oblique stroke in a stylised letter, S or Z respectively, projected onto the face of the tree.

Ozawa (1972) found that the between-tree variation in spiral grain angle for *Larix kaempferi* increased with distance from the pith, although the mean values were still quite small in the outermost growth rings compared to those found near the pith. The large between-tree variation means that in a few instances the grain angle may be sufficiently large (>6°) to present a risk of twist on drying. This is possible in both the very first few growth rings and in the last few growth rings of mature wood, as illustrated in Fig. 7.2.

Spiral grain is hard to detect. Visual assessment is not particularly reliable, as the slope of grain is not readily apparent. Slope of grain can be determined

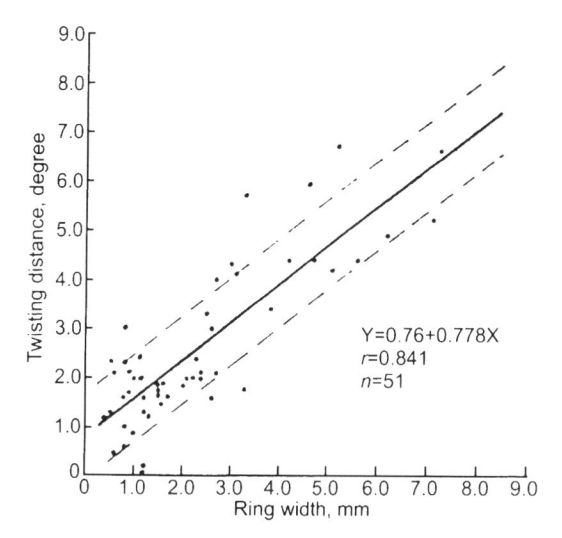

Fig. 7.2. Between-tree variation in spiral grain for *Larix kaempferi*, based on a study of 30 trees and sampling these at 4 m above ground level. (Ōzawa 1972)

from measurement on two perpendicular faces, traditionally using a sharp-pointed scribe; but this procedure is rarely used commercially. There have been efforts to develop tools more suited for industrial processes. General and localised slope of grain can be determined from the anisotropy of electrical capacitance (McDonald and Bendtsen 1986), or by focussing a laser beam on the surface and measuring the directional intensity of the backscattered light (Eastlin and Johnson 1993). The latter technique gives a direct value for the dive angle of the softwood tracheids with respect to the length of the board, the true angle of spiral grain. Both techniques are applicable to in-line grading, but have yet to be widely adopted.

Spiral grain, although permitted in some end-uses, limits the lumber's serviceability in others. In general, the limiting effect of spiral grain is seen in the twisting of the lumber, but it also affects its strength and stiffness. The most frequently quoted example is that of ladder rungs. For such components, Standards limit the slope of spiral grain to 1 : 20 (BSI 1990). The impact strength of clearwood is very sensitive to spiral grain. With board grades, spiral grain is not a visual defect and is not restrictive; however, there is an element of double-checking, as strict limitations are placed also on the extent of twist which can be attributed in part to spiral grain. For structural lumber, spiral grain affects both the strength and twist of members: the slope of the grain is limited to as little as 1 in 12 (4.8°) for the select structural grade under Canadian grading rules (NLGA 1979). Again, restrictions on twist apply which frequently overrule mere slope-of-grain considerations. Bowing of joists and arching of roof trusses are examples of structural problems arising from changes in moisture content influencing spiral grain and thus product performance (Tuomi and Temple 1975).

Spiral grain reinforces the effects of grain deviation in the immediate vicinity of knots in the two opposite quadrants as the natural grain inclination sweeps round the knot: the local grain angle decreases by corresponding amounts in the other two quadrants.

Spiral grain is most severe in juvenile wood and in the outermost wood of senescent old-growth trees. There are few generalised patterns, so it is important to appreciate the trends that apply to individual species. For example, with the yellow pines spiral grain is most severe near the pith in the upper logs of the stem (Hallock 1965).

Spiral grain is rarely uniform. If it were, it would present industry with far fewer problems. Difficulties arise because of the changing inclination of the spiral grain within the log and within boards cut from it. Since longitudinal shrinkage of lumber is sensitive to the grain angle, then differences in grain angle on either edge of a length of quarter-sawn lumber will result in differing longitudinal shrinkage on each edge and the board will show edgewise bending known as *crook* or *spring* on drying (Fig. 7.3). Crook is especially difficult to counter in a kiln as the weight of wood above is less able to prevent such distortion, unlike cupping, twist and bow, which can be restrained.

7.1.6 Juvenile and Mature Wood

In the 1950s the old-growth forests supplied most mills, and sawmillers could afford to be particular about wood quality. Because these huge logs had so much high-quality mature wood, relatively poor processing technology could be tolerated. With second-growth and plantation wood, the mill is confronted with smaller logs containing proportionately more juvenile wood which is far more susceptible to degrade on drying and subsequent processing. The concept of juvenile or corewood as a cylindrical column of wood enclosing the pith and running from the base to the top of the tree is broadly appropriate, although there are some gradual differences in wood properties at different heights up the tree.

In 1956, Jayne observed:

"The period during which this juvenile wood is produced varies in different trees and even in individuals of the same species, but in general, it is safe to assume that wood after about the fiftieth growth ring will possess the structure of mature wood."

Today, no serious commentator would insist that the juvenile wood zone extends to the 50th growth ring. The definition of juvenile wood is arbitrary and subjective. Zobel and van Buijtenen (1989, p. 83 ff.) emphasise that the transition from juvenile wood to mature wood occurs over several years, and there is a transition zone. The usefulness of defining such a transition zone is questionable, as the principal feature of juvenile wood is that its properties and characteristics approach outerwood values asymtotically. However, it is the

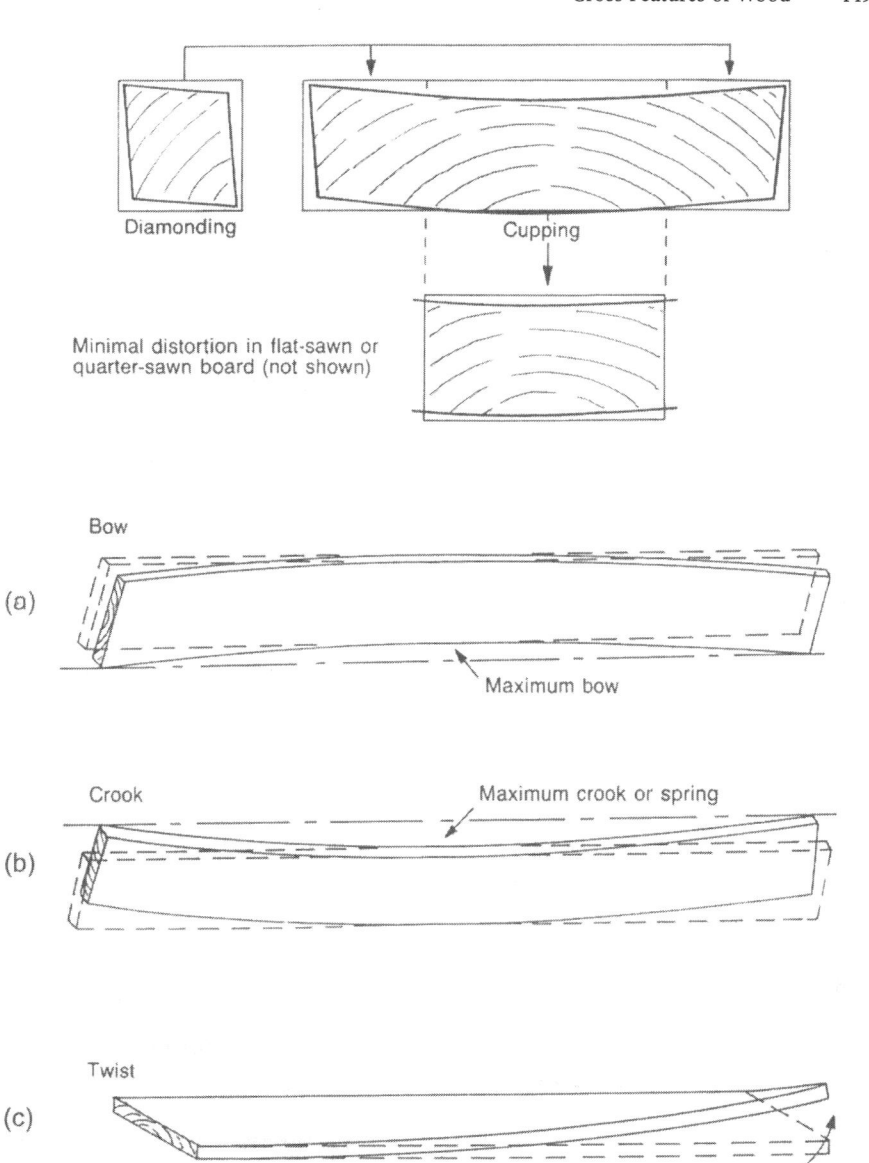

Fig. 7.3a–c. Warping of unrestrained lumber on drying. **a** Bow. **b** Crook or spring. **c** Twist. (Reproduced from Walker 1993, Primary Wood Processing, by kind permission of Kluwer Academic Publishers.)

initial *large change* in any wood property or characteristic with radius that largely identifies wood as lying in the juvenile zone (Fig. 7.4).

More generally, this juvenile zone is defined in terms of growth rings from the pith (typically 5–30 in number), rather than by its diameter (150 to

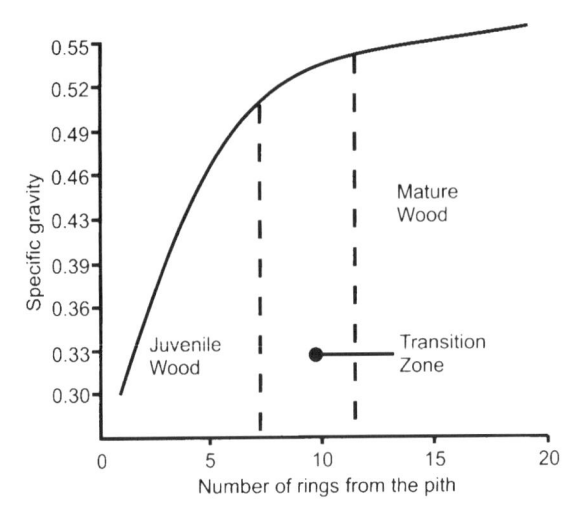

Fig. 7.4. Schematic representation of changes in wood quality with growth ring from the pith outwards for *Pinus* spp. (After Zobel and van Buijtenen 1989)

400 mm), as the latter is far more variable, being affected by genetic and environmental factors. From another viewpoint, size is what determines the commercial significance of juvenile wood, which is small in extent and insignificant in slow-growing, old-growth forests, but extensive and very significant in fast-growing plantations. In ideal circumstances, it is possible to grow millable trees in 15–25 years, while even in cool, temperate climates it is possible for trees to reach economic maturity in 45–50 years. With fast growth and a short rotational span, much of the stem volume can be juvenile wood, perhaps exclusively in a 15-year-old stand, declining to about one-tenth after 45 years.

Juvenile wood quality is often considered unsatisfactory, resulting in inferior solid-wood products. The deficiencies of juvenile wood in fast-grown pines are summarised in Table 7.3. In this table, the key features have been ranked subjectively in decreasing order of significance for the manufacture of solid-wood products. The characteristics that most prejudice the use of juvenile wood are the presence of abundant compression wood, the large microfibril angle, spiral grain and low density. These characteristics, taken in combination with large knots, pose problems in using the juvenile wood of fast-grown softwoods for high-quality lumber. It is the preponderance of juvenile wood in the fast-grown tree that influences lumber quality adversely, rather than the fast growth of the tree itself.

Although these deficiencies apply with particular force to the juvenile wood of the hard pines, the same trends are seen in the case of Douglas-fir (*Pseudotsuga menziesii*) and larch (*Larix* sp.). On the other hand, with firs (*Abies* spp.), spruces (*Picea* spp.) and hemlocks (*Tsuga* spp.), wood density is greatest near the pith and declines for a few growth rings before becoming constant or

Table 7.3. Undesirable features in juvenile wood of *Pinus* spp. as they affect solid-wood quality

Features	Effects
Above-average amounts of compression wood, especially in the butt log	Problems with crook, bow and low stiffness
Large microfibril angle in initial growth rings, especially in the butt log	Problems with crook, bow and low stiffness
Severe spiral grain in first few growth rings, especially above the butt log	Problems with twist
Above-average hemicellulose content	Accentuates any instability in lumber
High sapwood moisture content in young trees	Higher transport and drying-energy costs
Lower basic density	Potential for collapse during high-temperature drying

increasing modestly towards the cambium (Zobel and van Buijtenen, 1989, p. 107). As a group, hardwoods are far more uniform, with little change in density and fibre length both across and up the stem. With hardwoods, growth stresses, *brittleheart* and tension wood are likely to be the debilitating features of their lumber, which are not exclusive attributes of juvenile wood.

The gradual exhaustion of natural forest resources has forced industry to accept juvenile wood from plantations and second-growth forests. As its range and variation in properties become better understood, industry will find appropriate end-uses. There is enormous variability in wood quality between trees, and some juvenile wood, being of higher density, will be of better quality than some mature wood. It is the intrinsic properties of the individual piece of lumber that determines its quality.

The psychological significance of using juvenile wood should be appreciated. Historically, the industry of the Pacific Northwest of North America has used old-growth, and more recently second-growth, Douglas-fir (*Pseudotsuga menziesii*). In mature wood, the basic density is usually about 15 to 25% greater than that found on average in fast-grown plantations less than 30 years old which contain a very high proportion of juvenile wood. In the case of Douglas-fir (*Pseudotsuga menziesii*), the juvenile wood zone is relatively prolonged and changes only gradually compared with that in hard pines and *Pinus radiata*, where the zone is confined to a region close to the pith.

The differences between juvenile wood and mature wood, and between the butt and the top log are less great in hardwoods and the presence and properties of juvenile wood need not be considered separately. However, large *growth-stress gradients* are often present in young hardwood trees and, for uses other than for fibre production, there are strong reasons for prolonging the rotation age until the butt log diameter is at least 0.6 metres. Favoured fast-grown hardwoods are generally medium-density species which offer the broadest utility and produce short-rotation wood with a low extractive content. Of these eucalypts have been the most successful.

7.2 Intrinsic Features of Wood

7.2.1 Density

Density has long been regarded as the easiest and most reliable measure of intrinsic wood quality. Firstly, for the hard pines in particular, density normally increases with the ring number from the pith and so correlates well with all the other maturing wood properties and characteristics. Secondly, the shrinkage of wood is, to a first approximation, dependent on the basic density of the wood (Stamm 1964). Thus a plot of external volumetric shrinkage for a large number of hardwoods shows a reasonable correlation with basic density, as illustrated in Fig. 7.5.

The regression line drawn through the data corresponds to the relationship:

$$\alpha_v = X_{sp} \frac{\rho_b}{\rho_w}, \tag{7.1}$$

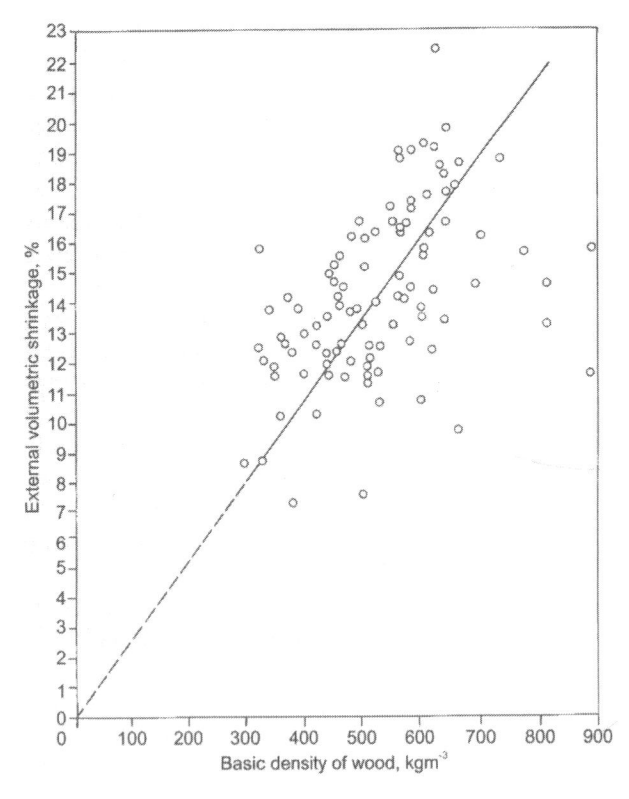

Fig. 7.5. The external volumetric shrinkage of 106 North American hardwoods from the green to the oven-dry state plotted against their respective basic densities. (Reproduced from Stamm 1964, Wood and Cellulose Science, reprinted by permission of John Wiley & Sons, Inc)

where α_v is the fractional volumetric shrinkage, ρ_b is the wood's basic density and ρ_w is the density of water. The parameter X_{sp} is the moisture content at the so-called *shrinkage intersection point*, which for these hardwoods has a mean value about 0.27 kg kg^{-1}. It is close to the fibre-saturation point, and it is often assumed that these two points are the same. Half the tested specimens deviate less than 11% from their predicted shrinkage value from this equation which explicitly links the external volumetric shrinkage to the loss of adsorbed water. There are exceptions; a partial explanation for this is that such woods are rich in extractives which take up part of the swollen cell wall that would otherwise be occupied by water, thus preventing the cell wall from shrinking as much as expected on drying (Stamm 1964).

If c is the mass of extractives (with a density of ρ_e) per unit volume of dry wood, of which a fraction x_{cw} is sorbed on the cell-wall material, then the bulking effect of these chemicals is $c\, x_{cw}/\rho_e$. It follows that the fractional volumetric shrinkage is smaller by this amount:

$$\alpha_v = X_{sp}\frac{\rho_b}{\rho_w} - \frac{c}{\rho_e}x_{cw}. \qquad (7.2)$$

Walker (1993) illustrates the effect of bulking by extractives with a worked example for wood having an unextracted bulk density of 530 kg m^{-3}, with 30 kg m^{-3} of extractives sorbed within the cell wall. The volumetric shrinkage to the oven-dry condition on the basis of Eq. 7.1 would be 15.0%, but if the presence of extractives were taken into account the estimate would fall to 12.9%, a reduction of one-seventh.

Kiln-drying needs to take account of both the overall wood density in a load and any variations within a single piece of lumber. Therefore, the within-ring variations, the within-tree variations and the between-tree variations in density are all significant.

Most species, apart from *Araucaria* sp. and diffuse-porous hardwoods, show distinct differences in wood density across the growth ring. This is primarily in response to seasonal climatic variations and the formation of latewood. The density variation across a growth ring far exceeds the density variation between trees. As an extreme case, Harris (1969) cites the contrast between latewood (870 kg m^{-3}) and earlywood (170 kg m^{-3}) in adjacent growth rings in the outerwood of a sample of Douglas-fir, *Pseudotsuga menziesii*. More usual within-ring and between-ring variations for temperate softwoods are shown in Fig. 7.6. The density profile across the growth ring is more uniform in juvenile wood than mature wood since in the former the latewood density is less and there is a lower percentage of latewood.

In Malaysia, the pines, *Pinus caribaea*, *P. merkusii* and *P. oocarpa*, produce little latewood during the first two to four growth layers, but latewood develops strongly thereafter, being characterised by frequent false rings. Even with subsequent latewood formation the ratio of earlywood to latewood density for these species is low, approximately 1 : 1.5, whereas the corresponding ratios for the pines, *Pinus radiata* and *P. taeda*, and Douglas-fir (*Pseudotsuga*

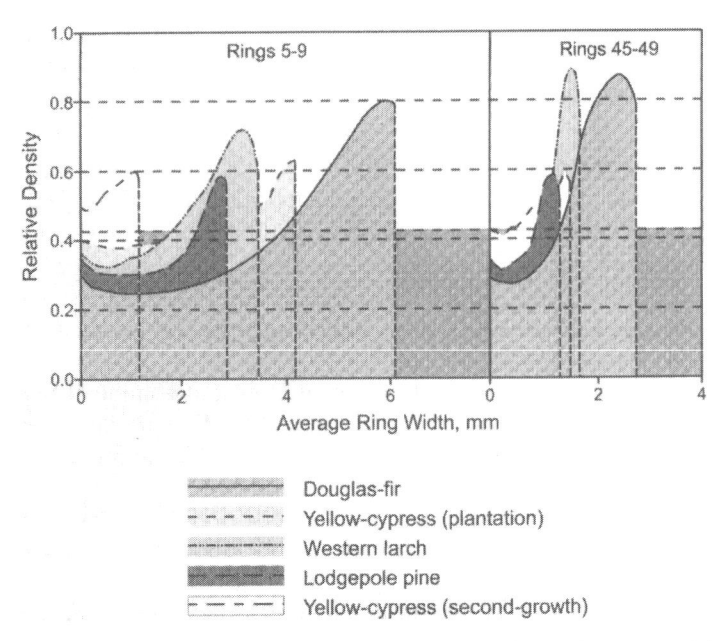

Fig. 7.6. Within- and between-ring variations in basic density for commercial softwoods growing in British Colombia. Data have been taken from rings 5–9 in the juvenile wood and from rings 45–49 in the mature wood. (After Jozsa and Middleton 1994)

menziesii) were found to be 1 : 1.8, 1 : 2.3 and 1 : 5.0, respectively (Harris 1973). On the basis of the moderate differences in density between earlywood and latewood, these tropical pines and the temperate *P. radiata* can be described as eventextured, while their wide growth rings would classify them as coarse-grained. With other softwoods, such as spruce (*Picea* sp.) and hemlock (*Tsuga* sp.), the differences in density between earlywood and latewood are comparatively small and the transition from earlywood and latewood is gradual. These species are desired on account of their uniformity of density.

Within-ring density variations can cause problems in drying for two reasons. The greater the within-ring density variation the greater the differential shrinkage at the ring boundary. This, in turn, may create sufficiently severe drying stresses to encourage delamination and internal checking (Booker 1994). Secondly, a low absolute density in earlywood can result in collapse on drying, especially at high temperatures (Booker 1996).

Within-tree density variations for the hard pines, Douglas-fir (*Pseudotsuga menziesii*) and some other, but by no means all, softwoods begin with low-density wood adjacent to the pith, which increases steadily in the juvenile zone before levelling off asymptotically within the mature wood. The mean basic density of the stem's cross-section decreases and its moisture content increases with distance up the stem; this merely reflects the steep, radial gradient of basic density in the vicinity of the pith and the proportionately greater amount of corewood higher up the stem. On the other hand, the general rule for true firs (*Abies* spp.), hemlock (*Tsuga* sp.) and spruce (*Picea* sp.) is for the basic density to decrease for the first few annual rings from the pith before levelling off or increasing moderately toward the cambium (Zobel and van Buijtenen 1989). Trends in ring density at breast height for a number of rapidly-grown, second-growth species are shown in Fig. 7.7.

The situation is more complex with hardwoods:

"All possible patterns of wood density variation appear in hardwoods. The middle to high density diffuse-porous hardwoods generally follow a pattern of low basic density near the pith and then an increase, followed by a slower increase or levelling off toward the bark. The low-density, diffuse-porous woods, such as Populus, seem to have a somewhat higher density at the pith, although some have a uniform density from pith to bark. The ring-porous hardwoods tend to have a high density at the centre, which decreases and then increases to some extent toward the bark" (Zobel and van Buijtenen 1989, p. 113).

Growth rate has little effect on the wood properties of diffuse-porous species. These have approximately the same proportion of vessels across the growth ring, regardless of the growth rate. On the other hand, growth rate has a noticeable influence on the density of ring-porous hardwoods, which usually produce denser wood when fast-grown: the volume of vessel tissue produced each year in a ring-porous hardwood remains constant regardless of the total growth during the growing season and therefore the wider the growth ring the smaller the proportion of vessel tissue.

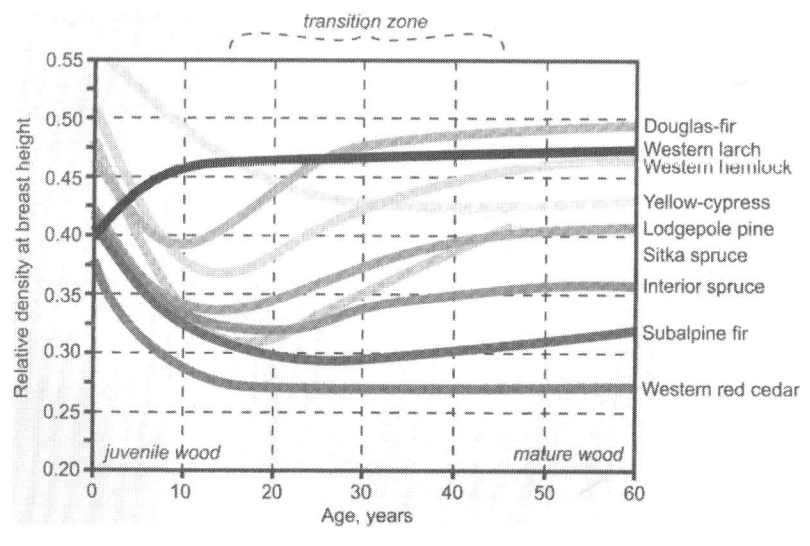

Fig. 7.7. Trends in basic density at breast height for commercial softwoods growing in British Colombia. (After Jozsa and Middleton 1994)

156   Lumber Quality

The within-tree variations exemplified by the hard pines and medium- to high-density diffuse-porous hardwoods have received particular attention as many important plantation species fall into these groups. As noted, with these species the juvenile wood is of lower density and poorer quality than the wood in the rest of the tree. In corewood, the S_2 layer is quite thin and so it is not unexpected that the lignin content is greater and the cellulose content less than that of outerwood.

Regardless of species or where the forests are established, the variation in wood properties between trees is very great. For *Pinus radiata* in New Zealand, the range of basic density within a typical stand is shown in Fig. 7.8. Where tree-improvement programmes emphasise high basic density, the between-tree variations will be reduced and the distribution will centre on the medium- to high-density range shown. Unfortunately, the variability of corewood is usually less than that of outerwood, so an equivalent increase in corewood density will be harder to achieve through forest management, in which case the within-tree variability may actually increase.

Another example of within-stand variability relates to Douglas-fir (*Pseudotsuga menziesii*). The mean basic density of the cross-section decreases on ascending the stem, but, because the distributions are so broad, the section density at the top of one tree may exceed the section density at the base of another tree. For a comparable sample, taken from the same aged trees on the same site, a between-tree variation in average cross-sectional basic density of 15 to 25 kg m^{-3} can be expected. High-density trees tend to have both more late-

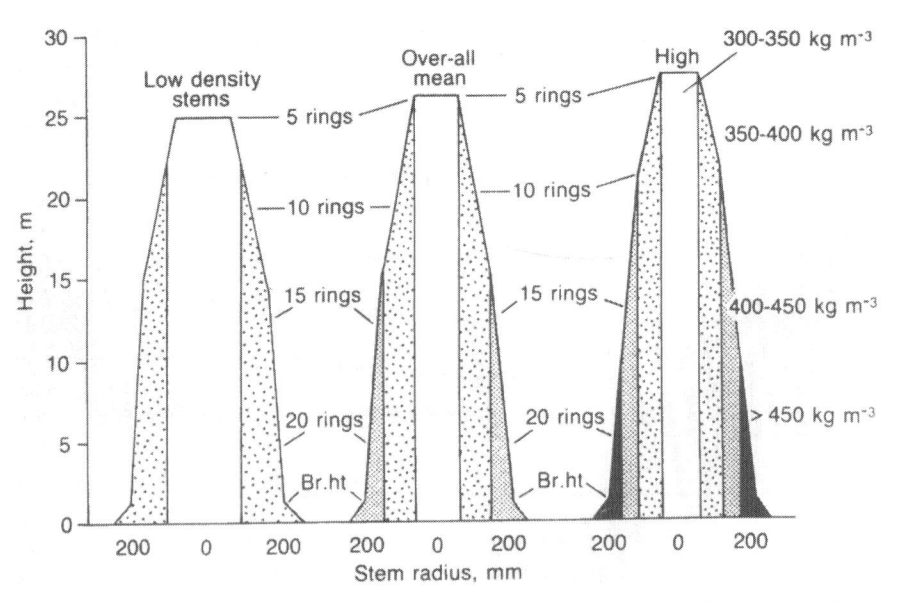

Fig. 7.8. Within-stand variation in wood density for 24-year-old *Pinus radiata* growing in the central North Island of New Zealand. (After Cown and McConchie 1983)

wood and higher average earlywood and latewood densities. These between-tree differences are assumed to reflect the high level of genetic variation within the population. Unfortunately, in the short term, only a 5% overall increase in wood density of Douglas-fir (*Pseudotsuga menziesii*) is likely from genetic improvement and this will do little to offset the effect of the shorter rotations envisaged in the Pacific Northwest (McKimmy 1986).

This raises the interesting observation: high-density juvenile wood of equivalent intrinsic quality to low-density juvenile wood is less stable: it will shrink more on drying and move more in service by virtue of its higher density. The same point can be made even more effectively when considering mature wood. Mature wood is far less prone to warp and instability because the intrinsic wood quality of mature wood is far superior to that of juvenile wood, and not because it is of higher density.

7.2.2 Collapse

Collapse occurs at moisture contents above fibre saturation. It is a physical flattening of the fibres induced by internal surface tension generated as water is removed from initially saturated lumens. Collapse thus takes place mainly at the beginning of the kiln schedule. Its presence is revealed by a washboard surface. It is liable to occur when low-density wood with a high moisture content close to saturation is dried at moderate to high temperatures. In the case of a collapse-susceptible species, *Sequoia sempervirens*, Siau (1995) estimates the compressive strength of the cell wall perpendicular to the grain to be 1.5 MPa for old-growth wood in the green state under kiln conditions. A capillary pressure of this amount corresponds to a radius of $0.1\,\mu m$, which is significantly larger than some of the pit openings, which may be as small as $0.01\,\mu m$. Some collapse on drying may be expected.

Innes (1995) developed a stress model of a eucalypt-wood fibre which is assumed to be a thick-walled cylinder, as such a fibre is typically round and has a ratio of outside diameter to wall thickness of order 5:1. The predicted stress profiles are calculated to be very sensitive to temperature. For example, the fractional change in radius, or *radial strain*, pulling in the fibre wall at 20 °C is predicted to be two-thirds of the nominal ultimate value of 0.02, but at 25 °C it is estimated to reach 95% of the ultimate value. Thus, Innes suggests that there is a *collapse-susceptible temperature* above which significant collapse is likely.

The traditional solutions to collapse are either to air-dry or kiln-dry at low temperatures (<45 °C), since at low temperatures the rigidity of the cell wall is greater and more able to counter the capillary forces, whereas at higher temperatures the rate of loss in strength is greater than the decline in capillary tension due to the reduction in surface tension. The alternative policy is to accept that collapse is going to occur on drying and to recover that collapse towards the end of the schedule by steaming, when the moisture content is around 15–20%. Conditioning unlocks the existing stresses on the cell walls to get them to spring back to their original shape. After conditioning, the recovered wood cells can be redried without risk of recollapse as the cells are no longer saturated and there can be no significant capillary-tension effect.

There is one specific example where no recovery is possible, and that relates to high-temperature drying of low-density softwoods. High-temperature drying occurs above the thermal softening point of lignin, and drying stresses are considerably relaxed by viscoelastic and mechanosorptive strains (as discussed in the following chapter) at kiln temperatures. On conditioning, there are insufficient stresses locked within the cell walls able to effect recovery.

In tests in drying the hardwood, *Nothofagus truncata*, Grace (1996) showed that there is no sharp collapse-susceptible temperature, but rather the steam-recovered shrinkage (which is assumed to represent collapse) develops rapidly with temperature from a small value at room conditions. In these tests, the apparent collapse strain increased 20-fold from 0.005 to 0.10 when the drying temperature was raised from 20 to 80 °C. Collapse-susceptible hardwoods are often either air-dried or predried in sheds at low temperatures before kiln-drying is attempted once the lumber is below fibre saturation.

In kiln-drying, there are two features which are strongly influenced by basic density: collapse and warp. Whenever kiln conditions favour collapse, then this will be accentuated by low density; and whenever kiln conditions favour internal checking at the growth-ring boundary, then this will be accentuated by large differences in basic density between earlywood and latewood.

7.2.3 Warp

The distortion of lumber on drying, or *warp*, results from anisotropic shrinkage of wood. The shrinkage of wood is different in the longitudinal, radial and tangential directions. The longitudinal (axial) shrinkage from the green to the oven-dry condition is at least an order of magnitude less than the transverse shrinkages. Tangential shrinkage is usually 1.5 to 2.5 times that of the radial shrinkage.

To a first approximation, the longitudinal shrinkage can be ignored in mature wood. The fractional volumetric shrinkage then becomes:

$$\alpha_v = \alpha_r + \alpha_t - \alpha_r \alpha_t, \tag{7.3}$$

where α_r is the radial shrinkage coefficient and α_t is the tangential one. Since the tangential coefficent is between 1.5 and 2.5 times greater than the radial one and if the cross-product is ignored as being negligibly small, then as an approximate rule-of-thumb

$$\alpha_v \approx 1.5 \alpha_t \approx 3 \alpha_r. \tag{7.4}$$

Lumber is not used in the oven-dry state. For many in-service uses the desired percentage moisture content falls within the range of 8–15%, and the shrinkage will be only a half to three-quarters of the oven-dry shrinkage value. The actual shrinkage strain (or fractional change in length) in the ith direction is given by

$$\varepsilon_i = \alpha_i(X_p - X_{sp}), \tag{7.5}$$

where X_p is the desired end-moisture content. Equation (7.5) follows the convention that shrinkage represents a negative change in length.

Carrington (1996), on investigating the shrinkage of disks cut from the central stem of an 8-year-old *Pinus radiata* tree, has found that the sectional shrinkage is also linear with moisture-content change. The sectional strain s could be related to the radial and tangential strains, with an average deviation of 16% from experimental values, by the expression

$$s = 1 - [1-\varepsilon_r]^2[1-\varepsilon_t]. \tag{7.6}$$

Equation (7.6) suggests that the value of s is about twice the tangential strain or four times the radial value.

The magnitude of the shrinkage in any direction depends on the *stiffness* of the microfibrils and the *rigidity* of the cell-wall matrix. In a standard beam-bending test, the stiffness in bending or the *modulus of elasticity* is estimated as

$$E_b = F_p L^3 / [4Bh^3\Delta], \tag{7.7}$$

where F_p is the load at the proportional limit, L is the span between the beam's supports, B is the breadth of the beam, h is the depth of the beam and Δ is the mid-point deflection of the beam under the load F_p. The rigidity is defined through the horizontal shear stress, which is a maximum at the mid-depth or neutral plane of the beam, falling to zero at the upper and lower surfaces. The maximum shear stress at this mid-plane is given by

$$S = 3F_R/2Bh, \tag{7.8}$$

where F_R is the load at the moment of failure.

The ratio of stiffness to rigidity, E/S, varies with species and wood density. In greenwood and wood at a moisture content of about 20%, the shear-stress modulus S is small, while the elastic modulus E increases with diminishing moisture content below fibre saturation (Keey and Keep 1999); thus, as wood dries, the E/S ratio will increase.

Barber and Meylan (1964) developed a shrinkage model based on this matrix-microfibril interaction. In its simplest form, the model considers only the properties and behaviour in the thick, dominant S_2 layer, where the microfibrils usually make an angle of 10° to 30° with the tracheid axis. The cell wall is considered to consist of an amorphous matrix of lignin and hemiceluloses, which can adsorb moisture thereby swelling, and strong, stiff cellulosic microfibrils, which neither adsorb moisture nor swell. In isolation, the matrix would shrink isotropically. However, the microfibrils restrain the matrix from shrinkage in the direction parallel to the microfibril axis, so forcing the wall to shrink excessively in the transverse plane to the microfibrils. Thus, whenever the microfibrils are aligned parallel to the cell axis, there will be minimal longitudinal shrinkage on drying. One counter-intuitive outcome of this model is that, when the microfibril angle is about 30°, the wood will swell or *expand* longitudinally on drying, but for angles greater than about 40°, the wood begins markedly to shrink in the longitudinal direction on drying.

This theory is supported by experimental observations. Kelsey (1963) showed that, for some wood for which E/S is large, longitudinal swelling can take place in the early stages of drying. Both juvenile and reaction wood show uncharacteristically large longitudinal shrinkage. Meylan (1968), on examining the longitudinal and tangential shrinage in juvenile wood of *Pinus jeffreyi*, where the microfibril angle can be greater than 40°, observed a longitudinal shrinkage of 7%, well in excess of the tangential shrinkage of the same samples, as illustrated in Figs. 7.9 and 7.10.

Shrinkage and shrinkage anisotropy have a profound effect on the management of lumber-drying practice. Lumber is cut to a particular, rough green-target size in the mill. This takes account of sawing variations, shrinkage to the end-moisture content, planing allowances and the percentage of undersized boards permitted, typically 0% or 5% (Brown 1986). The planing allowance does not directly affect kiln-drying unless boards are presurfaced green, as discussed in Chapter 11.

The total-process standard deviation, which takes account of both within- and between-board variation coming from each and all saws within the mill, and the minimum permitted green dimension allow one to calculate the desired mean green dimension, such as thickness. For example, if the saws are producing lumber to a standard deviation of 1 mm, then 95% of the lumber will lie within 2 mm of the average. Therefore, in order to meet a minimum

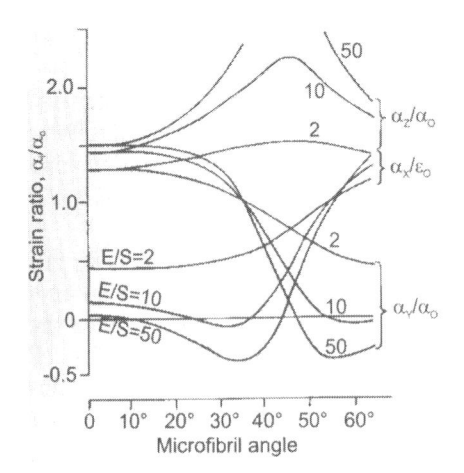

Fig. 7.9. Predicted longitudinal shrinkage of wood: the strain ratio of the longitudinal shrinkage coefficient α_l to the isotropic value α_o as function of the cellulosic microfibril angle, the stiffness E and the rigidity S. (After Barber and Meylan 1964)

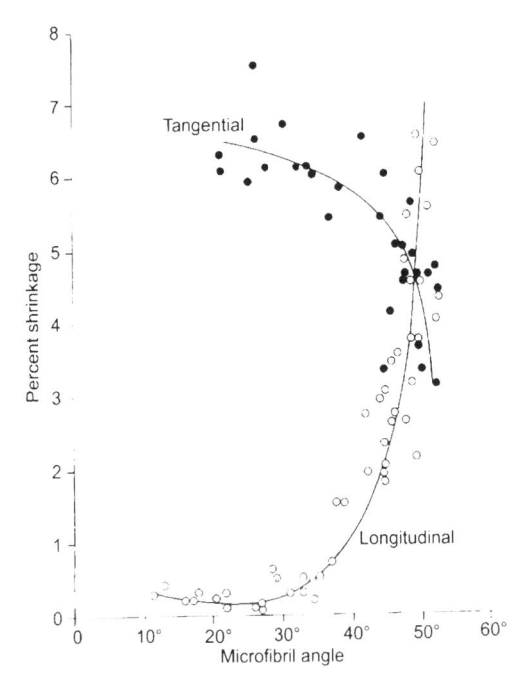

Fig. 7.10. Longitudinal and tangential shrinkage as a function of the microfibril angle. (Meylan 1968)

thickness of 50 mm, the mean thickness of all boards should be 52 mm, with 5% of the lumber in the stack lying outside the range 50 to 54 mm. As a consequence, the material in each layer of the stack will be of variable thickness. Any thin or undersized lumber is susceptible to warp and distort as it dries, thus enhancing the unevenness of the airflow in the kiln, as discussed in Chapter 9. Top-weighting of the stack with loads of order $1000\,\mathrm{kg\,m^{-2}}$ helps restrict twist and cupping, but will be less effective against the lateral distortion of lumber such as crook or spring.

Moreover, at the end of the drying schedule, normally there will be significant moisture-content variations in the stack, particularly if the airflow is not uniformly distributed across the length of the boards. This leads to variable shrinkage throughout the kiln load. In estimating the shrinkage in a kiln, it is assumed that the lumber is tangentially oriented in both thickness and width directions, since tangential shrinkage is invariably greater than the radial amount. Again, quarter-sawn lumber, which is potentially more susceptible to crook or spring, will shrink more in its thickness, and so will become progressively less subject to any physical restraint from weights on the stickered stack. In particular, different amounts of shrinkage will take place in a load with a species which shows large differences in density and wood quality in the juvenile zone. Similar problems would arise in the drying of mixed species.

The final factor to consider is variability in sticker thickness. Stickers should be manufactured from kiln-dried material and machined to standard widths and thicknesses. After repeated use, especially in high-temperature kilns, stickers should be discarded, since they become thinner having been subject to viscoelastic and mechanosorptive creep.

7.2.4 Reaction Wood

Environmental factors in the forest result in the production of *reaction wood*. Wind is the main cause of leaning stems, although poor root development, soil creep, and factors such as phototropism can play a part. Once a tree is leaning, enormous forces are necessary to straighten it. Not only must the stresses sustain the bending moment of the leaning tree, but they must eventually overcome the rigidity of the stem itself, forcing it to bend so that it regains an upright position. These stresses are generated by the formation of reaction wood in response to the lean. Even a large stem can be straightened given enough time for the stem to accumulate a large amount of reaction wood. These stresses, and the ultrastructural responses to them, are features of the wood which emerge on sawing and drying to plague the kiln operator.

In *Ginkgo biloba* and all softwoods, reaction wood is found on the underside of the stem. The tissue is called *compression wood*. In many but not all hardwoods, reaction wood forms on the upper side and the tissue is called *tension wood*. Both types of reaction wood act to correct the lean of the stem. During the formation of compression wood, the tracheids on the underside expand longitudinally, so pushing the stem up; whereas with tension wood, the fibres on the upper side shrink longitudinally to pull the stem up. In general, the stem is enlarged on the underside in compression wood and on the topside in tension wood. The presence of reaction wood can be deduced if the stem is somewhat elliptical as a consequence of increased growth in the reaction-wood zone, but reaction wood is not confined to obviously leaning trees. If the stem eventually straightens, the presence of reaction wood near the pith may not be suspected.

A high incidence of reaction wood near the pith is understandable, as a thin stem is misaligned easily and seeks to straighten itself. Instability, or a wandering leader, is a particular problem in fast-grown exotic species which generate reaction wood as the stem moves to counteract any lean. The lean can be overcorrected with reaction wood forming on alternate sides of the stem. Indeed, very vigorous growth in conifers can produce mild compression wood all round the stem, the same effect being noted after heavily thinning or fertilising stands on good sites. Fast growth and wide spacing encourage the formation of reaction wood, but this can be countered by removal of the worst-formed stems during a thinning operation. Also, tree breeders are able to minimise the production of reaction wood and its severity by selecting

Table 7.4. Characteristics and properties of reaction wood

Feature	Compression wood	Tension wood
Density	15–100% greater than in corresponding normal wood	10–30% greater than in corresponding normal wood
Physical features	Darker in colour and very hard	Darker in, among others, in eucalypts; fur has a silvery sheen in most temperate hardwoods. Develops a woolly surface when sawn longitudinally
Longitudinal shrinkage	1–5% or greater, compared with 0.1–0.3% in normal wood	0.5–1.5%
Warp on drying	Liable to warp badly	Can warp and liable to collapse
Strength	Comparable to normal wood, but strength does not reflect the higher density	Generally superior strength to normal wood

straight-stemmed trees with small, wide-angled limbs. The adverse effects of reaction wood are summarised in Table 7.4.

7.2.4.1 Compression Wood

In an elliptical cross-section of a tree, the growth rings are narrow on the upper face and much broader underneath, so that the pith is located nearer the upper face of the leaning stem. In the compression-wood zone, the wood is denser, with more thick-walled cells. Clear differences between normal tracheids and cells with severe compression wood are revealed by a scanning electron microscope (Fig. 7.11a). The compression-wood cells are more rounded and shorter than normal tracheids, with intercellular spaces between tracheids. The S_3 layer is always absent. The S_2 layer has an outer highly-lignified zone $S_2(L)$, containing about 70% lignin, and an inner less lignified zone (with about 40% lignin) in which there are deep helical cavities and ribs whose inclination correspond to that of the microfibrils (30–50°). The large microfibril angle in the S_2 layer means that the tracheids shrink more in the longitudinal direction on drying compared to normal wood (Fig. 7.11a). Compression wood can contain as much as 40% lignin compared to 30% in normal wood. The content of cellulose in normal wood is about 40–45%, but is only about 35% in compression wood, with a corresponding reduction in hemicellulose content from 30–35% in normal wood to about 25% in compression wood. Compression wood should not necessarily be regarded as abnormal tissue; rather it can be considered to be an extreme form of normal wood because there is a continuous gradient of characteristics from compression wood through normal wood to *opposite wood* on the other side of the stem.

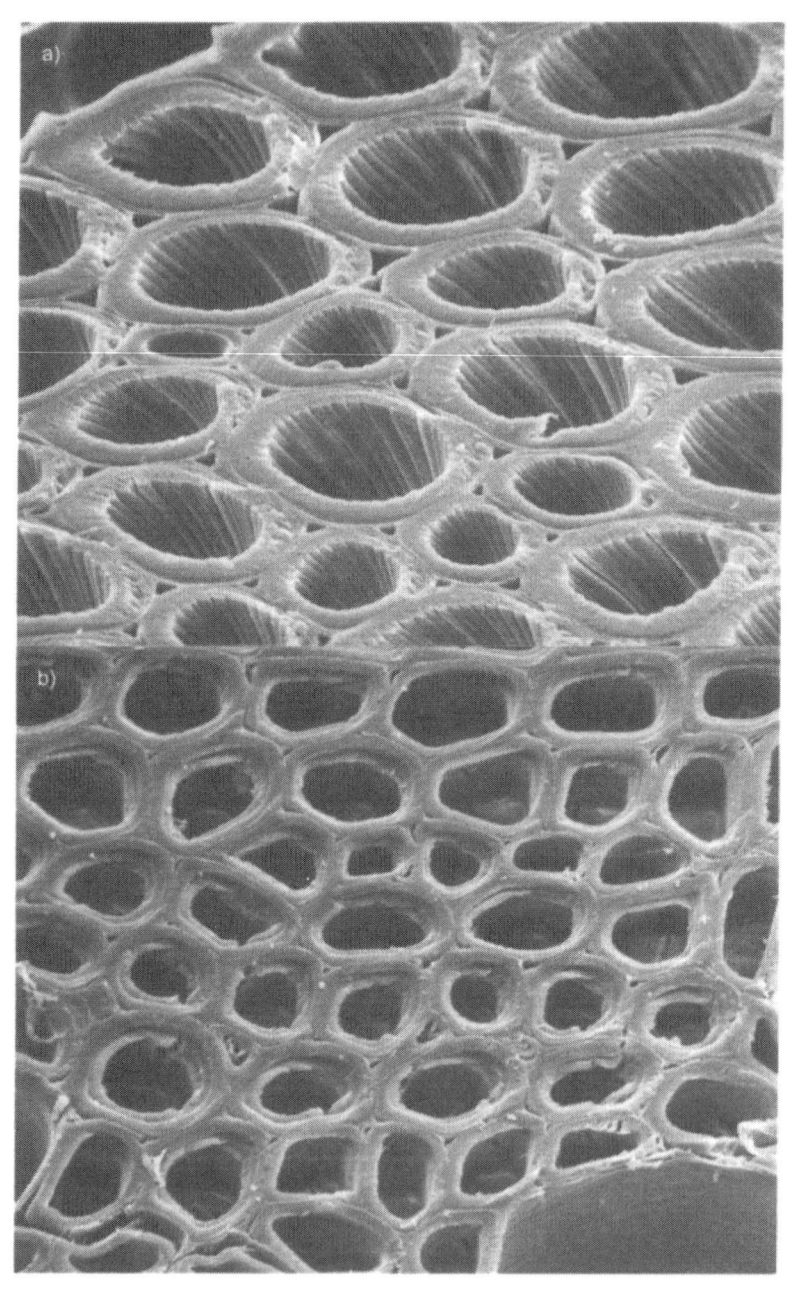

Fig. 7.11a,b. Cell-wall feature of reaction wood. **a** Compression wood. **b** Tension wood. (Courtesy of Dr. B.G. Butterfield)

Burdon (1975), in a study of 12-year-old *Pinus radiata* trees growing on four sites, estimated that 30–45% of the stems contained mild to severe compression wood, while Timell (1986, p. 4) mentions that stems have been found to contain on the average 15% compression wood, both in virgin forests of spruce in Canada and in plantations of Southern pines in the United States. Compression wood is usually most evident near the pith, suggesting that the young, thin stem frequently needs to reorient itself, to correct for a slight lean or to reposition its live crown to better capture the light in competition with surrounding trees. Compression wood is especially abundant in butt logs.

Gaby (1972), in a study of warp in southern pine studs, was able to pick out pieces with significant, severe compression wood. However, in that initial visual examination, he failed to detect the presence of mild compression wood, despite almost 30% of with-pith studs actually having significant amounts of mild compression wood in the first three annual rings for at least half of the stud length. Compression wood is an insidious problem.

Longitudinal shrinkage is greatest near the pith, and in the butt log. Not only is there an abundance of mild compression wood with its large microfibril angle near the pith, but even in normal wood the cellulose microfibril angle is greatest next to the pith, decreasing gradually with ring number. At a few growth rings from the pith, the incidence of compression wood is much reduced and the normal wood has a smaller microfibril angle. As already noted, Barber and Meylan (1964) and Barber (1968) provide a non-linear model describing longitudinal shrinkage in terms of the cellulose microfibril angle. In this model, severe longitudinal shrinkage only becomes serious when the microfibril angle exceeds about 30°. Harris and Meylan (1965) have provided experimental validation (Fig. 7.10), as have others, with Ying et al. (1994) offering a particularly detailed example. The significant feature of both the theoretical and experimental results is that a large microfibril angle is necessary to induce significant longitudinal shrinkage, and that this condition is found only in compression wood and in normal wood near the pith.

The amount of crook and bow is greatest in the butt log of softwoods for two reasons. First, the microfibril angle is higher at the base of the tree than it is further up the stem, and also because there appears to be more compression wood in the butt log. In these cases, the absolute microfibril angles will lie in the critical range (30°–50°). Secondly, the *gradient* in microfibril angle with distance from the pith will be steep. Both contribute to differential shrinkage in the longitudinal direction between opposite edges (crook) or faces (bow) of juvenile lumber.

7.2.4.2 Tension Wood

This is formed on the upper side of a leaning stem, although it is not present in all genera or species. Tension wood is less easy to identify than compression wood. In dry-dressed, temperate hardwoods, it can have a silvery sheen; in tropical woods, it appears as darker streaks; while in green sawn timber the fibres become pulled out, resulting in a woolly surface. Unlike compression wood, individual tension-wood fibres tend to be proportionately less heavily lignified than normal, principally because these cells are characterised by the presence of a gelatinous (G) layer which usually replaces the S_3 layer, although it may occur inside this layer or replace both S_2 and S_3 layers. The G-layer is unlignified and readily separates from the rest of the cell wall (Fig. 7.11b). The G-layer consists of crystalline cellulose whose microfibrils lie at a low angle (0–5°) to the axis of the fibre. Because of the presence of the gelatinous layer in tension-wood fibres, which consist of pure cellulose, the cellulose content of tension wood can be as high as 50–60%. The lignin content is correspondingly lower, namely about 15–20%, as is the hemicellulose content (about 30%). While tension-wood fibres can be found over a sizeable area of tissue, more usually they tend to be scattered amongst other fibres. Tension-wood fibres are often absent in latewood. Boyd (1985) has emphasised that, even in localised areas of severe tension wood, only a proportion of the fibres (ca. 30%) have the non-lignified G-layer that distinguishes them as tension wood, whilst a larger proportion of the fibres have thick, fully lignified cell walls. Vessels in tension wood are smaller than normal ones, and more sparsely distributed.

The longitudinal shrinkage of tension wood (0.5–1.5%), while not as great as that of compression wood, is nevertheless much higher than that of normal wood. As with compression wood, the other major problem is that the timber may collapse and warp owing to excessive and uneven shrinkage.

7.3 Processing Implications

7.3.1 Sawing Strategies

Timell (1986, p. 1803 ff.) provides a good introduction to sawmilling strategies to combat warp in juvenile wood and compression wood. The propensity to warp, which is intrinsic to the wood itself, can be ameliorated by appropriate cutting patterns at the headrig. This is especially true for fast-grown plantation species. The cutting pattern impinges on drying, whether stresses are released on sawing or the cutting results in subsequent drying stresses that are balanced within the board.

7.3.2 Warp on Drying

A number of wood characteristics are directly implicated in warp. These are compression wood, microfibril angle and spiral grain. The severity of warp is determined not so much by the presence of these characteristics or even by the magnitude of their effects; rather it is their *gradients* within and across the lumber that are so economically and physically destructive. In addition, and in contrast, basic density and hemicellulose content influence the magnitude of the effect. The hemicelluloses, being non-crystalline and strongly hygroscopic, are largely responsible for adsorption of water within the cell wall and this, in turn, will affect shrinkage on drying. Density has a similar effect, as the extent of shrinkage is directly proportional to the basic density of wood to a first approximation (Stamm 1964, p. 218 ff.).

The location within the tree from which the board has been sawn and the orientation of the growth rings within that board determine its manner of warping on drying, while the wood density and the final moisture content influence the severity of the distortion, as illustrated in Fig. 7.12.

Warp has become more of a problem in recent years as the availability of lumber from old-growth stands has declined and mills have become more dependent on logs from younger, second-growth forest and short-rotation plantations. Younger trees have proportionately more unstable juvenile wood.

Degrade from crook and bow is most severe in the butt log and near the pith (Hallock and Malcolm 1972; Haslett et al. 1991). This is because there is a steep, axial *shrinkage gradient* in this zone where the microfibril angle is changing over the range from 50° to 30°. The effect is far less significant further up the tree where the initial microfibril angle is less, and indeed, on average,

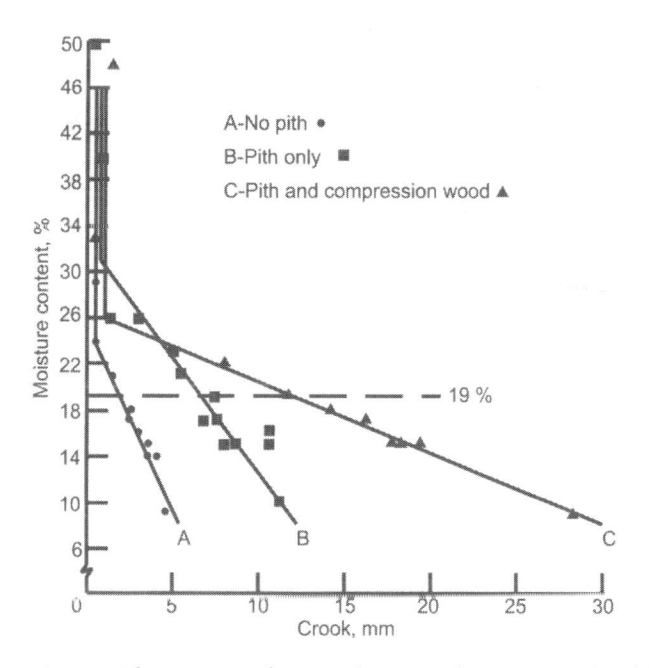

Fig. 7.12. The manner and extent of warp on drying. (Reproduced from Gaby 1972, Warping in Southern pine studs, reprinted courtesy of the Southern Research Station, USDA Forest Service)

lies below the critical value of 30°. Most crook and bow is thus expected to be found in the butt log, which, having the largest diameter, is the one that appeals most to sawmillers.

Conversely, spiral grain increases in severity up the stem (Cown 1992a), so twist is more of a problem in the upper logs (Hallock 1965; Haslett et al. 1991; Kano et al. 1964). Again, there appears to be a critical spiral grain angle, approximately 6°, above which twist becomes of economic significance. Figure 7.13 shows the average pattern of spiral grain in the case of *Pinus radiata*. Much of the wood close to the pith above the butt log will be susceptible to twist, as observed by Haslett et al. (1991). They noted that 54% of logs having small end-diameters between 150 and 300 mm, which were sawn into 100 × 50 mm lumber, failed to meet the national grade requirements for twist; whereas the failure percentage of larger logs, greater than 400 mm in diameter has fallen to 22%, but only because there was proportionately less corewood in the larger logs.

Earlier work with *Larix kaempferi* by Ozawa (1972), using material sampled at 4 m above ground level, would indicate that twist might be a problem in a few trees both in the first few growth rings and in the outermost wood. In this case, the grain angle exceeds the critical value of 6° in the juvenile wood and mature wood, with both kinds of wood having similar grain-angle gradients per unit distance.

This suggests an alternative answer to the question as to whether fast growth in plantation forestry reduces wood quality, because of the increase proportion of juvenile wood under these conditions. If the *gradient* in any property is crucial in accentuating warp, then fast growth should be viewed as a positive benefit.

To restate these points: crook and bow are most severe in the corewood of the butt log, where the microfibril angle is large and declining rapidly and where compression wood is frequently encountered; twist is most severe in the corewood of the upper logs, where spiral grain angles are large and changing rapidly.

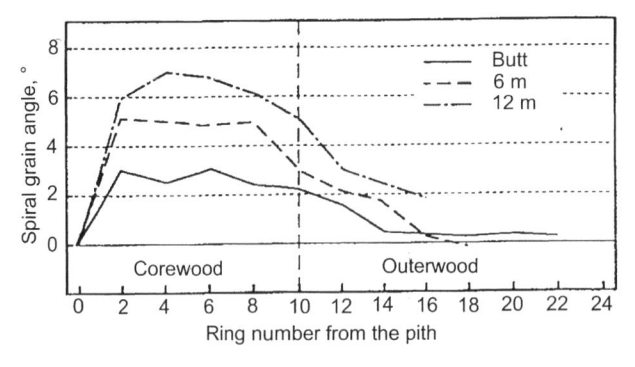

Fig. 7.13. A typical pattern of spiral grain in *Pinus radiata* in relation to radial and axial position in the stem. (After Cown 1992a)

Techniques, such as drying large-dimension lumber, both in the width and thickness, minimise warp because of the balancing effects due to symmetry about the pith and restraining effects of normal wood mixed with any compression wood. This is analogous to the saw-dry-rip (SDR) method of balancing growth stresses within full-width flitches. Both problems are likely to be relieved better by kiln-drying at high temperature, arising from the beneficial plasticisation of moist lignin. If the lumber is subsequently ripped, or dressed heavily on one face, any residual stresses will be released and the pieces will warp; however, the worst of the residual initial stresses should have been relieved during the high-temperature schedule, so that any warp should be kept within limits.

8 Stress and Strain Behaviour

Drying wood generates strains and thus stresses in a board. If the rate of drying is relatively slow and the extent of moisture-content change small, essentially only elastic strains and viscoelastic creep occur. However, if the drying is faster, mechanosorptive behaviour develops as the moisture content is reduced. Kiln-drying is done as fast as possible within limits imposed by an excessive buildup of strain and stress, with likely appearance of checks or splits in the boards. These checks arise because the strains and stresses have exceeded failure criteria in local regions of the lumber. The prediction of these mechanical changes in a transverse direction is therefore one prerequiste for the optimisation of lumber-drying schedules.

Strain is the fractional change in linear dimension that results from an applied stress. Most approaches to assessing stress and strain levels in lumber on drying have treated the lumber as a continuum material and have modelled different components of strain using constitutive equations. Around 200 cells are required to approximate an infinite continuum (Harrington 1997, pers.comm.), which compose an entity only a little smaller in diameter than the thickness of lumber boards that are dried. Nevertheless, this continuum approach will be used as the basis for discussing strain development in this chapter.

These components of strain are commonly given as follows:

$$\varepsilon = \varepsilon_s + \varepsilon_i + \varepsilon_{ve} + \varepsilon_{ms}, \tag{8.1}$$

where ε_s = shrinkage strain; ε_i = instantaneous strain; ε_{ve} = viscoelastic strain; and ε_{ms} = mechano-sorptive strain.

Some workers also include a thermal strain to take account of the thermal expansion of wood and the loss of bound moisture as the temperature is raised (Chen et al. 1997a,b). For convenience, Eq. 8.1 is sometimes written in strain-rate form as follows (a point to be noted in interpreting the literature):

$$\dot{\varepsilon} = \dot{\varepsilon}_s + \dot{\varepsilon}_i + \dot{\varepsilon}_{ve} + \dot{\varepsilon}_{ms}. \tag{8.2}$$

Typical strain behaviour for a piece of lumber, which is subjected to a constant stress while being dried, is shown in Fig. 8.1.

Instantaneous strain (ε_i) arises as soon as the lumber is loaded, and stays constant as long as the stress is kept at a constant value. *Viscoelastic strain* (ε_{ve}) is the time-dependent movement of the wood in the absence of any change in moisture content. As soon as moisture is removed from the lumber, other contributions to the strain arise. If the moisture content of the lumber falls below

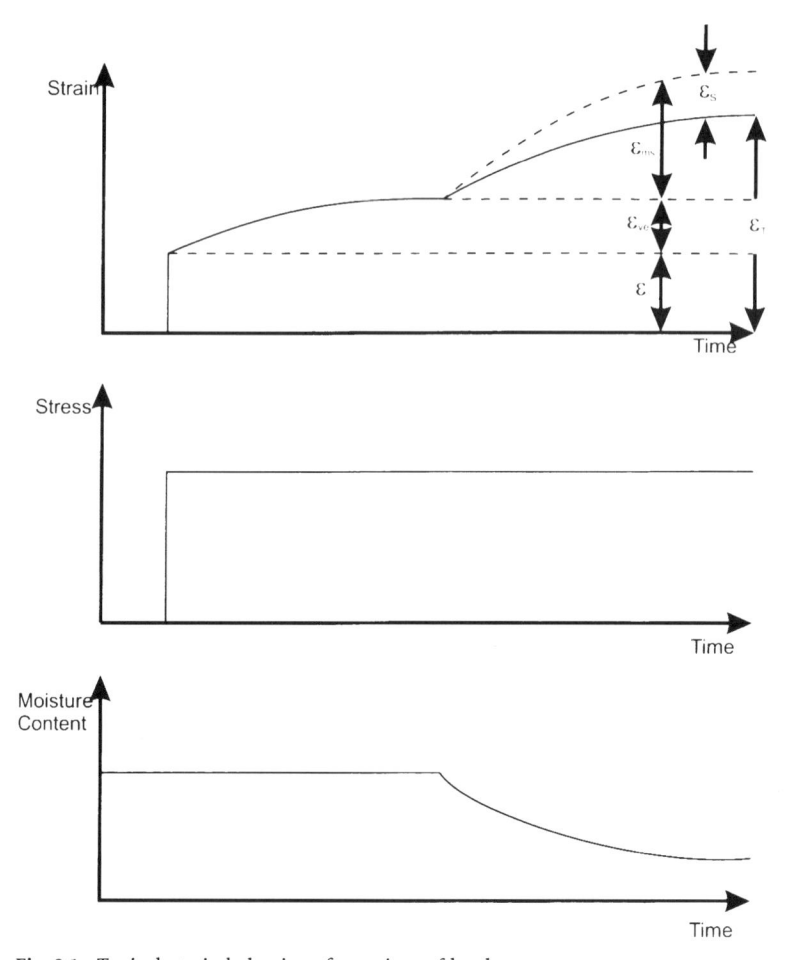

Fig. 8.1. Typical strain behaviour for a piece of lumber

the fibre-saturation point, the lumber shrinks, and this *shrinkage strain* (ε_s) reduces the length of the piece of lumber in the transverse direction. *Mechanosorptive strain* (ε_{ms}) also occurs in response to changes in moisture content, allowing the wood to stretch more than would occur due to instantaneous and viscoelastic strain. Viscoelastic and mechanosorptive strains both depend on time while the wood is drying and its moisture content is changing. Both these strains are sometimes described as *creep* (e.g. Salin, 1992).

A physical constraint on the stresses which develop within a board is that the sum of the internal forces within the board must be zero, so the integral or sum of the stresses through the thickness of the board must also be zero. This feature is often called the requirement for *internal* or *static equilibrium*:

$$\int_{\text{whole thickness}} \sigma \, dx = 0 \qquad (8.3)$$

or in finite-element form,

$$\sum_{\text{whole thickness}} \sigma \, \Delta x = 0. \tag{8.4}$$

8.1 Mechanical Analogues

These analogues have been used by many authors to describe the development of strains in wood, and a useful review of these models in the context of drying is given in Hasatani and Itaya (1996). The analogues include elastic spring and viscous dashpot elements, together with combinations which include these elements in series or in parallel, such as Maxwell and Kelvin models. These models have two elements, a linear spring and a linear dashpot. In the Maxwell model, the two elements are connected in series; whereas in the Kelvin model, the two are connected in parallel. When a Maxwell and a Kelvin model are connected in series, this combination is called the standard linear or Burgers model (Gittus 1975). These various models are illustrated in Fig. 8.2.

8.1.1 Elastic Element

This may be represented by a spring element, as shown in Fig. 8.2a. From Hooke's law, the relationship between the strain (ε) and the imposed stress (σ) is given by

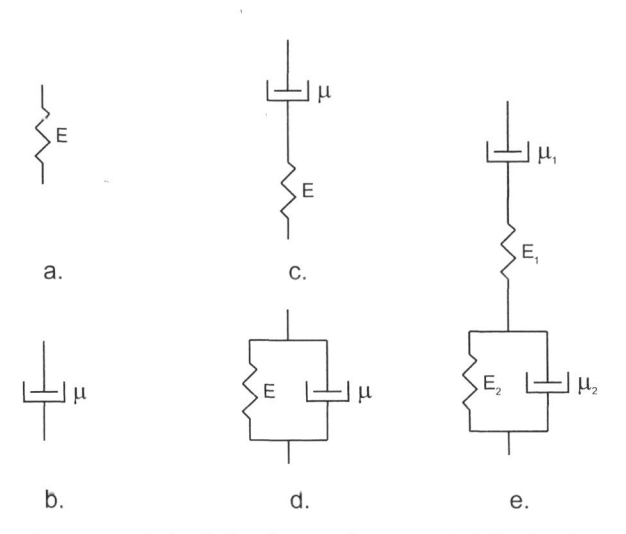

Fig. 8.2a–e. Mechanical analogues of stress-strain behaviour in wood. a Spring element. b Viscous element. c Maxwell model. d Kelvin model. e Burgers model

$$\varepsilon = \frac{\sigma}{E}, \tag{8.5}$$

where E is the *modulus of elasticity*. It is sometimes known as the *stiffness*.

8.1.2 Viscous Element

This element is frequently represented by a dashpot, as illustrated in Fig. 8.2b. The linear version of this element has a constant Newtonian viscosity (μ), so that the relationship between the rate of strain ($\dot{\varepsilon}$) and the imposed stress ($\dot{\varepsilon}$) is

$$\dot{\varepsilon} = \frac{\sigma}{\mu}. \tag{8.6}$$

8.1.3 Maxwell Model

This model consists of spring and dashpot elements in series, as shown in Fig. 8.2c, so that the overall strain consists of a component from each element:

$$\varepsilon = \varepsilon_{spring} + \varepsilon_{dashpot}, \tag{8.7}$$

where ε_{spring} is the strain in the spring, and $\varepsilon_{dashpot}$ is the strain in the dashpot. The rate of strain is given by

$$\dot{\varepsilon} = \frac{\dot{\sigma}}{E} + \frac{\sigma}{\mu}. \tag{8.8}$$

Hence, for constant stress, the relationship between strain rate and imposed stress is simply that of the viscous dashpot element, which is sometimes written in the form

$$\dot{\varepsilon} = \frac{\sigma}{\mu} = m\sigma. \tag{8.9}$$

In integrated form, this equation becomes

$$\varepsilon = \frac{\sigma}{E} + \frac{t\sigma}{\mu}. \tag{8.10}$$

8.1.4 Kelvin Model

As depicted in Fig. 8.2d, there are spring and dashpot elements in parallel, so each element supports the applied stress, and the stress-strain relationship is given by

$$\sigma = E\dot{\varepsilon} + \mu\dot{\varepsilon} \tag{8.11}$$

Sometimes it is presented in the integrated form for constant stress:

$$\varepsilon = \frac{\sigma}{E}\left[1 - \exp\left(-\frac{Et}{\mu}\right)\right]. \tag{8.12}$$

8.1.5 Burgers Model

Figure 8.2e illustrates a combination in series of the Maxwell and Kelvin model known as the Burgers model. Hasatani and Itaya (1996) give the relationship for the strain at constant stress in this case as follows: –

$$\varepsilon_{ve} = \frac{\sigma}{E_1} + \frac{t}{\mu_1} + \frac{\sigma}{E_2}\left[1 - \exp\left(-\frac{E_2 t}{\mu_2}\right)\right]. \tag{8.13}$$

8.2 Shrinkage

Shrinkage occurs in wood because the diameters of the fibres dwindle when bound moisture is lost from the cell walls. No shrinkage occurs above the fibre-saturation point. The amount of linear shrinkage has invariably been found to be proportional to the difference between the local moisture content and the local value of the moisture content at fibre saturation at the same temperature:

$$\varepsilon_s = \alpha(X_{FSP} - X) \text{ for } X < X_{FSP}. \tag{8.14}$$

The volumetric shrinkage is approximately equal to the volumetric moisture loss from the cell wall.

To measure the shrinkage strain, Oliver (1991) recommends that 1-mm-thick samples, for which the moisture content distribution is virtually uniform, should be dried at room temperature to about 12% moisture content. Typical linear shrinkage strains for many Australian eucalypts in the radial direction is about $0.04\,\text{mm}\,\text{m}^{-1}$ (Oliver, 1991). Kininmonth and Whitehouse (1991) gave the transverse shrinkage strains to the oven-dry condition for radiata pine (*Pinus radiata*) as 0.072 (tangential) and 0.034 (radial), while, for sitka spruce (*Picea sitchensis*), Yokota and Tarkow (1962) have reported strains of 0.075 (tangential) and 0.043 (radial).

Free shrinkage in *Pinus radiata* was also investigated by Keep (1998). For the larger, 5-mm-thick samples used in her tests, the moisture content appeared to have reached equilibrium while the sample was continuing to shrink. It was also observed that the final shrinkage was less than expected from previous work. The ratio of the time for the sample to reach the equilibrium moisture content and the time for shrinkage to be complete increased with temperature, suggesting that moisture diffusion increases more rapidly with temperature than relaxation of moisture stress. Hence, as the temperature is raised, stress equilibrium takes a proportionately longer time to reach than moisture equilibrium does.

8.3 Instantaneous Strain

When lumber is loaded rapidly under tension, bending or compression, the resulting instantaneous strain is usually proportional to the imposed stress up to a certain point called the *proportional limit*. In the region below the limit, Hooke's law may be used to relate the stresses and the strains; thus

$$\varepsilon_i = \frac{\sigma}{E} \quad \text{or in rate form} \quad \dot{\varepsilon}_i = \frac{\dot{\sigma}}{E}. \tag{8.15}$$

where E is the modulus of elasticity or stiffness. Above this region, non-linearity occurs. However, there may not always be a sharp transition between fully elastic behaviour and elastoplastic deformation, as illustrated in Fig. 8.3 by Carrington's data (1996) in the tensile testing of *Pinus radiata* samples across the grain. Below fibre saturation the wood becomes less yielding, as the plasticising effect of moisture in the cell walls diminishes with its loss. Comparative tests under ambient conditions and a dry-bulb temperature of 90 °C, show that, at the higher temperature, the wood is more plastic and fails at a lesser *ultimate strain*.

In kiln-drying, only transverse stresses are of interest, with tensile stresses being of concern as being potentially more damaging than compressive stresses. Oliver (1991) suggests that surface checks appear at a transverse strain of 0.02, indicating that this value might be an appropriate failure criterion in kiln-drying for Australian eucalypts. Carrington's data (1996), shown in Fig. 8.3, imply that a single-valued criterion might not be adequate. Chen et al. (1997a), in calculating failure-strain envelopes during the high-temperature drying of a sapwood board of *Pinus radiata*, illustrate the varying susceptibility to failure as the temperature and moisture content change.

The term *compliance J* is also used to describe the ratio of strain to stress (ε/σ). The modulus of elasticity is the inverse ratio of stress to strain, which is reported more normally, because most mechanical-testing machines are controlled strain-rate devices, and the strain is therefore a controlled input. The compliance is sometimes reported, however, because it is the ratio of an effect or outcome of drying (strain) to a cause or input (stress).

8.3.1 Linear Loading

Oliver (1991) suggests that the proportional limit for many species, in the absence of further information, should be taken as a strain of 0.005.

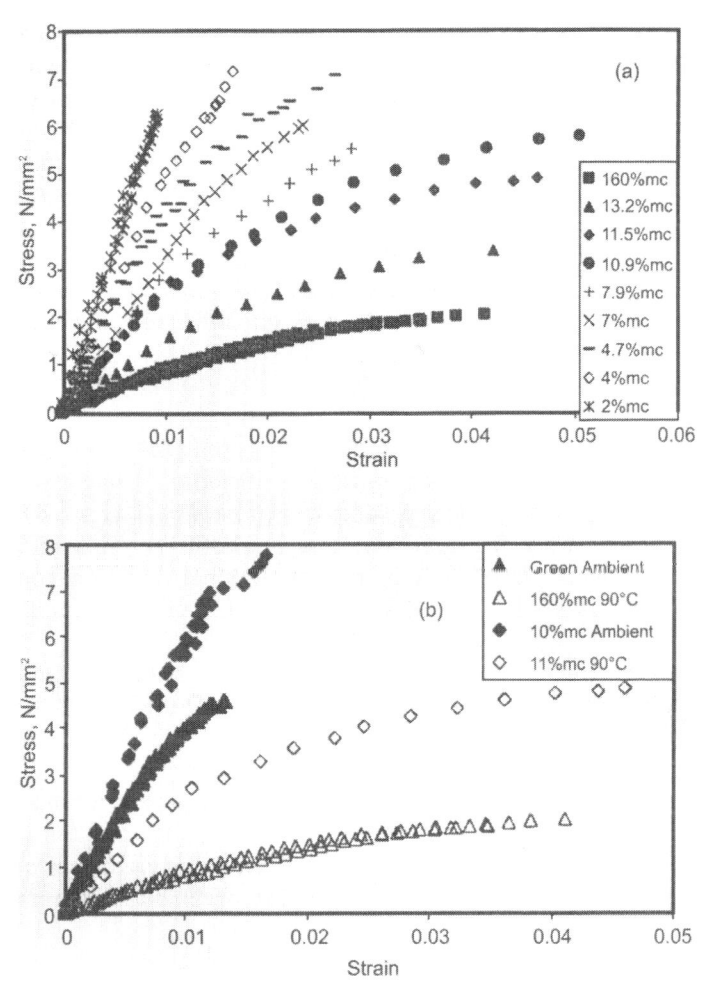

Fig. 8.3a,b. Tensile stress-strain curves for *Pinus radiata* sapwood. **a** Variation with percentage moisture content (mc) at a dry-bulb temperature of 90 °C. **b** Comparison of behaviour with air-dried and green wood at ambient conditions and at 90 °C. (Carrington 1996)

Carrington's data (1996), presented in Fig. 8.3, are not inconsistant with this value, although at moisture contents below 10% *Pinus radiata* sapwood appears to be elastic to a strain of 0.01. Reported failure strains vary from 0.038 for green sapwood to 0.009 at 2% moisture content at a temperature of 90 °C. By contrast, for longitudinal tension tests, Dinwoodie (1981) suggests that the limit of proportionality is approximately 60% of the ultimate stress, while for longitudinal compression the limit is about 30–50% of the ultimate value. Walker (1993), however, indicates that the proportional limit for axial stress is about 80% of the ultimate value.

For this linear part of the stress-strain curve, Oliver (1991) indicates that the slope (the modulus of elasticity E) varies roughly linearly with temperature to become very small at a temperature close to 100 °C, although the effect of temperatures above 80 °C is uncertain due to lack of data. Keep (1998), on the other hand, shows that the transverse elastic modulus for green *Pinus radiata* sapwood only falls from 219 MPa at 25 °C to 90 MPa at 90 °C, as shown in Table 8.1.

Ellwood (1954) suggests that the ratio of longitudinal elastic modulus at zero moisture content (E_o) to that in the green state (E_g) is related to temperature T (°C) for Tasmanian eucalypts (*Eucalyptus* spp.) and for American beech (*Fagus grandifolia*) by the equation,

$$E_g = E_o(0.475 - 0.0038T). \tag{8.16}$$

This linear relationship is also consistent with the findings of Palka (1973) and Wu and Milota (1995). For a range of softwoods including Douglas-fir (*Pseudotsuga menziesii*), spruces (*Picea* spp.) and Scots pine (*Pinus sylvestris*), Palka (1973) has found that the fractional decrease in elastic modulus with increase in moisture content was $0.016X$ in the longitudinal direction and $0.032X$ in the radial and tangential directions (where X is the dry-basis moisture content).

The effect of moisture content on the modulus of elasticity has been found to be significant only below about 20% moisture content (Dinwoodie 1981; Oliver 1991; Carrington 1996; Keep 1998). Keep's data for sapwood *Pinus radiata* under tensile loading in the transverse direction are given in Table 8.2. The data illustrate that, as bound moisture is removed, the wood becomes stronger but fails at a smaller ultimate strain.

Oliver (1991) suggests that this feature is due to the increasing influence of bound water on the stiffness of the lumber structure, when the moisture

Table 8.1. Summary of mechanical parameters determined for green *Pinus radiata* sapwood (Keep 1998)

Temperature (°C)	25	90
MOE (MPa)	219	90
Failure strain (%)	4.83	3.80

Table 8.2. Summary of mechanical parameters under tensile loading determined at 20 °C for *Pinus radiata* sapwood in the transverse direction. (Keep 1998)

Property	Green	20% emc	15% emc	8% emc
Elastic modulus (MPa)	219	294	486	669
Failure strain (%)	4.83	3.34	1.70	0.80
Ultimate stress (MPa)	4.76	4.61	5.18	6.20

becomes more strongly bound to the cell walls at moisture contents below $0.02\,\text{kg}\,\text{kg}^{-1}$. In this low moisture-content range, the effect of moisture content on the modulus is virtually linear (Dinwoodie 1981, Oliver 1991). For example, Oliver gives the following equation for the longitudinal modulus of Australian eucalypts:

$$E = E_o - 500(E_o - E_g)X, \tag{8.17}$$

where X is the moisture content.

This linear dependence of the modulus of elasticity on the moisture content follows a similar trend to that found by Palka (1973) and Wu and Milota (1995). Palka found that the fractional decrease in modulus was 0.02 (i.e. 2%) for every 1% increase in moisture content for the radial and tangential directions, and 0.01 for the same increase in the longitudinal direction. Even when the moisture content has been changed cyclically, the modulus of elasticity has been found to be a unique function of moisture content by Hearmon and Paton (1964).

The effects of both temperature and moisture content have been combined by Oliver (1991) to give the expression:

$$(E - E_g)[E - E_o + 500(E_o - E_g)X] = \frac{E_g^2}{1000} \tag{8.18}$$

for the longitudinal modulus of eucalypts. Oliver (1991) indicates that the radial and tangential moduli of elasticity are around 10 and 5% of the longitudinal modulus, respectively. This relative stiffness in the transverse directions has an important bearing on the extent to which a drying board can cope with the developing stresses.

On the basis of the similarity between the exponential decrease in both modulus of elasticity and heat of sorption with increasing moisture content, the following correlation was proposed by Keey and Keep (1999) for the relationship between the modulus of elasticity and the moisture content below the fibre-saturation point for *Pinus radiata*:

$$[\text{MOE}/\text{kPa}] = 900 \exp(-17X) + 90, \quad X < X_{FSP}. \tag{8.19}$$

The value for this modulus appeared to be uniform above the fibre-saturation point, indicating that the infilling of tracheids with free water has no effect on the plasticity of softwood fibres.

The cell-wall structure is divided into several layers, each having a characteristic fibre angle. The microfibril angle of the dominant S_2 layer has been found by Cave (1978) (for *Pinus radiata*) and Hunt and Shelton (1987) (for *Pinus ponderosa*) to be of major significance in determining the elastic modulus of the wood. Small, acute angles result in small compliances, so the elastic modulus of the wood is greater.

If the strain is used as the failure criterion, then Langrish et al. (1997) observe that the value of the modulus of elasticity, while affecting the predicted stresses, does not affect the predicted strains. For illustrative purposes, this feature can be seen by considering only instantaneous elastic and shrinkage strains, together with a one-dimensional moisture-content distribution through the thickness of a lumber board.

If any one layer i could move independently of all other layers (Fig. 8.4), its shrinkage strain ε_{si} could be well approximated by the linear expression of equation (8.14), written here in the form:

$$\varepsilon_{si} = \alpha M_i, \tag{8.20}$$

where the moisture change is given by $M_i = X_{FSP} - \min(X_i, X_{FSP})$. The stress σ_i that would be required to keep the layer at its original length can be calculated from Hooke's law:

$$\sigma_i = E_i \alpha M_i, \tag{8.21}$$

where the modulus of elasticity E_i is also dependent on temperature and moisture content. However, none of the layers can move independently, as any board will keep its general shape as shrinkage occurs. Therefore, the internal forces must be calculated in relation to the average shrinkage $\bar{\varepsilon}_s$ of the whole board. From the requirement for internal force equilibrium, the sum of these internal forces must be zero, a fact which can be expressed in terms of the finite slices by the following expression:

$$\sum E_i (\bar{\varepsilon} - \alpha M_i) \Delta x_i = 0. \tag{8.22}$$

From this equation, the average shrinkage strain of the board $\bar{\varepsilon}_s$ is calculated to be:

$$\bar{\varepsilon}_s = \frac{\sum E_i \alpha M_i \Delta x_i}{\sum E_i \Delta x_i}. \tag{8.23}$$

Here, variations in the modulus of elasticity tend to cancel each other out. The strain which is imposed on each layer to keep it at the same length as the lumber board as a whole may be estimated from:

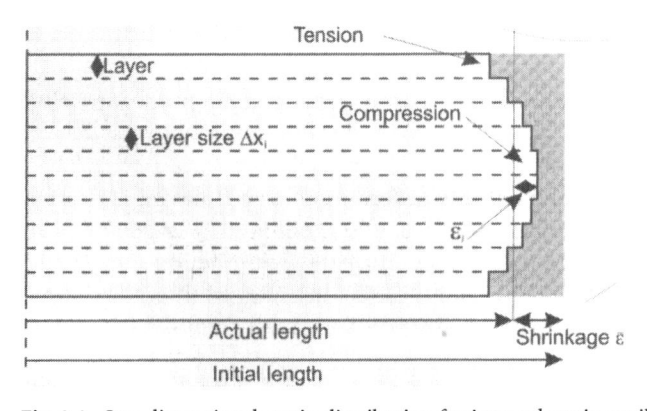

Fig. 8.4. One-dimensional strain distribution for internal static equilibrium

$$\varepsilon_i = \bar{\varepsilon}_s - \alpha M_i. \tag{8.24}$$

Again, the strain imposed on each layer is not dependent on the modulus of elasticity.

8.3.2 Non-Linear Loading

Above the proportional limit, Oliver (1991) proposed the use of the following equation to relate the stresses and strains:

$$\sigma_u - \sigma = (\sigma_u - \sigma_y)\exp[-k(\varepsilon_i - \varepsilon_y)], \tag{8.25}$$

in which σ is the local stress, σ_u is the local value of the ultimate stress (taken as $0.01E$ by Oliver), σ_y is the local value of the stress at the proportional limit (defined by the value of E and the strain at the proportional limit, 0.005), ε is the local value of the instantaneous strain and ε_y is the strain at the proportional limit. The value of k in Eq. (8.25) was chosen by Oliver (1991) to be tangent to the linear part of the stress-strain curve at the proportional limit, and he suggested that the value of k could be estimated as 200 for most eucalypts. Further, Oliver (1991) showed that the local slope to the stress-strain curve in the non-linear region was given by

$$E = k\,|\sigma_u - \sigma|, \tag{8.26}$$

in which the absolute value of the stress difference is used because the modulus is always taken as positive.

Bodig and Jayne (1982) suggested a more elaborate fitting equation to describe the instantaneous response to applied stress above the proportional limit:

$$\varepsilon = \frac{\sigma}{E}\left[1 + k_I n_I (\sigma - \sigma_y)^{n_I - 1}\right], \tag{8.27}$$

where k_I and n_I are constants. The authors claim this is a general expression for all species and wood-based materials.

8.3.3 Unloading

The process of drying for a lumber board involves periods during which the stress and strain increase in the lumber with increasing time (loading) and those in which the stress decreases (unloading). When unloading occurs, Oliver (1991) shows different stress-strain curves for loading and unloading, with the unloading section being taken as a straight line with a slope equal to that in the linear part of the loading curve, as illustrated in Fig. 8.5.

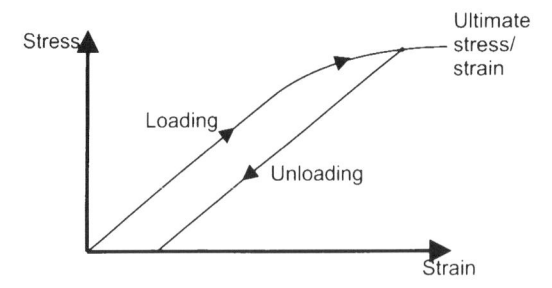

Fig. 8.5. Typical stress-strain curve for rapid loading of lumber, showing loading and unloading behaviour. (Oliver 1991)

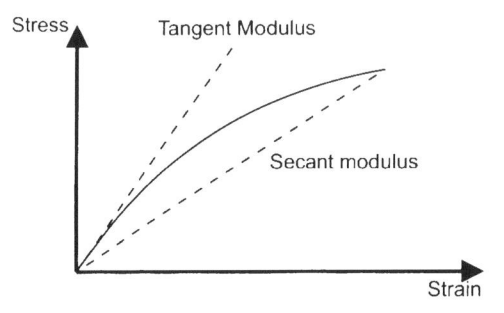

Fig. 8.6. Typical stress-strain curve for slow-loading of wood samples, showing different interpretations of the modulus of elasticity. (After Dinwoodie 1981)

8.3.4 Slow-Loading Tests

In slow-loading tests, over periods of several minutes, no proportional limit is generally observed, and the stress/strain curve is concave downwards with a curvilinear shape, similar to those depicted in Fig. 8.3. When determining the modulus of elasticity, the traditional treatment of these curves has been either to take the tangent to the curve through the origin (*the tangent modulus*) or to connect the failure point to the origin (*the secant modulus*), with the tangent modulus being the commoner, as illustrated in Fig. 8.6 (Dinwoodie 1981). These slow-loading tests almost certainly include some viscoelastic and mechanosorptive components of strain, and so may not represent only the instantaneous component of strain.

Ugolev (1976) has used a power-law relationship to fit this type of slow strain-rate behaviour for beech (*Fagus sylvatica*) and birch (*Betula* sp.), as follows:

$$\sigma = \sigma_b \left(\frac{\varepsilon}{\varepsilon_b} \right)^m \tag{8.28}$$

where σ_b, ε_b and m are fitted parameters.

8.4 Viscoelastic Strain

Wood is a material which shows both elastic and viscous aspects of stress-strain behaviour, including the phenomena of creep (an increase in strain with time under a constant stress) and relaxation (a decrease in strain with time for a constant stress). These components of strain are time-dependent. Under constant stress, the type of behaviour shown in Fig. 8.7 is typical, with the strain curve being divided into primary (the initial non-linear response), secondary (the long, nearly linear section), and a short tertiary section just before the wood fractures divided (Bodig and Jayne 1982). The tertiary section is not shown in Fig. 8.7 because it is normally very short in duration compared with the primary and secondary sections, and typically, it is not a useful operating region when drying lumber boards.

Alignment and rotation of the polymer components (cellulose, hemicellulose, etc.) of wood (Rice and Youngs 1990), uncoiling and recoiling of molecules, and occasional hydrogen-bond breaking and reforming (Bodig and Jayne 1982) have all been proposed as mechanisms for viscoelastic strain.

Both mechanical analogies for the process of viscoelastic strain development (such as the Kelvin model) and an empirical approach (the Bailey-Norton equation) have been used.

8.4.1 Mechanical Analogues

Various theoretical descriptions of viscoelastic behaviour have been proposed, based on combinations of elastic springs and viscous dashpots, as described in Section 8.1. For Tasmanian eucalypts, Oliver (1991) indicates that the shape of the curve of creep (viscoelastic) strain against time is close to that described by the Kelvin model, and is described by

$$\varepsilon_{ve} = \frac{\sigma}{E} A [1 - \exp(-k_1 t)]. \tag{8.29}$$

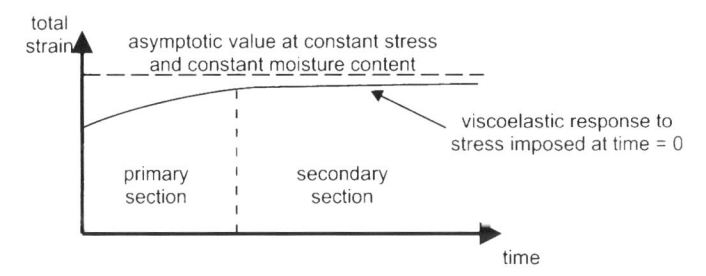

Fig. 8.7. Typical viscoelastic response for wood under load

The strains given by this equation follow the observation that the rate of viscoelastic strain decreases as the amount of strain increases. The parameters in Oliver's model are non-linear functions of temperature and moisture content, in agreement with the finding of Kingston and Clarke (1961) that the use of linear elements in the Kelvin model resulted in poor agreement between their test results and model predictions for beams of mountain ash (*Eucalyptus regnans*).

For eucalypts, Oliver (1991) suggests that the parameter A is given by:

$$A = \frac{1}{b_1 + \sqrt{[k_2 - a_2(R_o - R_s|\sigma/\sigma_u|) + b_1^2]}}, \qquad (8.30)$$

in which

$$b_1 = 0.5(a_2 + R_o - R_s|\sigma/\sigma_u|) \qquad (8.31)$$

and

$$a_2 = 0.04 R_o \qquad (8.32)$$

$$R_o = 105.6 - 3.46T + 0.0319T^2 \qquad (8.33)$$

$$R_s = 60 - 1.99T + 0.019T^2, \qquad (8.34)$$

where T is the temperature of the wood in degrees Celsius.

In Eq. (8.29), the instantaneous strain is (σ/E), so taking the ratio of viscoelastic strain to the instantaneous strain ($e = \varepsilon_{ve} / \varepsilon_i$) gives

$$e = A[1 - \exp(-k_1 t)], \qquad (8.35)$$

so that differentiation gives

$$\dot{e} = k_1 A \exp(-k_1 t) = k_1(A - e), \qquad (8.36)$$

and since e is equal to $\varepsilon_{ve} / \varepsilon_i$, further differentiation gives

$$\frac{1}{\varepsilon_i}\frac{d\varepsilon_{ve}}{dt} - \frac{\varepsilon_{ve}}{\varepsilon_i^2}\frac{d\varepsilon_i}{dt} = k_1\left(A - \frac{\varepsilon_{ve}}{\varepsilon_i}\right). \qquad (8.37)$$

The value of k_1 was not specified. This equation allows the change in viscoelastic strain (ε_{ve}) to be related to the change in the instantaneous strain (ε_i).

Hardtke et al. (1996) used the following version of the Kelvin model to described viscoelastic strain for European beech (*Fagus sylvatica*):

$$\varepsilon_{ve} = D^{-1}(\sigma - E_{ve}\varepsilon_{ve}) \qquad (8.38)$$

where D was taken as 10 MPa and E_{ve} was equated to the modulus of elasticity for instantaneous strain.

Salin (1989) used an equation based on the Burgers model for describing viscoelastic strain in Scots pine (*Pinus sylvestris*) and Norway spruce (*Picea abies*), but found, like Rice and Youngs (1990), that this component of strain could not fully account for the observed stress levels in the wood. Salin suggested that neglect of the mechanosorptive strain was responsible for the discrepancy.

8.4.2 The Bailey-Norton Equation

The Bailey-Norton equation is a power-law model which has been used frequently to describe the viscoelastic response of a material under a constant applied stress (Schniewind and Barrett 1972; Wu 1993; Wu and Milota 1995). This equation is commonly used in the field of material science (Kraus 1980) and has the following form for constant imposed stress:

$$\varepsilon_{ve} = A\sigma^q t^n, \tag{8.39}$$

where A, q and n are parameters. Ranta-Maunus (1993) used the following version of the equation for the total strain (ε) under constant stress, for Norway spruce (*Picea abies*): –

$$\frac{\varepsilon}{\varepsilon_i} = 1 + 0.06 t^{0.27}, \tag{8.40}$$

where ε_i is the instantaneous strain, and the time (t) is measured in hours. For southern pine, Wu (1993), Milota and Wu (1994) and Wu and Milota (1995) have used the following equation for the total compliance (ε/σ):

$$\frac{\varepsilon}{\sigma} = \frac{\varepsilon_i}{\sigma} + k_C t^{n_C}, \tag{8.41}$$

in which the time (t) is measured in minutes, and the parameters k_C and n_C for creep strain were fitted to the following functions of temperature. In tension, these coefficients are:

$$k_C(T) = 11.8 \times 10^{-5} \tag{8.42}$$

$$n_C(T) = 0.363 + 0.00237 T \qquad R^2 = 0.94, \tag{8.43}$$

where the temperature (T) ranges from 32.2 to 82.2 °C and R^2 is the correlation coefficient. (There was no significant correlation with temperature for the coefficient k_c). In compression, these coefficients were determined to be

$$k_C(T) = 7.05 \times 10^{-5} + 1.62 \times 10^{-6} T \qquad R^2 = 0.83 \tag{8.44}$$

$$n_C(T) = 0.323 + 0.00207 T \qquad R^2 = 0.76. \tag{8.45}$$

Keep (1998) measured the viscoelastic behaviour of *Pinus radiata* over the temperature range from 20 to 140 °C, and produced the following correlations for the material parameter values under a stress of 0.65 MPa:

$$\text{At } X = 0.05\, kg\, kg^{-1}: k_C(T) = 11.8 \times 10^{-5} + 4.05 \times 10^{-8} T \qquad R^2 = 0.80 \tag{8.46}$$

At $X = 0.15 \, kg \, kg^{-1}$: $k_C(T) = 2 \times 10^{-4} \exp(0.0165T)$ $R^2 = 0.81$ (8.47)

At $X = 0.20 \, kg \, kg^{-1}$: $k_C(T) = 2 \times 10^{-5} \exp(0.0572T)$ $R^2 = 0.61$. (8.48)

Evidence of a glass transition was detected at the highest temperature used. For a test performed at 140 °C with an equilibrium moisture content of 5%, the strain behaviour exhibited much greater plastic behaviour than that at lower temperatures. This was felt to be due to thermal softening of the lignin which binds the microfibrils together in a more brittle matrix when the temperature is lower. The glass-transition temperature for biomaterials is known to decrease monotonically with increasing moisture content (Mujumdar 1997), and thus the transition temperature for green wood would be expected to be significantly lower than that reported by Keep (1998) for much drier wood. However, Riley et al. (1999) observe from tests on compressive strains that the progressive softening over the temperature range 50–140 °C may not be simply related to the glass-transition points of the constituent lignin and hemicellulose. They suggest that the plastic limit may be a function of stress level and temperature.

The decrease in the viscoelastic strain rate with time at constant stress is known as *strain-hardening*. It corresponds physically to the stretching of the wood-cell ultrastructure, with progressively less deformation being possible as the polymer chains become more strongly aligned in the direction of the applied stress. Hence, the rate expression given by the Kelvin equation seems a reasonable process-engineering approach to modelling this phenomenon. A detailed analysis of Keep's viscoelastic data for *Pinus radiata* has shown that the two-parameter Kelvin model fitted most of the data sets best (Haque 1998).

Nevertheless, the Bailey-Norton equation is widely used, even though it is unclear how it explicitly represents this physical process of stretching the ultrastructure. Moreover, in practice, the use of this latter equation requires some involved reasoning, as will now be explained. In order to calculate increments in viscoelastic strain over specific time intervals, Wu (1993) has discussed two concepts, the strain-hardening principle and the time-hardening principle. At a particular time t, suppose that the stress is σ_1 and the amount of viscoelastic strain is ε_{ve1}, but that, over a time step Δt, the stress changes to σ_2. In Fig. 8.8, two viscoelastic strain curves are shown for constant stress levels σ_1 and σ_2 respectively from the start. With the strain-hardening principle (Fig. 8.8a), the new strain is calculated by taking a strain increment for a duration Δt which starts on the strain curve for the stress level σ_2 at the starting strain ε_{ve1}. With the time-hardening principle (Fig. 8.8b), the new strain is calculated by taking a strain increment for a duration Δt which starts on the strain curve for the stress level σ_2 at time t. Wu (1993) indicates that the time-hardening principle underestimates the strains but is easier to apply, so this principle was used in his work (Wu 1993; Milota and Wu 1994; Wu and Milota 1995).

For stress reversal (a change from tension to compression and vice versa), Wu (1993) used the following procedure, which is illustrated in Fig. 8.9. On

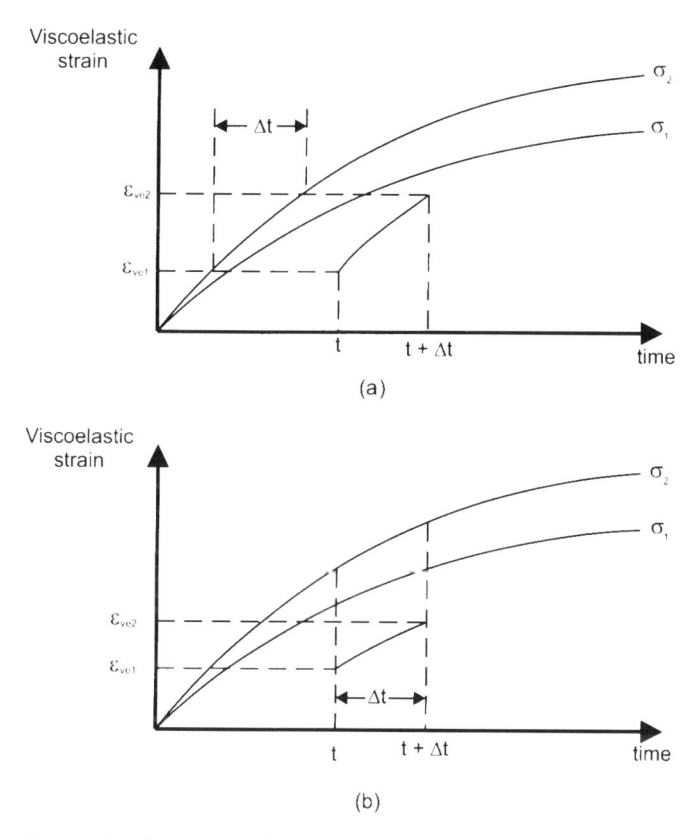

Fig. 8.8a,b. Illustration of the strain-hardening and time-hardening principles. **a** Strain-hardening principle. **b** Time-hardening principle

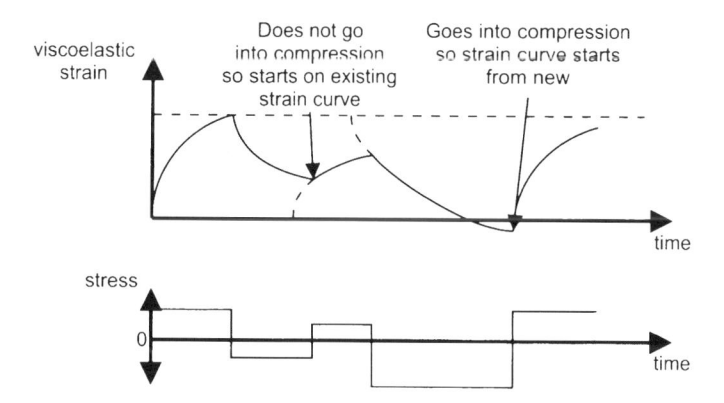

Fig. 8.9. Illustration of Wu's procedure (1993) for stress reversal

changing from tension to compression, all previous strain history is assumed to be lost, so the specimen is assumed to behave like a new (and unstrained) specimen. Any change back to tension is considered to retain the effects of the strain history from the point of compression, if the total strain was tensile at the point of changing back to tension. If the total strain was compressive at this point, the material after the change back to tension is assumed to behave like a new specimen again.

8.5 Mechanosorptive Strain

This component of the total strain occurs in response to simultaneous mechanical loading and changes in moisture content. Early work (Armstrong and Kingston 1960, 1962; Schniewind 1968; Ranta-Maunus 1975) established that the amount of mechanosorptive strain depends on the size of the change in moisture content and is not significantly affected by its duration or the rate at which it is attained.

Most workers have found that the mechanosorptive effect occurs only below the fibre-saturation point (together with shrinkage). However, Wu (1989) discovered a strong mechanosorptive effect at all moisture contents for Tasmanian eucalypts.

A suggested mechanism for mechanosorptive strain is hydrogen-bond breaking and reforming (Hunt 1982), with movement of moisture increasing the rate at which this process occurs. In polymers whose chains are distorted in three dimensions with shapes determined by adjacent water molecules, movement of water allows such chains to form lower-energy configurations under the influence of applied stress (Oliver 1991).

8.5.1 Qualitative Observations

The first detailed qualitative description of mechanosorptive strain was given by Armstrong and Kingston (1960). They identified three types of mechanosorptive strain:

- *Case* 1: Decreasing the moisture content ($\Delta X < 0$) increases the mechanosorptive strain.
- *Case* 2: Increasing the moisture content ($\Delta X > 0$) to a value *below* the previous highest moisture content decreases the mechanosorptive strain.
- *Case* 3: Increasing the moisture content ($\Delta X > 0$) to a value *above* the previous highest moisture content increases the mechanosorptive strain.

8.5.2 Quantitative Analysis

The first quantitative model for the development of mechanosorptive strain was proposed by Leicester (1971a, b) for messmate stringybark (*Eucalyptus obliqua*) and Ranta-Maunus (1975) for five species:

$$\varepsilon_{ms} = m\sigma\Delta X \tag{8.49}$$

While the model was developed using continuum-mechanics theory, it may also be viewed as being analogous to the Maxwell model for viscoelastic strain (Eq. 8.9).

Qualitatively, Ranta-Maunus (1975) observed two types of behaviour. The first kind, which he called *birch-like*, was characterised by different values for the parameter m, depending on the three types of mechanosorptive strain as identified by Armstrong and Kingston (1960). The second type of behaviour, called *spruce-like*, was characterised by only one value of the parameter m. For example, analysis of the data of Armstrong and Kingston (1960) for a birch-like material, hoop pine (*Araucaria* sp.), by Ranta-Maunus (1975) gave values for the parameter m in Table 8.3.

The numerical values shown in this table are consistent with the qualitative behaviour reported by Armstrong and Kingston (1960). In case 1, with both m and ΔX negative, mechanosorptive strain will be predicted to increase with time, while in case 2, with m and ΔX of opposite sign, mechanosorptive strain will be predicted to decrease. Finally, in case 3, with both m and ΔX positive, mechanosorptive strain will be predicted to increase.

For drying lumber, case 1 (decreasing moisture content) seems most relevant, but during reconditioning, steaming and solar (cyclic) drying, the moisture content at the surface of lumber boards will increase, so the other two cases are also relevant in kiln-seasoning practice.

This description has some attractiveness, since only one parameter (m) needs to be fitted to experimental data. Hence, the approach has been adopted by Oliver (1991); Wu and Milota (1995); Mauget and Perré (1996), and many others. Values for the parameter m, when the change in moisture content is converted into units of kg moisture (kg dry wood)$^{-1}$, show a wide range of

Table 8.3. Values of the parameter m for hoop pine (*Araucaria cunninghamii*) in the equation of Ranta-Maunus (1975)

Case	Moisture-content change	Symbol used by Ranta-Maunus (1975) for m	Numerical value for hoop pine (*Araucaria cunninghamii*), MPa^{-1}
1	$\Delta X < 0$	a	−0.02
2	$\Delta X > 0$	a^+	−0.01
3	$\Delta X > 0$ above previous level	a^{++}	0.02

values. The parameters in Table 8.3 for hoop pine (*Araucaria cunninghamii*), from Ranta-Maunus (1975), have values of +0.02 to −0.02 MPa^{-1}. Oliver (1991) used values for the constant m of $330/E_g$ Pa^{-1} and $60/E_g$ Pa^{-1} for Tasmanian eucalypts when the shrinkage strain and the instantaneous strain have, respectively, the same sign and the opposite sign (E_g is the elastic modulus for green lumber, Pa). Wu and Milota (1995) fitted values for the parameter of +0.073 MPa^{-1} for tension and −0.23 MPa^{-1} for compression in the drying of Douglas-fir (*Pseudotsuga menziesii*). Mauget and Perré (1996) used a single value of 0.33 MPa^{-1} for European beech (*Fagus sylvatica*). A range of values for wetting and drying have been given by Ranta-Maunus (1993) for some softwoods, including pines and spruces, at different stress levels, temperatures and ranges of moisture contents and for radial and tangential directions. The ranges of values for drying and wetting are [−0.04 MPa^{-1} to −0.24 MPa^{-1}] and [+0.03 MPa^{-1} to −0.09 MPa^{-1}], respectively.

In a series of papers, Hunt (1982, 1984) and coworkers (Hunt and Shelton 1987; Hunt and Gril 1996) challenged the assumption that the viscoelastic and the mechanosorptive components of strain can be modelled separately. In his 1982 paper, Hunt used a Kelvin model (spring and dashpot in parallel) to represent the "recoverable creep" (viscoelastic strain), as distinct from the mechanosorptive creep (or strain), which was modelled by a parallel spring, dashpot and pinned-slider element (Fig. 8.10). In this representation, the spring gives a force for recovery after the removal of load, the pinned slider limits the compliance, and the dashpot controls the recovery rate. Removal or replacement of the pins represents the breaking or remaking of hydrogen bonds. Rules were stated for removing the pins (case 3 of Armstrong and Kingston, 1960) and replacement (case 2), but it was admitted that the explanations for these rules were not clear. In 1987, Hunt and Shelton went further and suggested that it was not certain at that time whether the two processes of viscoelastic and mechanosorptive strain were basically the same or separate phenomena. In 1996, he still continued to argue that the two processes are essentially the same.

Hunt (1984) performed experiments with beech (*Fagus sylvatica*) to measure creep compliance (both mechanosorptive and viscoelastic) at three different stresses (6, 9 and 12 MPa). The percentage moisture contents of the samples were cycled between 4 and 20%. A useful outcome of the results from

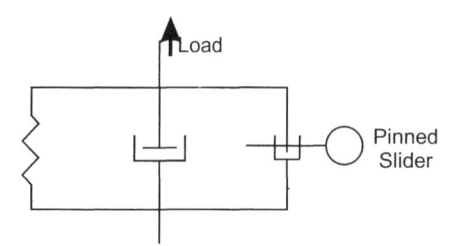

Fig. 8.10. Pinned-slider model of Hunt (1982) for mechanosorptive strain

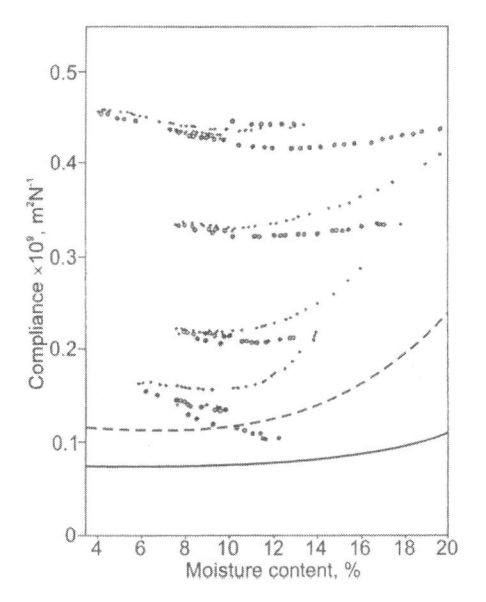

Fig. 8.11a,b. Observed and predicted compliance curves for mechanosorptive strain. **a** Observed for stress level 12 MPa, beech (*Fagus sylvatica*). **b** Predicted from the model of Ranta-Maunus (1975). (After Hunt 1984)

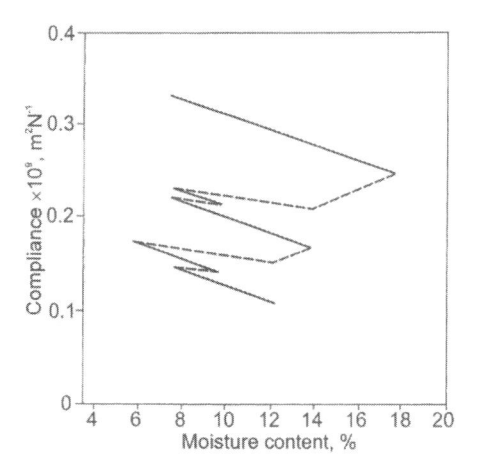

these tests was a critique of the model of Ranta-Maunus (1975), highlighting a number of points which are discussed below.

1. *Mechano-sorptive compliance is virtually independent of stress level.* The model of Ranta-Maunus (1975) implies that the compliance (ε_{ms}/σ) should be independent of stress level and equal to $m \, \Delta X$. At the low stresses in Hunt's experiments, the stress indeed appeared to have little effect on the compliance. For example, at a stress of 6 MPa, the maximum compliance was $0.45 \times 10^{-9} \, m^2 N^{-1}$, while at 9 MPa and 12 MPa the maximum compliances were $0.53 \times 10^{-9} \, m^2 N^{-1}$ and $0.45 \times 10^{-9} \, m^2 N^{-1}$.

2. *Shapes of compliance curves.* Figures 8.11a and 8.11b (taken from Hunt 1984) show the observed compliance with varying moisture content at a stress

of 12 MPa and the compliance estimated from the model of Ranta-Maunus (1975) if all the parameters a^-, a^+ and a^{++} were constant. The shapes of the curves are significantly different from those predicted with constant values of the parameters.

3. *Compliance threshold.* Below a compliance level of around $0.15 \times 10^{-9} \, m^2 N^{-1}$, all changes in moisture content caused an increase in compliance. No such threshold was predicted by the model of Ranta-Maunus (1975).

4. *Values of parameters in model.* Above the compliance threshold, drying ($\Delta X < 0$, case 1) caused an increase in mechanosorptive strain below a moisture content of $0.01 \, kg \, kg^{-1}$ (a^- negative), as expected from the model of Ranta-Maunus (1975). However, the mechanosorptive strain decreased with drying (a^- positive) at higher moisture contents. Likewise, while adsorption ($\Delta X > 0$, case 2) caused a decrease in mechanosorptive strain below moisture contents of $0.11 \, kg \, kg^{-1}$ (a^+ negative), as would be expected from the model of Ranta-Maunus (1975), it caused an increase in mechanosorptive strain at higher moisture contents (a^+ positive). Hence, from comparing the predicted and observed compliance curves, the values of the parameters a^+ and a^- do not appear to have constant values.

Also, from the observed compliance curves, there did not appear to be a sharp change in behaviour when the lumber was wetted to a moisture content above the previous highest level (case 3). This behaviour differs from the model of Ranta-Maunus (1975), in which the parameters a^+ and a^{++} are usually significantly different. The parameters for hoop pine (*Araucaria cunninghamii*) given in Table 8.3 are typical of the data presented by Ranta-Maunus, with the two parameters a^+ and a^{++} differing by a factor of two.

5. *Minima in compliance curves.* During each drying or wetting stage at compliance levels greater than the compliance threshold, there was a minimum in the observed compliance against moisture content curve at a moisture content of $0.10 \, kg \, kg^{-1}$ during desorption and $0.11 \, kg \, kg^{-1}$ during adsorption. The model of Ranta-Maunus (1975) predicted a minimum at the transition from case 2 (a^+) to case 3 (a^{++}) behaviour (Fig. 8.11b), but not at these moisture contents, which were well within the range of previous moisture contents. Hence, from the model of Ranta-Maunus (1975), this range of moisture contents would not have been expected to trigger this transition.

These findings suggest that a model such as that of Ranta-Maunus (1975), which does not predict an upper limit on mechanosorptive compliance when the moisture content is cycled up and down (as was observed by Hunt 1984), is unlikely to be appropriate for evaluating drying schedules in which the moisture content of the lumber increases at any point in the schedule. Such increases will be found during solar (intermittent) drying and during reconditioning or steaming at the end of conventional drying schedules.

Ranta-Maunus (1993) has acknowledged some of these features, including the point that the parameter m in his 1975 model is not necessarily constant and is different for drying and wetting. Nevertheless, the 1975 model still predicts no inherent limit on compliance due to cyclic wetting and drying, and the 1993 paper still quotes constant values for wetting and drying of pines (*Pinus* spp.) and spruces (*Picea* spp.).

In further work, Hunt (1989) presents an equation of the following form for unloaded specimens:

$$\frac{\varepsilon}{\varepsilon_\infty} = \left[1-\exp\left(-\frac{t}{\tau_1}\right)\right] + K\left[1-\exp\left(-\frac{t}{\tau_2}\right)\right]\exp\left(-\frac{t}{\tau_2}\right), \qquad (8.50)$$

where τ_1 is the characteristic time for the initial sorption and τ_2 is that for relaxation. This equation could be interpreted in terms of a modulus of elasticity for creep (E_c) as follows:

$$\varepsilon = \frac{\sigma}{E_c}\left[1-\exp\left(-\frac{t}{\tau_1}\right)\right] + K\left[1-\exp\left(-\frac{t}{\tau_2}\right)\right]\exp\left(-\frac{t}{\tau_2}\right). \qquad (8.51)$$

The equation has some similarity in form to that for the Kelvin model:

$$\varepsilon = \frac{\sigma}{E}\left[1-\exp\left(-\frac{Et}{\mu}\right)\right], \qquad (8.12)$$

where μ/E is equivalent to a time constant τ.

Typical values for the characteristic times with thin (0.5-mm-thick) specimens of Scots pine (*Pinus sylvestris*) were 60 s to 500 s for initial sorption (τ_1) and 1.8 to 11 h for relaxation (τ_2), but the characteristic times for the change in moisture content were around 1 hour, so the implication is that the development of mechanosorptive strain occurs in two steps, one involving relatively rapid surface sorption and the other occurring over a longer time scale. This conclusion mirrors the observation that sorptive behaviour itself is a two-step process. At high humidities and at zero load there was an initial rapid expansion on sorption followed by a slow relaxation. The characteristic times were similar for desorption and sorption, and the times for dimensional changes, either with an applied tensile stress or without one, form almost a single population.

Salin (1992) assessed the quality of drying for 75 × 150 mm, commercial-length boards of Scots pine (*Pinus sylvestris*) over 14 trials in a pilot-scale dryer. An average of 93 boards were used in each trial, and different drying schedules were also used. "Value loss" due to checking was assessed at the end of each trial. He made the somewhat surprising finding that there was a strong correlation between the observed value loss and the predicted maximum stress from a strain model which included only elastic and shrinkage behaviour. He also indicated that pure viscoelastic strain was only a small part of the overall strain. Further, when the mechanosorptive model of Ranta-Maunus (1975) was used, the correlation between value loss and the predicted maximum stress was very poor. He concluded that the uncertainty introduced by this model made the overall model "*unusable for prediction of quality loss due to checking*". With cyclic changes in moisture content, as found in solar (cyclic) drying schedules and reconditioning, the model of Ranta-Maunus (1975) predicts that the deformation may increase indefinitely. Such predicted behaviour is unrealistic. Salin therefore suggested that it should be a necessary condition that any model for mechanosorptive strain should predict a dwindling rate of mechanosorptive strain with an increase in the amount of mechanosorptive strain itself. He proposed that the simplest approach to take account of this requirement would be an equation of the following kind, which is analogous to the Kelvin model for viscoelastic strain:

$$\dot{\varepsilon}_{ms} = m(\sigma - E_{ms}\varepsilon_{ms})\frac{dX}{dt}, \qquad (8.52)$$

where E_{ms} is a constant related to the maximum mechanosorptive strain achievable for a given stress. This equation reduces to that of Ranta-Maunus (1975) if $E_{ms} = 0$. The use of Salin's equation is also consistent with the observations of Hunt (1984) and Hunt and Shelton (1987) that there is an upper limit on the amount of mechano-sorptive compliance (ε_{ms}/σ), which here would correspond to $1/E_{ms}$.

Hardtke et al. (1996) used the following version of this type of approach to describe mechanosorptive strain for European beech (*Fagus sylvatica*):

$$\dot{\varepsilon}_{ms} = D^{-1}(\sigma - E_{ms}\varepsilon_{ms})\frac{dX}{dt}, \qquad (8.53)$$

where D was taken as $0.1\,MPa$ and E_{ms} was taken as half the modulus of elasticity for instantaneous strain.

Martensson and Svensson (1996) noted several features of mechanosorptive strain behaviour observed in tests with constant stress and at constant strain. For experiments conducted with constant stress:

- during desorption, the mechanosorptive strain rate is highest at the start and decreases during drying;
- during absorption, the behaviour is less clear, but some increase in mechanosorptive strain occurs;
- the higher the stress level, the larger the mechanosorptive strain rate; and
- the mechanosorptive effect is less in compression than in tension.

For constant-strain experiments:

- when drying is interrupted by a wetting period, the mechanosorptive strain is zero at the start of this wetting period, and for the rest of the drying period, the mechanosorptive strain rate is negative; and
- when drying starts again after re-wetting, the mechanosorptive strain is less than it was during the first drying period.

They proposed the following equation for mechanosorptive strain:

$$\dot{\varepsilon}_{ms} = \kappa\sigma\,|\dot{\varepsilon}_s| - \frac{\beta}{s_{max}}(\varepsilon - \varepsilon_s)|\dot{\varepsilon}_s| \qquad (8.54)$$

where κ is a coefficient with units of (m^2N^{-1}), β is a dimensionless constant and s_{max} is the total shrinkage from the green to the dry condition, but the values of the parameters which they used for Norway spruce (*Picea abies*) were not given. A feature of the equation is that mechanosorptive strain is only predicted to occur below the fibre-saturation point, when shrinkage takes place.

8.6 Relative Magnitude of Strain Components

8.6.1 Elastic and Other Strains

For black spruce (*Picea mariana*), Lin and Cloutier (1996) predicted maximum stress levels over a 120-h drying time using a model which included viscoelastic strains which were twice those predicted by a model which included only elastic strains. Perré (1996a) has predicted maximum stresses from an elastic strain model that are twice those from a strain model including mechanosorptive strains for the drying of Norway spruce (*Picea abies*) over a 50-h drying time. For lumber of the same species, Ranta-Maunus (1993) has shown viscoelastic strain compliances which were of the same order as those for instantaneous strains with data from pines (*Pinus* spp.), spruces (*Picea* spp.) and hinoki (*Chamaecyparis obtusa*).

Predicted viscoelastic and mechanosorptive strains which were both up to 50% of the elastic strains were presented by Doe et al. (1994) during the drying of Tasmanian eucalypts. Similar results were predicted by Ganowicz and Muszynski (1994), but the type of lumber being analysed was not specified.

8.6.2 Viscoelastic and Mechanosorptive Strains

The comparative importance of mechanosorptive strain relative to viscoelastic strain is the subject of some debate. If viscoelastic and mechanosorptive strain arise because of the same process, as suggested by Hunt and Gril (1996), then it seems reasonable to use essentially the same equation to describe both processes. The Kelvin-type model of Salin (1992) allows an observed limit on compliance ($1/E_{ms}$) to be fitted and hence gives physically realistic behaviour in this respect. Salin (1992) considered that even a strain model which included only pure elastic strain would be better suited to the optimisation of drying schedules than one which included a poor approach for the mechanosorptive model. Further, Hunt and Gril (1996) suggest that the effect of moisture change is to alter the rate at which the combined viscoelastic and mechanosorptive strain occurs. This total time-dependent strain will be called *combined creep* from now on. If Hunt's suggestion is valid, then the use of a rate constant, equivalent to the term $m(dX/dt)$ in Salin's model, would appear to be justified. Such a rate constant should behave according to the observed limits, which are that the strain rate should have a constant (low) value for no change in moisture content (corresponding to "normal" viscoelastic strain), and the strain rate should increase as the rate of moisture change increases, in keeping with past experience.

On the other hand, Oliver (1991) has argued strongly that viscoelastic and mechanosorptive strain are different processes, on the grounds that (at least, for Tasmanian eucalypts):

- the ratio of mechanosorptive to instantaneous strain ($\varepsilon_{ms}/\varepsilon_i$) is almost linear with moisture content;
- above the fibre-saturation point, the slope of the line relating ($\varepsilon_{ms}/\varepsilon_i$) to moisture content is nearly linear with temperature; and
- the time constants are very different for viscoelastic and mechanosorptive strain (mechanosorptive strain occurs almost as rapidly as the change in moisture content).

However, these observations are not necessarily inconsistent with the suggestion of Hunt and Gril (1996) that the adsorption or desorption of water causes the rate of change of the combined creep strain to increase, since a characteristic feature of most experiments to measure viscoelastic strain is that changes in moisture content are prevented. In experiments to measure mechanosorptive strain, on the other hand, the moisture content is deliberately altered.

Some workers (Dahlblom et al. 1994; Carlsson et al. 1996; Ormarsson et al. 1996) neglect viscoelastic strain on the grounds that they consider it to be a small component of the overall strain. However, Milota and Wu (1994) suggest that viscoelastic (creep) strain accounts for as much as 30% of the stress-induced strain, depending on the time required for drying, and has a significant effect on the maximum stress levels. They found good agreement between the measured net strains at the surface, just below the surface, and the centres of 50 × 190 mm boards of Douglas-fir (*Pseudotsuga menziesii*) and those predicted by a strain model which included all components of strain. The viscoelastic component was modelled using the Bailey-Norton equation, and the mechano-sorptive component was treated using the model of Ranta-Maunus (1975).

8.7 Solution Procedures

8.7.1 One-Dimensional Analysis

Time-dependent strains (such as viscoelastic and mechanosorptive ones) are dependent on the stress levels in a non-linear way, unlike the free-shrinkage strain (which is independent of the stress level) and the instantaneous strain, which often has a relatively simple and virtually linear relationship with the stress level. The time-dependent strains may be treated in a number of ways, including the approach of Ferguson and Turner (1996) who follow Zienkiewicz and Cormeau (1974) in taking the stress level for each time step as that at the start of the time step, thereby avoiding the need for an iterative procedure. Other workers (Wu 1993; Milota and Wu 1994; Wu and Milota 1995) have treated the viscoelastic strain in this way (with the stress level at the start of the time step), but for the mechanosorptive strain (which is frequently larger in magnitude), the stress level was taken as that halfway through the timestep, and so iteration was required to solve for the stresses and strains. Internal static equilibrium (Eqs. 8.3 and 8.4) must be obeyed, and it is also usual to assume that plane sections retain their planar shape.

8.7.2 Two-Dimensional Analysis

Starting with the consideration of an isotropic material, with an elastic modulus E and Poisson's ratio v, it is possible to relate the stresses $[\sigma]$ and the strains induced by them $[\varepsilon]_i$ for the situation inside lumber, which is one of so-called plane strain, since the shrinkage in the longitudinal direction is small. Plane stress is not realistic, since the longitudinal displacement is not locally free. The equation relating the instantaneous stresses and strains is given by:

$$\begin{bmatrix} \sigma_x \\ \sigma_y \\ \tau_{xy} \end{bmatrix} = \frac{E(1-v)}{(1+v)(1-2v)} \begin{bmatrix} 1 & \frac{v}{1-v} & 0 \\ \frac{v}{1-v} & 1 & 0 \\ 0 & 0 & \frac{1-2v}{2(1-v)} \end{bmatrix} \begin{bmatrix} \varepsilon_{xi} \\ \varepsilon_{yi} \\ \gamma_{xyi} \end{bmatrix}, \qquad (8.55)$$

where σ_x and σ_y are the normal stresses within the cross-section of the lumber board and τ_{xy} is the shear stress. The corresponding components of instantaneous strain are ε_{xi}, ε_{yi} and γ_{xyi}.

As shown by Fergusson and Turner (1996), the strains arising due to thermal expansion are negligible compared with the strains due to free shrinkage. The total strains in the lumber may be calculated using standard techniques such as the initial strain method (Zienkiewicz and Taylor 1989), in which free shrinkage, viscoelastic and mechanosorptive strains are treated as initial strains.

More sophisticated analysis is necessary if the development of stress and strain in the radial-longitudinal and tangential-longitudinal planes is to be assessed (Dahlblom et al. 1994; Perré 1996a). This analysis must consider three components of normal strain and three of shear strain, with six corresponding stress components.

8.8 Experimental Apparatus

It is possible to test wooden beams which are loaded in bending (Leicester 1971a, b) to determine parameters in viscoelastic and mechanosorptive strain models for drying, but in this case the stress field through the board thickness is non-uniform, varying from compression to tension. Since the magnitude of these strains and their development with time are dependent on the magnitude and sign of the stresses, the results from these tests are more difficult to interpret than those from tests in pure tension.

On the other hand, Oliver (1991) suggests that compression and bending tests are rather more representative of stress and strain behaviour in seasoning than tension tests, in which minor defects can cause failure. However, lumber boards usually fail during drying due to tensile stresses; whereas, on loading a beam, only about half of the bent beam is exposed to tensile stresses, which vary from zero at the mid-plane of the beam to a maximum at the lower surface of a beam. Oliver further indicates that any minor defect during seasoning modifies the local stress field without affecting the overall stress field. Therefore, to measure the rupture stress and strain, a bending test might have some advantages over tension tests. Nevertheless, it is difficult to appreciate how a minor defect in a bent beam, unless it is in an area of the beam subject to compression (and hence of less interest for the purposes of drying than the tensile portion), is less important in terms of measuring the modulus of elasticity, for example, than in a tensile test.

A relatively simple experimental apparatus has been described by Muszynski and Olejniczak (1996) for the measurement of wood properties (instantaneous, shrinkage, viscoelastic and mechanosorptive strains). Three end-matched samples of Scots pine (*Pinus sylvestris*) were placed in a chamber in which the temperature and humidity were carefully controlled (Fig. 8.12). The first sample was placed under a constant tensile load, enabling the total and instantaneous strains to be estimated. The second sample was free to deform without restraint, enabling the free shrinkage strain to be measured under the same temperature and humidity conditions as the first sample. The strains of both these samples were recorded on a data-acquisition computer. The third sample was weighed continuously to monitor the drying rate, an important factor in determining the mechanosorptive strain. The mass of this sample was also continuously recorded, and the drying rate of this sample was assumed to be the same as those of the two other samples. Parameters in Maxwell and Kelvin models for viscoelastic strain, and in the models of Ranta-Maunus (1975) and Salin (1992) for mechanosorptive strain, were estimated by least-squares fitting. Keep (1998) used a similar experimental procedure for her tensile creep tests.

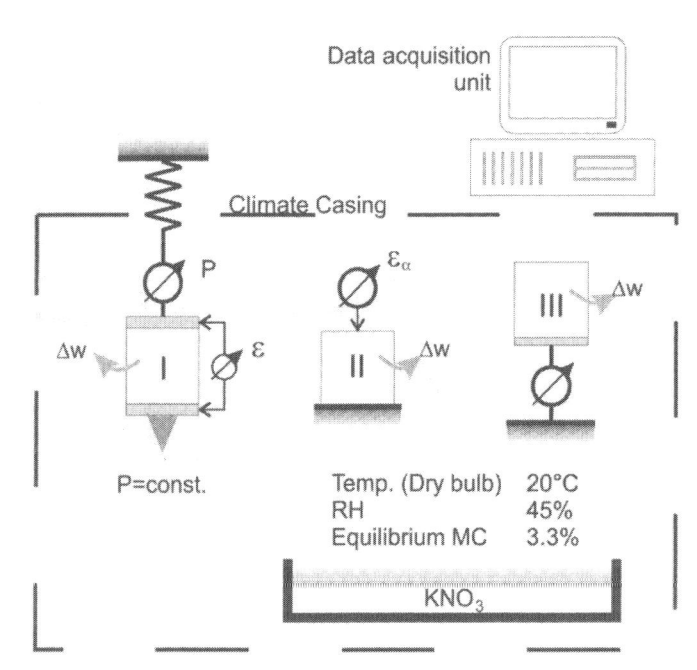

Fig. 8.12. Schematic diagram of an apparatus to measure instantaneous strain, free shrinkage and weight loss. (Muszynski and Olejniczak 1996)

Another approach, illustrated in Fig. 8.13, has been suggested by Brandao and Perré (1996). This is based on asymmetrical convective drying performed on a board with three faces sealed with an impermeable coating (for example, on both edges and the lower face). Due to the asymmetrical distribution of moisture content and temperature, the board becomes cupped during the test, and the deflection can be used, together with a two-dimensional stress-strain analysis, to "back-calculate" mechanical properties. One uncertainty in the application of this method is that the stress levels inside the board will be very uneven, so the approach amounts to curve-fitting to existing mechanical models.

8.9 Applications

There is a link between this foregoing stress-strain analysis and Grace's observation (1996) for a hardwood *Nothogagus truncata* that flat-sawn boards are more difficult to dry than quarter-sawn boards. With flat-sawn boards, both the growth rings and the tangential shrinkage are essentially parallel to the wide faces of the boards, whereas with quarter-sawn boards radial shrinkage dominates. Tangential shrinkage tends to be larger than the radial shrinkage

200   Stress and Strain Behaviour

Classical convective drying

A ─────────── A'   Initial state and first drying period: growth stress only

Second drying period: Tensile stress close to the exchange surfaces

End of drying: stress reversal

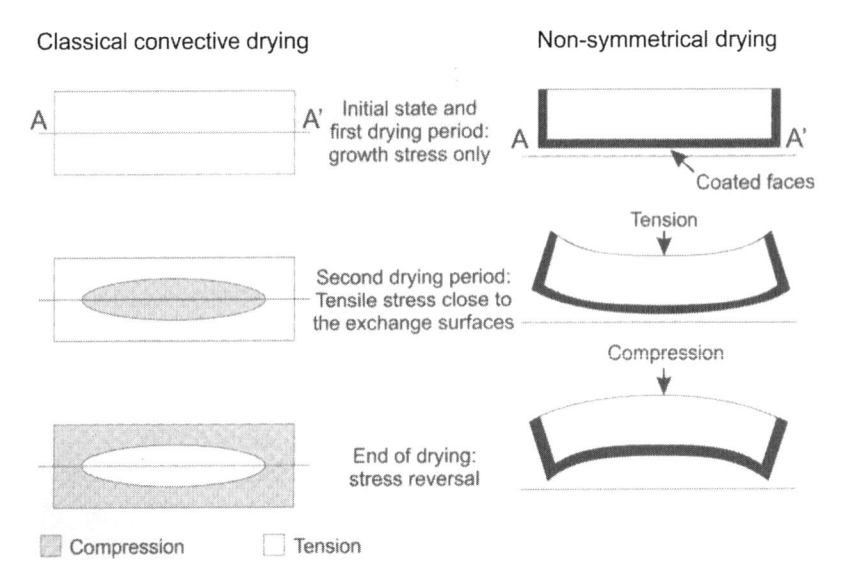

▨ Compression   ☐ Tension

Non-symmetrical drying

A ─────────── A'
↘ Coated faces

Tension ↓

Compression ↓

Fig. 8.13. Board-cupping test to determine mechanical properties of wood. (Brandao and Perré 1996)

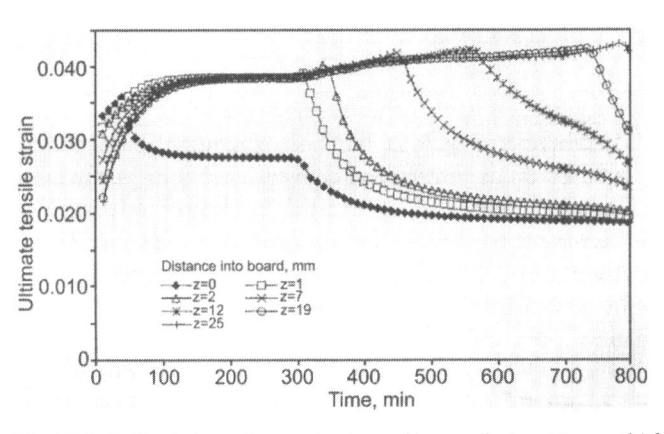

Fig. 8.14. Estimated maximum strain profiles on drying 50-mm-thick sapwood *Pinus radiata* boards at 120 °C (dry-bulb) and 70 °C (wet-bulb). (Chen et al. 1997a)

for a given extent of drying, and thus the maximum strain in a flat-sawn boards is likely to be greater than that in a quarter-sawn board when drying to the same moisture content. Since the failure strain is taken to be the limiting factor on drying, this criterion is more likely to be exceeded when drying flat-sawn boards than quarter-sawn ones.

Chen et al. (1997a) used the planar stress model to estimate the profile of the maximum tensile strain during the high-temperature drying of sapwood *Pinus radiata*. These profiles, reproduced in Fig. 8.14, show that the maximum strain, which indicates the failure envelope, remains at about 0.04 mm mm^{-1} for much of the drying before dwindling to a value between 0.02 and 0.03 mm mm^{-1} towards the end of drying except for wood at the surface. There, it falls more rapidly to the lower limit as the wood dries out preferentially and becomes less plastic.

In the high-temperature drying of permeable softwoods, significant tensile stresses are induced near the surface as the evaporative front withdraws into the core of the lumber board. These associated strains are relaxed by the combined creep. The enhanced softening of the material at the kiln temperatures means that such wood subjected to high temperatures may not necessarily fail due to the relief effect of these creep strains. Chen et al. (1997b) show that if creep strains are ignored, then the wood in the outer one-tenth of the 50-mm board is likely to have failed under commercial drying conditions of 120 °C dry-bulb temperature and 70 °C wet-bulb temperature. The effect of the creep strains is to reduce the peak tensile stresses to one-third of the value if creep were absent. The local mechanosorptive strain is strongly related to the rate at which the moisture content is changing, and this will be greatest about the irreducible-saturation condition when a sharp change in the mechanism of moisture movement through the wood occurs. Finally, the overall board shrinkage is influenced markedly by the viscoelastic and mechanosorptive strains: these creep strains are predicted to reduce the board shrinkage from about 3.5 to 2%.

The predicted stress profiles, illustrated in Fig. 8.15, when both viscoelastic and mechanosorptive strains are included show how these strains relax the developing stresses. The tensile stress at the surface rises rapidly to a maximum value of 1.2 MPa as the surface tries to shrink over the much wetter core, and peaks successively at progressively deeper positions in the board as the evaporative zone withdraws from the surface layer. This value may be compared

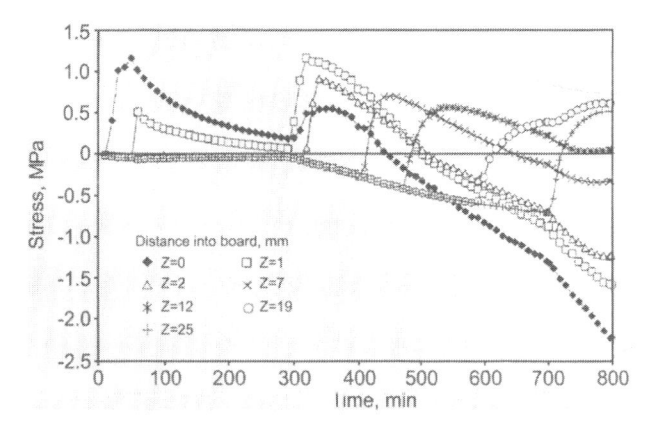

Fig. 8.15. Estimated stress profiles generated during the drying of 50-mm-thick sapwood *Pinus radiata* boards at 120 °C (dry-bulb) and 70 °C (wet-bulb). (Chen et al. 1997b)

with the tensile-failure data of Hinds and Reid (1957) for *Pinus radiata*, who report a value of 1.9 MPa for failure in the tangential direction when the wood is green and one of 3.3 MPa when dry. As the board dries out, stress reversal takes place, but the larger compressive stresses are considered to be less dangerous than tensile ones which tend to tear the material apart. A maximum compressive stress of -2.3 MPa is predicted to arise at the surface at the end of drying, which may be compared with the failure value of -5.9 MPa for the dry wood in compression reported by Hinds and Reid (1957).

In a further paper, Chen et al. (1997c) use the planar stress model to investigate alternative stress-relief policies to cooling and steam-conditioning the softwood stack. These strategies included; simple cooling under a cover, simple steaming without an intermediate cooling process, and intermittent drying and conditioning throughout the schedule itself. For the simple method of cooling, a proper wrapping of the stack would be essential to ensure adequate stress relief. Should the intermediate cooling step after drying be omitted, a slower rate of moisture pick-up and stress relief may result during the subsequent steaming period. Intermittent drying and conditioning during the schedule results in a reduced moisture content after the steaming step. However, Chen et al. (1997c) suggest that, in terms of stress relief of internal checking, the benefit of intermittent drying and conditioning may not be marked as that which might be found with a less permeable species.

9 Airflow and Convection

The flow of air inside a kiln provides a means of keeping the lumber at the required environmental conditions of temperature and relative humidity besides being a carrier for the evaporated moisture. A uniform airflow through the stack is one prerequisite for reaching uniformity in the dryness of the boards at the end of the schedule.

In the drying of impermeable species and at low temperatures, the drying is controlled by the slow diffusional mechanisms within the wood. The effective diffusion coefficient, however, rises rapidly with temperature. Luikov (1966), for example, states that the moisture-diffusion coefficient in the tangential direction in a softwood varies with absolute temperature as T^{10}. This dependence suggests that the diffusion coefficient rises fivefold in raising the temperature from 60 to 120 °C. Over the same temperature range, the external mass-transfer coefficient for the evaporation from the wood surface increases by only one-third, so the moisture movement within the wood becomes less dominant in restricting the overall drying process than does the moisture-vapour transfer in the air. Thus, the convection is much more significant in the drying of permeable softwoods at relatively high temperatures. This is particularly so when superheated steam replaces air in the ultrahigh-temperature seasoning of species such as *Pinus radiata*.

Perré and Moyne (1991) demonstrated the importance of the external heat and mass-transfer processes in high-temperature seasoning by the simulation of the drying of spruce (*Picea* sp.). They found that all of the unbound moisture was removed from the lumber after 20 h within a total drying time of 50 h. During the drying of unbound moisture, the external mass-transfer process will be controlling the moisture-loss rate since the internal mass-transfer resistance is small when unbound moisture is present. This result suggests that the external resistance is a significant factor for approximately 40% of the total drying time.

There have been suggestions that this situation could be exploited by reducing the air velocity in the kiln after the lumber reaches the *critical point* when the drying process no longer remains unhindered by the moisture movement within the wood. This so-called *Hi-Low* strategy has the attraction of saving on fan-energy consumption. However, Riley and Haslett (1996) and Nijdam and Keey (1996) have shown that external convection still plays an important role in controlling the drying process for a permeable species such as *Pinus radiata*, so that a *Hi-Low* strategy for this softwood is likely to lead to much longer drying times. Any savings in fan-power costs are likely to be outweighed by the higher costs associated with the lower throughputs with these longer times.

While the external convection may the affect the total time to dry diffusion-limited hardwoods, the external mass-transfer coefficients influence the moisture-content profiles near the wood surface. For a major portion of the drying time, the moisture contents in a zone near the wood surface are likely to range between fibre-saturation and the kiln-equilibrium values, as the kiln tests undertaken by Grace (1996) with boards of hard beech (*Nothofagus truncata*) demonstrate. In turn, the development of such a reduced-moisture zone will influence the developments of drying stresses and phenomena such as case-hardening, limiting the severity of the drying schedule which might be used. Therefore, consideration of the airflow and the convection in a kiln is important in the drying of both hardwood and softwood lumber.

9.1 Airflow in a Batch Kiln

Although there are differences in detail between the designs of kiln manufacturers, a standard lumber kiln is essentially a large, box-shaped chamber fitted with overhead fans for circulating the air and heating coils for maintaining the temperature at the set level. The humidity is controlled by means of opening vents in the kiln's roof, thus governing the amount of humid air that returns to the fans to be recirculated through the kiln again.

The lumber is stacked externally in a rectangular pile on a low, flat-bed trolley, with the rows of boards separated by wooden *stickers* of uniform thickness to provide duct-like spaces for the kiln air to flow through. Alternatively, the timber may be stacked in separate packages on bearers and loaded into the kiln by a fork-lift vehicle. The stack is squared off as far as possible to avoid having ragged inlet and outlet faces, thus providing a uniform resistance to the airflow across the length of the stack. Since the boards are stacked with their long sides facing the incident airflow, this dimension is called the *stack length*, and the dimension in the airflow direction is the *stack width*. After stacking, the lumber is railed into the empty kiln, and the drying schedule is started. Typically, kilns may have a single track for a 2.4-m wide stack, or be double-tracked to yield a 4.8-m wide stack. A cross-sectional view of a typical box kiln is shown in Fig. 9.1.

The lumber may also be dried in a progressive manner, when truckloads of the stacked boards are moved through a long chamber. In this case, the drying schedule is maintained by varying the temperature and relative humidity conditions with distance along the chamber rather than with time as in a box kiln. However, the airflow pattern through the stack of lumber is essentially the same.

To obtain as uniform an airflow as possible, the plenum spaces at each side of the stack have to be sufficiently wide, and manufacturers have geometrical

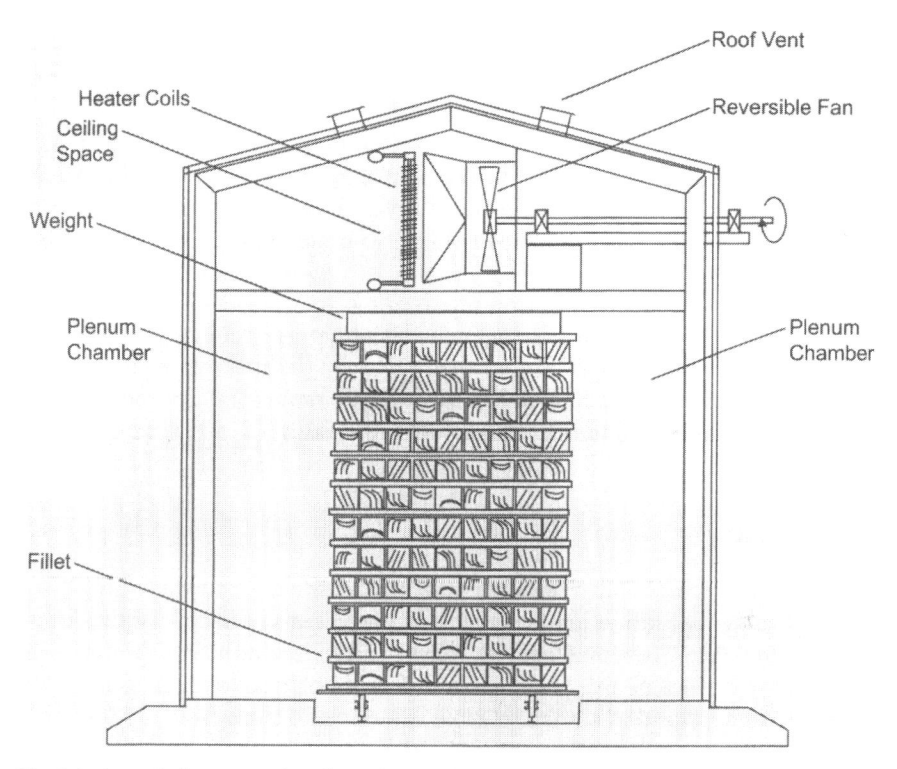

Fig. 9.1. A vertical cross-section through a single-tracked box kiln

norms for the size of the stack relative to the kiln itself, as well as the amount of sticker air-space relative to the plenum spaces themselves. The kiln has an internal ceiling and is fitted with end baffles or curtains to direct the airflow though the stack. The overhead fans circulating the air have to overcome the pressure drops accompanying the changes in air direction, the losses at the inlet and outlet faces of the stack and the pressure loss in forcing the air through the passageways in the lumber.

9.1.1 Velocity Distributions over a Kiln

Kröll (1978) reports the air distribution in a batch dryer which resembles the arrangements of a box-type lumber kiln in a number of respects. The described dryer contained 15 shelves and was fitted above an internal ceiling with both a heat exchanger and a fan revolving at 2000 r.p.m., pushing the air though the spaces between the shelves at a mean rate of $2.52\,\mathrm{m\,s^{-1}}$ with a coefficient of variation of 0.67. In the fifth gap from the top, the velocity was $5.5\,\mathrm{m\,s^{-1}}$, over twice the average, while in the top gap there was a backflow of about

−0.5 m s^{-1}, with the air streaming back towards the inlet in the opposite direction to the main flow.

However, the provision of a streamlined flow restriction at the 180° bend after the heating coils at the end of the false ceiling and bullnose-shaped distributors at the entrance to the shelved section substantially reduced the extent of the airflow maldistribution. The latter provision is impracticable in lumber kilns, but modification of the airflow issuing from the roof space is feasible by the use of suitable guides.

While the airflow maldistribution reported by Kröll (1978) may be extreme for a modern lumber kiln, existing kiln designs may yield a non-uniform airflow through well-stacked lumber loads, as may be inferred from the kiln audits reported by Nijdam and Keey (1996). They report the closeness of the final moisture-content to the desired value as the ratio

$$m = \frac{X_e - X_f}{X_o - X_f}, \qquad (9.1)$$

where X_e is the actual end-moisture content, X_f is the desired value and X_o is the initial value. The kiln-wide mean of m of the 16 runs reported for two kilns varied from −0.067 (overdrying) to +0.340 (underdrying), with a standard deviation for various boards taken from different positions as high as 0.055 in the drying of sapwood at a relatively low velocity of 3.5 m s^{-1}.

As a standard to achieve, Haslett (1998) recommends that the coefficient of variation for the air velocity across the outlet face of the stack should not exceed 0.12 at specified velocities between 4.5 and 8 m s^{-1} for the drying of *Pinus radiata* under conditions when the dry-bulb temperature is greater than 90°C.

9.1.2 Geometrical Considerations

Industrial rules of thumb generally equate the ceiling-space height to the plenum-space width and to the combined sticker-spacing height. Arnaud et al. (1991) have studied the influence of kiln geometry on the velocity distribution down a stack of lumber. They concluded that any flow maldistribution may be reduced by designing a kiln so that the ratio of the plenum-space width to the ceiling-space height above the stack should be at least equal to unity. If this condition is not met, the right-angled turn in the airflow from the ceiling space into the more restricted plenum would result in significant velocity gradients, with the flow directed towards the kiln wall. As Nijdam (1998) notes, when the plenum spaces become relatively wide, these regions act as infinite reservoirs, with minimal velocity changes and thus, from Bernoulli's theorem, limited pressure variations. Consequently, the airflow through the stack becomes uniform under these conditions. Another arrangement that yields an essentially similar situation is to make the kiln stack sufficiently wide that the pressure drop through the stack is much greater than elsewhere in the kiln. However, this solution is constrained by the practical limitations of the maximum possible output pressure from an axial-flow fan and the extra fan power involved with very wide stacks.

Nijdam (1998) investigated the effect of variations in plenum width by studying the behaviour of a hydraulic kiln model. Halving the plenum-space width relative to the combined sticker-spacing height from a ratio of 1.325 to 0.615, as illustrated in Fig. 9.2, resulted in a greater flow maldistribution down the stack, with a peak velocity appearing closer to the top. When the plenum-space width was the same as the combined sticker-space width, the peak velocity appeared about one-third the way down the simulated stack over a range of equivalent air velocities from 1 to 4.2 m s^{-1}, with a vertical profile similar to that reported by Kröll (1978) for the batch dryer mentioned in the previous section.

The ratio of the highest to the lowest between-board velocities varied from 1.5 to 1.8 over the velocity range tested in Nijdam's experiments. Because his kiln model is effectively shorter in width than the full-scale prototype, these variations are overestimated in relation to actual kiln operation since the shorter width of the model reduces its influence in distributing the inflow more evenly. Nevertheless, the reported variations are indicative of the likelihood of such maldistribution of the airflow in practice.

A feature of the tests at the highest simulated air velocities (of order 8 m s^{-1}) was the appearance of a vortex just after the right-angled bend at the top of the inlet plenum, with the fluid being directed towards the kiln wall, as shown in Fig. 9.2. Such a vortex impedes the flow into the upper sticker spaces and, if well-developed, would induce a reverse airflow, as observed by Kröll (1978). The influence of this vortex on the inlet-air velocities to the stack appears to become more extensive as the relative width of the plenum chamber narrows.

9.1.3 Pressure Drops over Kiln Sections

The pressure drop ΔP over a section in the kiln, such as a bend, can be conveniently expressed in terms of the kinetic energy of the airflow or the *velocity head*:

$$\Delta P = K \left[\frac{\rho u^2}{2} \right], \tag{9.2}$$

where K is a *kinetic-loss coefficient*. For air flowing at 10 m s^{-1}, a unit velocity head (or $K = 1$) corresponds to a pressure loss of about 50 Pa or 0.05% of an atmospheric pressure.

The kinetic-loss coefficient K is normally derived from experiments. In some cases, K may be found theoretically. For a 90° bend, with an inlet cross-section of S_1 and an outlet cross-section S_2, Bernoulli's theorem yields

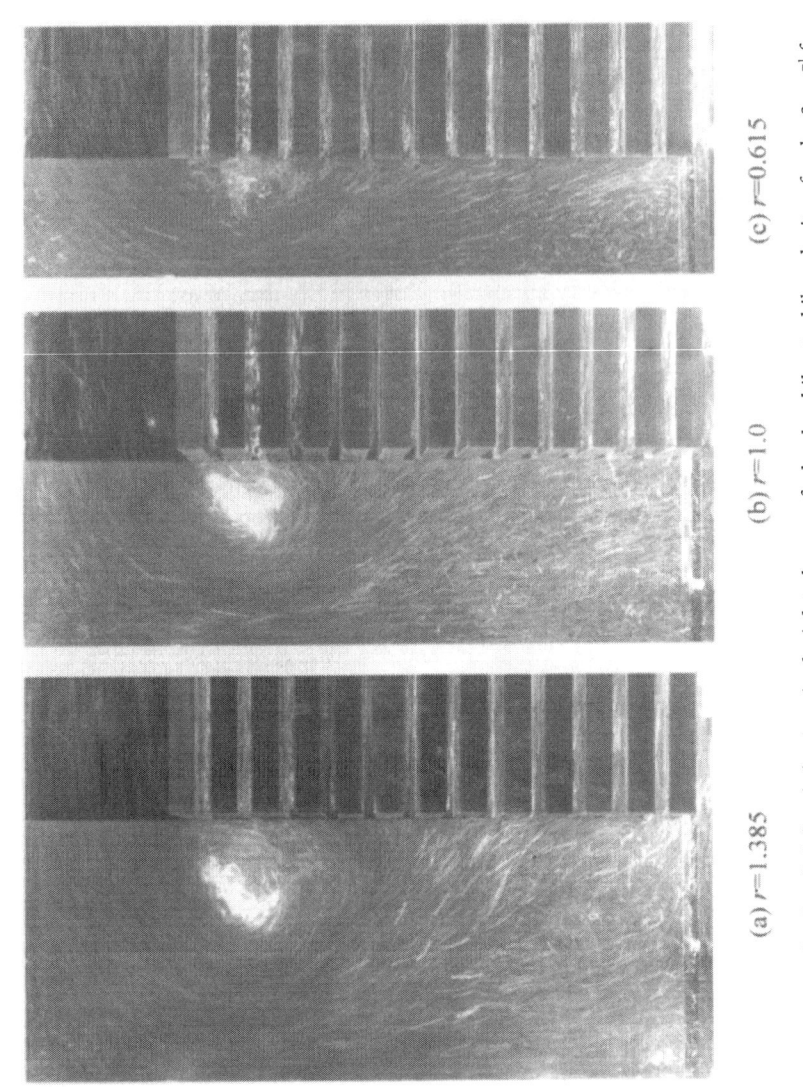

Fig. 9.2a–c. Simulated airflow behaviour in the inlet plenum of a lumber kiln at a kiln velocity of order 8 m s^{-1} for various ratios r of plenum width to combined sticker-spacing height: a 1.385, b 1.0, c 0.615. (Nijdam 1998)

$$K = \left[1 - \frac{S_1}{S_2}\right]^2. \tag{9.3}$$

For a considerable enlargement in which there is a gradual change in cross-sectional area,

$$K = \alpha \cdot \left[1 - \frac{S_1}{S_2}\right]^2, \tag{9.4}$$

where $\alpha \leq 0.15$. For the pressure drop around an obstacle, such as a heating tube, Kröll (1978) gives the coefficient K as

$$K = \alpha \cdot \left[\frac{S_1}{\varepsilon (S_1 - S_0)} - 1\right]^2, \tag{9.5}$$

where S_1 is the unimpeded upstream cross-section and S_0 is the projected cross-section of the obstacle; ε is the wake factor which takes a value between 0.8 and 1 for generally-round bodies, and the coefficient α is about 1.

Kröll (1978) also gives values for the kinetic-loss coefficient K for various right-angled bends, as reproduced in Table 9.1. If the inner sharp bend is replaced by a radiused one, there is a reduction in the pressure loss. This suggests that sharp bends and edges should be avoided in kiln design. Such edges provide places for flow detachment and subsequent eddying as noted in flow over blunt slabs discussed in Chapter 3.

The pressure drop through the airways in the stack depends primarily on the relative dimensions of the stack width L and the *characteristic cross-dimension* of the sticker spacing D, defined as the ratio of the cross-sectional area of the sticker space to its periphery. (Some authorities define the characteristic cross-dimension as four times this quantity, so that this size for a circular tube is equal to its actual diameter, and is referred to as the *hydraulic-mean diameter*). The kinetic-loss coefficient is given by

$$K = f \cdot \frac{L}{D}, \tag{9.6}$$

Table 9.1. Kinetic-loss coefficients K for flow around bends (After Kröll 1978)

B_2/B_1	R_1/B_1	R_2/B_1	K
1	0	1	2.7
1.5	0	1	1.28
1.5	0.25	1	0.98
2	0	1	1.38

B_1 is the upstream diameter; B_2 is the downstream diameter, R_1 is the inside-bend radius and R_2 is the outside-bend radius. A right-angled inside bend corresponds to $R_1 = 0$.

where f is known as the *friction factor* which is a function of the Reynolds number based on the characteristic cross-dimension, ($Re = uD/\nu$), and depends also on the roughness of the surface being swept by the air movement. However, the sublimation from a naphthalene-coated board was affected little by a 2-mm-deep slot cut into the surface in tests reported by Kho et al. (1989), and these workers concluded that even rough-sawn boards had insufficient roughness to influence the mass transfer. By the analogy between skin friction and mass transfer, we would expect that surface roughness in wood also has an imperceptible bearing on the pressure drop through the stack.

A number of expressions for f may be found in handbooks, while some authorities report $f/2$ as the *Fanning friction factor*. One of the simplest expressions is the Blasius equation (1913) which was derived for the shear stress τ in hydrodynamically smooth pipes under turbulent conditions:

$$\frac{\tau}{\rho u^2} = \frac{f}{2} = 0.0396 \, Re_p^{-0.25}. \tag{9.7}$$

The Reynolds number here, Re_p, is based on the pipe diameter. For developed laminar flow, the friction factor f takes a value $64/Re_p$.

Langrish and Keey (1996) have estimated the pressure drop across a kiln stack from first principles using computational fluid dynamics (CFD). They estimate the kinetic-loss coefficient to be 2.7 for flow of air at $5 \, m \, s^{-1}$ through a stack 2.4 m wide with sticker-spacing height of 20 mm. This coefficient is about two to three times the loss cited for the restricted right-angled bends noted in Table 9.1. Therefore, the flow resistance of the stack itself may not always completely dominate, thus rendering the flow more sensitive to the irregularities caused by the plenum design, as demonstrated in Nijdam's experiments. The pressure-drop predictions agree well with experimental data taken from measurements over a small stack of lumber in a laboratory kiln up to velocities of $7 \, m \, s^{-1}$ (Wu 1989).

In a kiln working at 90°C, the pressure drop across a kiln stack, 2.4 m wide with a 20-mm sticker spacing, would be about 75 Pa at a kiln-air velocity of $5 \, m \, s^{-1}$. The pressure drop would be greater in a high-temperature kiln which would be fitted with heating tubes offering less free space for the air recirculating through them.

When stacks are built from boards of random lengths, the boards are normally placed flush at one side but will be staggered at the other end. Preferably, the stacks are aligned at each side, leaving gaps in particular rows within the stack. These geometries produce complex flow patterns through the stack. Salin and Ohman (1998) evaluates the pressure drop through rows of 50 × 150 mm boards, separated by a shrinkage space of 5 mm and stacked with a sticker thickness of 25 mm for an air velocity of $3.0 \, m \, s^{-1}$. For a row of n boards, the pressure drop is estimated at $(5.02 + 0.99n)$ Pa. Should alternate boards be missing, then the pressure drop becomes $(1.13 + 2.44n)$ Pa. For a stack, 16 boards wide (corresponding to a 2.4-m-wide stack), the pressure drop for the rows with missing boards is nearly twice that for full rows. In practice, the pressure drop across the stack would reach a uniform value, and the air velocities over the rows would redistribute to compensate.

9.1.4 Stack-Velocity Distribution

If the pressure drop across the kiln is known, the theoretical velocity distribution down the kiln may be estimated from momentum balances. Since the kiln stack is relatively long, sidewall effects are minor and the airflow may be considered one-dimensional through the stack.

A momentum balance over a small volume element in the inlet plenum thus yields

$$\frac{\partial P_1}{\partial z_1} + \frac{2\rho u_1}{\alpha_1}\frac{\partial u_1}{\partial z_1} + \frac{f\rho u_1^2}{2D_C} = 0, \tag{9.8}$$

where P_1 is the local pressure, z_1 is the vertical distance from the top of the plenum in a downwards direction, u_1 is the air velocity down the plenum, α is a momentum-correction factor to take account of the disruption of the lumber-side boundary layer caused in the traverse over the sticker spaces and D_c is hydraulic mean diameter of the inlet-plenum chamber. A similar equation holds for the outlet-plenum space.

An overall momentum balance across a plane located at any distance down the lumber stack gives

$$\frac{\partial(P_1 - P_2)}{\partial z} + \beta\rho u\frac{\partial u}{\partial z} - \frac{f\rho u^2}{D} = 0, \tag{9.9}$$

where subscript 1 stands for the inlet plenum and 2 for the outlet space. In Eq. (9.9), u is the horizontal velocity in the sticker space, D is the associated characteristic cross-dimension and the friction factor f is a function of the Reynolds number based on this dimension. The coefficient β is an overall momentum-correction factor defined by

$$\beta = 2\cdot\left[\frac{1}{\alpha_1} - \frac{1}{\alpha_2}\right]. \tag{9.10}$$

Bajara and Jones (1976) have found experimentally a value of 1.55 for β for pipe manifolds with circular cross-sections.

The foregoing equations were solved by Nijdam (1998) to show how the velocities in the sticker spaces can vary down the stack. Figure 9.3 illustrates his computations for various values of the ratio of the plenum-space width to the combined sticker-spacing height. When this ratio is one-half, the air velocity in the sticker space varies down the stack from 7 to $4.2\,\mathrm{m\,s^{-1}}$ at a nominal

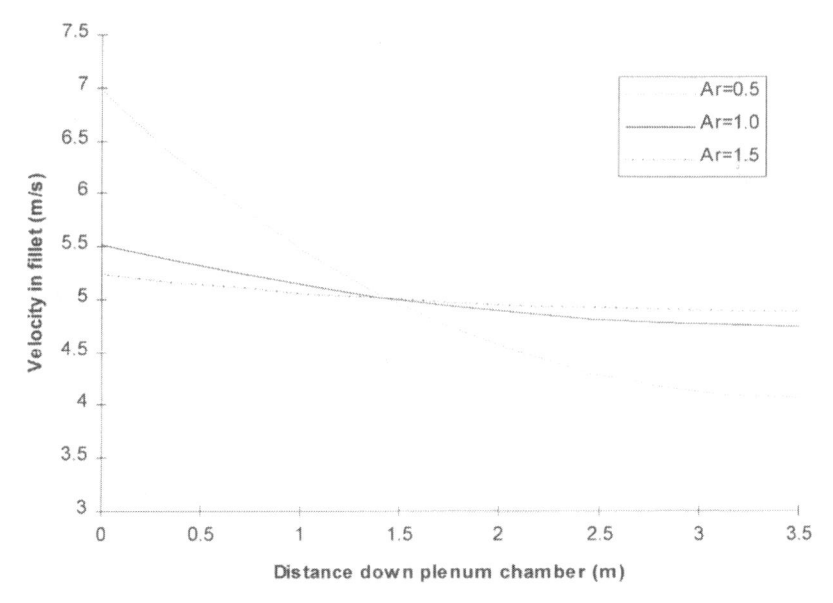

Fig. 9.3. Computed sticker-space velocities in a kiln at a mean kiln velocity of 5 m s^{-1} for various ratios Ar of plenum-space width to combined sticker-spacing height. (Nijdam 1998)

velocity of 5 m s^{-1}, but the variation is small once the plenum-space width and the combined sticker-space height have the same extent. Since these calculations assume that the air velocity entering the plenum space is uniformly distributed over the whole cross-section, these results represent the best-possible situation. In practice, the right-angled bend at the top of the plenum will cause some flow irregularities, and in extreme cases a vortex may arise, unless very careful attention is paid to the kiln design at this point. Thus velocity variations down the kiln are to be expected in most cases.

9.2 Flow between Boards

The sticker spacing provides a set of pathways of rectangular cross-section for the air to flow through the stack. These pathways are ostensibly similar to rectangular ducts, but there is a significance difference, as pointed out in Chapter 3. These "ducts" do not have continuous walls, but are built up from boards butted up to each other along their long sides. The air thus flows over a series of wall discontinuities at board-width intervals. The boards may not butt up perfectly, and small gaps may open up during kiln-drying. Even gaps as small as 1 mm wide will influence the flow, as the numerical simulations of the flow about such a gap indicate (Langrish et al. 1992, 1993).

Lee (1990) performed a series of smoke experiments to visualise the flow over a layer of short boards in a wind tunnel. The test boards were stacked in a way similar to that in a commercial kiln. The sample boards were 100 mm wide, 25 mm thick and 225 mm long. To follow the smoke trails in the tunnel, very low air velocities had to be used: 0.2, 0.4 and 0.6 m s^{-1}. However, wider gaps between adjacent boards than the separation expected in a commercial kiln were employed, in the range of 3 to 10 mm, so that dynamic similarity of the motion could be retained in the tests compared with that in an actual kiln. A significant feature of the airflow pattern was that eddies periodically formed within these small gaps between adjacent boards. After formation, such eddies then moved upward, deforming and finally escaping over the downstream board into the sticker duct. The period of oscillation for the formation of the eddies and their upward movement was observed to be in the range of 0.5 to 10 s, appearing to shorten with increasing air velocity and decreasing gap space.

Lee's results imply that, with kiln velocities in the range of 2 to 6 m s^{-1}, this periodicity would take place with gaps as small as 0.3 to 1 mm, which are highly likely to be present in practice, even with careful stacking. Corroboration of this phenomenon may be found in the numerical simulation of the flow, assumed to be time-dependent and turbulent, by Langrish and colleagues (1992, 1993), whose calculations showed the same features as Lee (1990) observed. A period of 2 s was estimated for the eddy motion at an air velocity of 0.6 m s^{-1} and a board side-gap of 10 mm; a further simulation for a velocity of 5 m s^{-1} and a side-gap of 1 mm gave a similar predicted period of oscillation, confirming the principle of dynamic similarity and the applicability of Lee's results to kiln operations. A simulation of the eddy motion is shown in Fig. 9.4.

Wiedemann et al. (1989) have measured the velocity and turbulence intensity at 11 positions in the 20-mm board-row spacing between two layers of lumber in a stack over a 2-min period. High levels of turbulence were observed at positions 2.3 and 17.8 mm from the surface of the lower board, suggesting that an eddy was present just behind the leading edge, as noted by others (Danckwerts and Anolick 1962; Sørensen 1969). Further evidence of turbulence caused by board side-gaps is provided in tests reported by Wu (1989), who investigated the effect of a 5-mm-wide gap on the turbulence intensity at positions 1, 2 and 3 mm over the surface of a stack which was four boards wide and placed inside a wind tunnel. The turbulence intensity was found to be consistently higher over the gap than at positions over the adjacent boards.

The flow regimes for typical spacing between the boards are usually beyond the laminar region for air velocities over 1 m s^{-1}. Above this velocity, the flow is transitional, in which the heat and moisture transfer is very sensitive to small change in the inlet-flow conditions. However, turbulent conditions are more likely with kilns when the airflow is disturbed periodically with a switchover in rotational direction of the fans and when charged with lumber stacks that

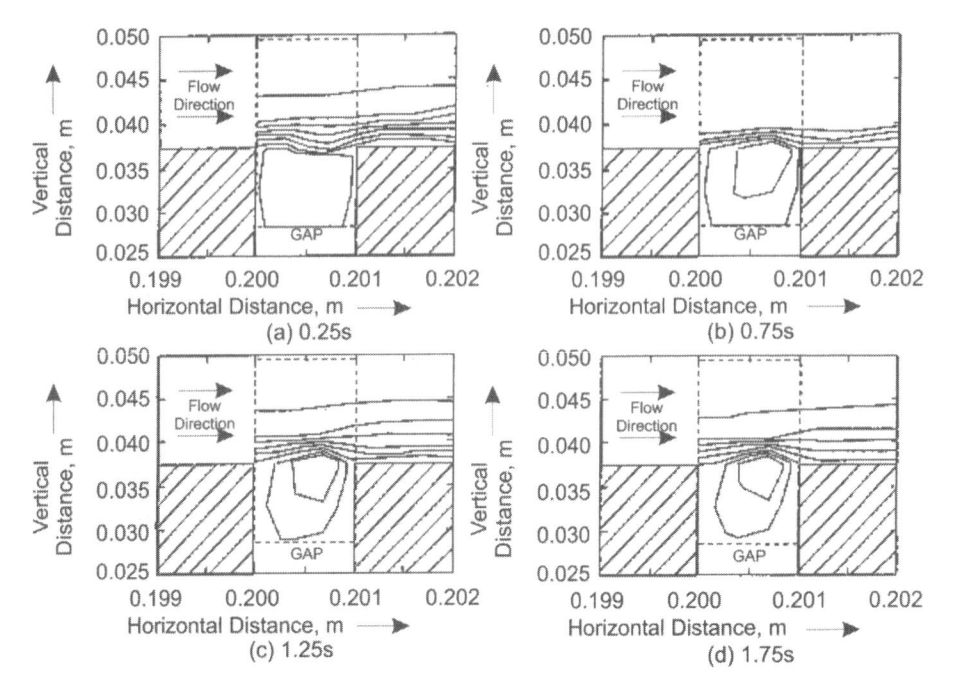

Fig. 9.4. Computational fluid-dynamic simulation of the motion between adjacent boards. (Reprinted from Langrish et al. 1993, Chem Eng Sci 48: 2219, with permission from Elsevier Science)

produce local eddies near the gaps between adjacent boards. Fully turbulent flow occurs above Reynolds numbers of 10 000, which correspond typically to kiln velocities over 4 m s^{-1}.

Stack widths are relatively short for the flow to become fully developed, however; that is the velocity profile across each sticker spacing no longer changes with distance along the airflow direction. In the entrance region, the profiles of velocity, temperature and humidity change markedly with distance. At a Reynolds number of 2000, Massey (1990) gives the entrance lengths for laminar flow as approximately 114 hydraulic diameters and 50 hydraulic diameters for turbulent flow. For a spacing between boards of 25 mm, these distances are 5.7 and 2.5 m, respectively, which are of the same order as the stack width of double-track (4.8 m) and single-track (2.4 m) kilns, respectively. However, the entrance lengths in commercial kilns are likely to be less at the higher operational Reynolds numbers normally employed. Further, Ashworth (1980) suggests that the greatest changes happen over the first 20 hydraulic diameters, which is still a substantial fraction of the stack width. The significant extent of the entrance region was also noted in the work of Gilliwald and Tschirnisch (1967). They measured the developing velocity profile between two planks at a spacing of 40 mm and a Reynolds number of 9000. The velocity profile was still developing at a distance 0.54 m from the inlet.

9.3 Convection in Kilns

The complex air motion inside the spaces between the board rows has an important bearing on the ease with which the evaporated moisture can be conveyed away to the outlet. Not only does the convected moisture reduce the drying potential of the air as it passes through the stack, but the air motion induces a saw-toothed profile of moisture-transfer rates due to the series of discontinuities where the adjacent boards butt, as discussed in Chapter 3. Superimposed on these variations are any irregularities induced by the maldistribution of the air as it enters the stack from the inlet-plenum chamber.

A more detailed consideration of the influence of the kiln conditions of velocity, temperature and humidity on the kiln-drying process is considered in the subsequent Chapter 10. In the following sections, we confine the discussion to the influence on the convection from a lack of uniformity in the airflow through the stack and from irregularities in stacking.

9.3.1 Airflow Maldistribution

The arrangement of axial-flow fans in the roof space and the placement of the heating coils ensures that the airflow is almost uniformly distributed as it approaches the end of the roof space to turn into the plenum chamber. Thus, velocity variations across the length of the stack are likely to be less than variations down the stack, particularly if the plenum width is restricted. The drying characteristics of the lumber kiln then is essentially similar to the behaviour of a set of smaller kilns in parallel, each being circulated with a uniform airflow but different from its neighbours. On this basis, Nijdam and Keey (1996) investigated the influence of airflow maldistribution on the drying of the lumber stack.

The extent of drying in each of these smaller kilns depends upon the *number of transfer units* (*NTU*), $\zeta = \beta \phi_M a L/G$, where β is the mass-transfer coefficient, ϕ_M is the humidity-potential coefficient, a is the boards' surface area in unit kiln volume and G is the mass flowrate of air over unit kiln cross-section. Normally, the value of ζ lies between 0.4 and 1. The mass-transfer coefficient is a function of air velocity, as discussed in Chapter 3:

$$\beta \propto u^n, \tag{9.11}$$

where the exponent takes a value of about 0.7. A comparison of two zones in a kiln with different air velocities, u_1 and u_2, yields

$$\frac{\zeta_1}{\zeta_2} = \left[\frac{u_1}{u_2}\right]^{n-1}. \tag{9.12}$$

The relationship between the moisture-loss rate and the number of transfer units is described more fully in Chapter 10. The extent of drying can be measured by the fractional approach to the final specified end-moisture content X_f, as given by Eq. 9.1:

$$m = \frac{X - X_f}{X_o - X_f} \quad (9.13)$$

Some computed results using audits from two kilns drying *Pinus radiata* under low-temperature schedules are listed in Table 9.2. Airflow maldistribution was suspected in these kilns.

A feature of these results is that the kiln-wide mean values of the moisture-content approach are negative, indicating that the load as a whole is overdried. Despite this overdrying, some boards in the kiln had yet to reach the specified end-moisture content, as the largest values for m are positive. Thus, merely extending the drying schedule is no solution to the problem of airflow maldistribution.

Another interesting feature arising from the kiln audits was the underdrying found when using a low kiln-air velocity of $3.5\,\text{m}\,\text{s}^{-1}$ in kiln A. In one test with sapwood boards, the final kiln-wide moisture-content approach was as high as 0.340, with some boards only half-dried ($m_{max} = 0.518$). This result incidentally shows the importance of using an adequate kiln-air velocity when drying a very permeable species, such as sapwood *Pinus radiata*.

9.3.2 Board Irregularities

Irregularities in the stack of lumber boards may be of two kinds:

- the boards may not be of even thickness everywhere, resulting in surface "steps" in the sticker spaces;

Table 9.2. End-moisture contents for two kilns drying *Pinus radiata* under low-temperature schedules (Nijdam and Keey 1996)

	Kiln A		Kiln B	
	Heartwood	Sapwood	Heartwood	Sapwood
Mean air velocity, $\text{m}\,\text{s}^{-1}$	6.9	6.9	5.0	5.0
NTU, ζ	1	1	0.95	0.95
m: kiln-wide mean	−0.067	−0.021	−0.065	−0.020
m: maximum	0.071	0.042	0.038	0.021
m: minimum	−0.133	−0.035	−0.120	−0.033
Standard deviation	0.034	0.011	0.035	0.011

- the ends of the stack may not be squared off if random-length boards are piled, giving variations in the resistance to the airflow besides providing ends which may cause eddy formation.

9.3.2.1 Uneven Thickness of Boards

Kho et al. (1989) measured the rate of sublimation of from a smoothed, recessed surface that was filled with naphthalene and placed in a board array within a pilot-plant kiln. The experiments were carried out at three air velocities from 3 to $7\,\mathrm{m\,s^{-1}}$. Using different heights for adjacent boards, as would be found in an imperfectly stacked kiln, was found to affect the local mass-transfer coefficients to varying extents. A lower leading board was found to have a greater influence on the front portion of the downstream sample board compared with a lower trailing board for the same height difference. When the sample board was lower than the adjacent ones, enhanced local mass-transfer coefficients were noted over virtually the whole length of the sample board, and a recession of 3 mm produced a greater enhancement than a 5-mm gap in the streamwise direction. This result is corroborated in a computational fluid-dynamic (CFD) analysis of a the flow about a board recessed 3 mm with respect to its neighbours (Langrish 1994). The predicted time-averaged streamlines close to the surface are shown in Fig. 9.5. The position of the stagnation point, about 20 mm from the leading edge of the recessed board, corresponds to a peak in the profile of the estimated mass-transfer coefficients in the naphthalene-sublimation studies of Kho et al. (1990), as illustrated in Fig.3.10 in Chapter 3.

For various reasons, including the reduction of the final moisture-content variations at the end of the schedule, Haslett (1998) recommends that the coefficient of variation for the board thickness should be under 0.04 for successful high-temperature drying of *Pinus radiata*.

9.3.2.2 Stack Ends

Salin (1996) has discussed the significance of the board layout at the ends of a stack where alternate boards are sometimes missing when random lengths

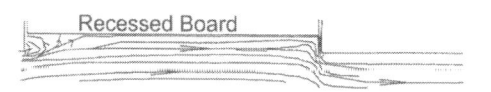

Fig. 9.5. Predicted time-averaged streamlines for the flow near a 100-mm-wide board recessed 3 mm from its neighbours. (Langrish 1994)

are stacked, with some combined board lengths being shorter than others. In such cases, some boards will be at least one board-width away from each other in the airflow direction. Any gap in the array is likely to generate substantial turbulence, and the transfer coefficients downstream from the gap may be expected to be higher than those for isolated boards.

Aligned or staggered configurations may be used in the end-stack sections. For stacks with gaps, Salin (1996) suggests that the transfer coefficients from all board surfaces for the second and subsequent rows are increased by 15% with an aligned configuration, and by 25% when these boards are staggered over those found for the boards in the first row. This statement is based on experience with bundles of circular heat-transfer tubes, for which Incropera and DeWitt (1990) claim that the heat-transfer coefficients in the second row of an inline array is from 30 to 60% higher, and those in the third row 50 to 70% higher, than those in the first row facing the flow. However, the reasons for choosing lower enhancements with inline boards compared with staggered arrangements are not given by Salin (1996).

9.4 Bypassing

Lumber kilns are normally fitted with baffles or curtains to prevent air from bypassing the stack. Some leakage around the sides of the stack is inevitable, however. Air can also bypass though any unblanked spaces between bearers. In the stacking of random-length boards, any gaps can provide alternative passageways for the flow of air. Significant bypassing reduces the kiln capacity. Nijdam and Keey (1996) note that with 100% bypass, with equal quantities of air passing through the stack as around it, the extent of drying is only 75% of that achieved if all of the delivered air had gone though the stack.

Langrish and Keey (1996) have used a CFD program to assess the airflow when there are gaps 100-mm wide around the outside of a stack and the stack edges may be well-aligned or ragged. With stacks having well-aligned edges, bypassing of order 10 to 25% is likely: whereas with ragged-edged stacks, bypassing of 100% is predicted. A bypass fraction of 40% of the air discharged by the fans is not uncommon in practice.

Even with well-sealed stacks, with minimal outside gaps, these various results suggest that stacks that have ragged edges or missing board lengths are likely to lead to more uneven drying and a loss in kiln capacity than when well-aligned stacks are dried.

9.5 Kiln Economics

The kiln-air velocity influences the economics of the kiln-drying of permeable timbers markedly. As the air velocity through the stack is increased, both the

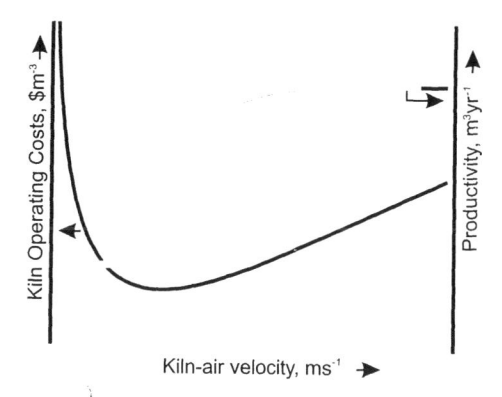

Fig. 9.6. Drying costs C ($\$\,m^{-3}$) and kiln productivity ($m^3\,yr^{-1}$) for kiln-seasoning softwood lumber

drying rate and the kiln productivity increase at the cost of extra fan power. Eventually, this extra cost outweighs the additional economic benefits in added kiln capacity, and the cost of drying per unit load of lumber rise. These trends are illustrated in Fig. 9.6.

Shubin (1990), by way of example, estimates the optimal kiln velocity, when operating costs are at a minimum, at $2.5\,m\,s^{-1}$ for drying $19 \times 250\,mm$ lumber boards of pine (*Pinus* sp.) from a moisture content of $1.2\,kg\,kg^{-1}$ to $0.1\,kg\,kg^{-1}$. This is a velocity much lower than current (1998) norms of $5–10\,m\,s^{-1}$ for drying softwood lumber, and may reflect the cost structures of the former USSR, with perhaps a relatively low value of capital costs compared with operating costs.

Figure 9.6 illustrates another feature. The kiln productivity increases linearly with kiln velocity as long as the external convection controls the drying rate. The optimal kiln velocity appears just below the value when the influence of the wood itself in hindering the rate of drying begins to play a role. This critical velocity is temperature-dependent, becoming higher as the kiln temperature is raised. At high kiln velocities, say above $10\,m\,s^{-1}$, there would be little gain in productivity by increasing the air velocity further unless very high temperatures were being employed.

The drying of impermeable timbers will exhibit no such economic minimum air velocity, since the drying is now controlled by moisture-movement processes within the wood itself. Under these conditions, the air velocity is chosen to ensure that the kiln conditions and wood temperature are maintained at their set values.

10 Kiln Operation

As noted in the earlier chapters, the air, on passing through a lumber stack in a kiln, gains in moisture content and loses heat, with a consequential fall in temperature. Thus the conditions change progressively throughout the kiln in the airflow direction. This change is inevitable even if the wood were perfectly homogenous and the airflow totally uniform across the face of the stack. While such ideal conditions do not pertain, the analysis of them, however, does provide a benchmark of kiln performance against which an actual kiln operation can be compared. Further, the non-linear interaction between humidity and temperature levels in controlling the convective transfer of both heat and mass means that some of the profiles of moisture-content change and drying rates with both time and distance in the kiln vary in a way that is not entirely intuitive at first sight, particularly when a large fraction of the humid outlet air is returned to the inlet face of the stack. Analysing the ideal case helps us to understand the appearance of such profiles.

The general theory of batch drying has been explored by Ashworth (1977) and is described in some detail by Keey (1978). These descriptions are based on the concept of the *characteristic drying curve* which provides a useful means of simplifying the drying kinetics, as outlined in Chapter 4. The essential requirement is a single batch-drying curve obtained with a wood sample of the same thickness as the lumber under consideration and undertaken under constant drying conditions of temperature and humidity. Ideally, these conditions should lie within the expected span of conditions in the kiln.

The characteristic drying curve yields a function $f(\Phi)$, where Φ is the normalised, volume-averaged moisture content defined by

$$\Phi = \frac{\overline{X} - X_e}{X_{cr} - X_e}, \tag{10.1}$$

where X is the moisture content, X_e is the equilibrium value under the prevailing relative humidity and X_{cr} is the so-called critical value marking the end of drying entirely controlled by the moisture-vapour transport in the air. For hardwoods and the heartwood of softwood, the critical moisture content may be taken as the initial green value, but in the drying of sapwood of permeable softwood species there is a short, distinct *constant-rate period*, independent of moisture content, before the drying rate falls as moisture is lost.

The *evaporation flux* j_V, or rate of drying over unit exposed surface, is given by

222   Kiln Operation

$$j_v = f(\Phi) \cdot \beta \, \phi_M (Y_W - Y_G), \qquad (10.2)$$

where β is the mass-transfer coefficient, ϕ_M is the humidity-potential coefficient, Y_W is the wet-bulb humidity and Y_G is the local value in the bulk of the air in contact with the surface. This formulation of the problem has the advantage in analysing kiln-wide behaviour in that the influence of the wood itself, $f(\Phi)$, the board and kiln geometry, β, and the process conditions, $\phi_M (Y_W - Y_G)$, are lumped into separate terms. Moreover, the effect of kiln-air velocity is contained only in the mass-transfer coefficient.

The theory may also be simplified by assuming that the heat transfer takes place much more quickly than the mass transfer. Tests monitoring the response of a small wool-felt pad to a step change in the inlet temperature of the air slowly filtering through reveal that the response in moisture-content change takes place after a period 250-fold greater than the time the air dwells in the bed, whereas the temperature change at the outlet is almost instantaneous (Nordon and Banks 1973). The extensiveness, or *number of transfer units NTU*, of a single-tracked lumber kiln is also relatively small, normally 1 or less, and thus the kiln will have the same general transfer characteristics as the wool pad of these tests. Even a double-tracked kiln will have much fewer transfer units than those dryers, typically silos for the slow drying of agricultural grains, which are sufficiently extensive for both waves of heat and mass transfer to be significant. Thus the drying behaviour in kiln-seasoning can be analysed by considering only the moisture transfer.

10.1 Drying under Constant External Conditions

In the limit, when the fraction of the humid air leaving the outlet face of the stack being returned by the fans to the inlet is very high and the stack width is relatively narrow, the drying conditions in the kiln are almost uniform. This limiting behaviour thus provides a starting point for considering kiln-drying.

A load of dry mass F having an exposed area A will lose over a time interval dt an amount of moisture given by

$$j_v A dt = -F d(X - X_e). \qquad (10.3)$$

Krischer and Kast (1978) suggest that it is convenient to define a fictitious *ideal drying time* t_f by analogy with Eq. (10.3):

$$j_v^o A t_f = F(X_{cr} - X_e), \qquad (10.4)$$

where j_v^o is the evaporation flux at the critical moisture content X_{cr}. This ideal drying time would be the time needed to reduce the load's moisture content from its critical value to the equilibrium amount if the drying were to continue at the unhindered rate j_v^o.

We may then define a relative drying time θ_f as

$$\theta_f = \frac{t}{t_f} = \frac{j_v^o A}{F(X_{cr} - X_e)}. \qquad (10.5)$$

Division of Eq. (10.3) by (10.34), and incorporating the non-dimensional quantities Φ for moisture content, $f(\Phi)$ for the drying rate and θ_f for the drying time, yields the equation

$$f(\Phi) \cdot d\theta_f = -d\Phi. \qquad (10.6)$$

Equation (10.6) can be solved if the function $f(\Phi)$ is known. For hardwoods and the heartwood of sapwood the function $f(\Phi)$ may be equated to Φ to a first approximation. This implies that the drying kinetics is first-order and the moisture-diffusion coefficient is independent of moisture content. Under these conditions, the time to dry to a product moisture content of Φ_p is given by

$$\theta_f = -\ln \Phi_p \qquad (10.7)$$

on noting that the critical moisture content is the initial green value. Many empirical correlations of lumber drying take the form of Eq. 10.7.

With the more permeable sapwood of softwood species, f may be approximated as $\Phi^{1/2}$, with the initial characteristic moisture content Φ_o somewhat greater than 1. For *Pinus radiata* for example, representative values lie between 1.25 and 1.5, but individual boards may lie outside this range. Equation (10.6) then yields for the drying time to Φ_p:

$$\theta_f = \Phi_o + 1 - 2\Phi_p^{1/2}. \qquad (10.8)$$

Although Eq. (10.8) predicts a finite time to remove all the moisture above equilibrium, unlike Eq. (10.7), this difference is believed to be a mathematical quirk and not to reflect any physical effect.

By way of example, on drying the heartwood of a softwood to an end-moisture content Φ_p of 0.15, Eq. (10.7) yields an estimate of 1.90 for the relative time of drying θ_p, whereas Eq. (10.8) for the sapwood (with $\Phi_o = 1.50$, and $\Phi_p = 0.05$) gives a value of 2.05 for time θ_p to reach the same final moisture content (in kg kg^{-1}), reflecting the differences in times needed to dry these wood types. Sapwood stacked in a kiln would take relatively even longer because of the higher moisture pick-up in the stack, as discussed in the following section.

10.2 Drying under Variable External Conditions

Under normal kiln-drying conditions, the temperature and humidity will vary throughout the stack. The fall in temperature and the rise in humidity as the air flows through the stack progressively reduces the maximum attainable drying rate from the boards.

Consider the moisture transfer from a short zone of length dz in the airflow direction through the stack, as illustrated in Fig. 10.1. Suppose the fans deliver a mass flowrate of dry air of G over unit sticker-duct area in the kiln. If the board thickness is b and the sticker spacing is s, the free cross-sectional area for the airflow is $s/(b+s)$ and the cross-section of the kiln blocked by the boards is $b/(b+s)$. Over a short time interval dt the local average moisture content falls by dX and the humidity in the air rises by an amount dY_G.

A moisture balance over this infinitesimally short zone yields

$$-dX[\rho_b b/(b+s)]dz = dY_G[G]dt. \tag{10.9}$$

The loss of moisture may also be related to the evaporation flux j_V:

$$-dX[\rho_b b/(b+s)]dz = j_v[a \cdot dz]dt, \tag{10.10}$$

where a is the exposed surface of the boards per unit volume in the kiln and is equal to $2/(b+s)$. On introducing Eq. (10.2) and putting Eq. (10.9) and (10.10) into partial differential form, one obtains

$$-\rho_b \frac{b}{(b+s)} \frac{\partial X}{\partial t} = G\frac{\partial Y_G}{\partial z} = f(\Phi)\beta \, \phi_M a \, (Y_W - Y_G), \tag{10.11}$$

which may be put into non-dimensional quantities for algebraic convenience:

$$\frac{\partial \Phi}{\partial \theta} = \frac{\partial \Pi}{\partial \zeta} = -f(\Phi)\Pi, \tag{10.12}$$

where Π is the fractional humidity potential normalised with respect the inlet potential ($\Pi = [Y_W - Y_G]/[Y_W - Y_{G_0}]$), the relative distance ζ is a modified NTU or extensiveness ($\zeta = \beta\phi_M az/G$) and the relative time θ (or extent) of drying depends upon the NTU and the capacity of the air to pick up moisture:

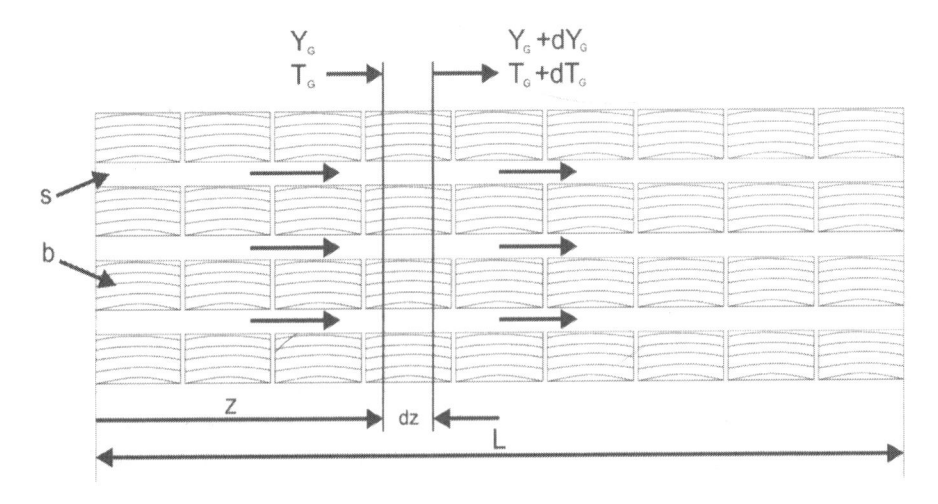

Fig. 10.1. Moisture transfer over an infinitesimally short zone in a batch lumber kiln

$$\theta = \frac{\beta \phi_M az}{G} \cdot \frac{G}{\rho_b(X_{cr} - X_e)} \cdot t. \tag{10.13}$$

The humidity potential can be eliminated from the pair of Eq. (10.12) to yield an expression for the local drying rate:

$$\frac{\partial \Phi}{\partial \theta} = \frac{\partial}{\partial \theta}\left[-\frac{1}{f(\Phi)} \cdot \frac{\partial \Phi}{\partial \zeta}\right] \tag{10.14}$$

Normally, Eq. (10.14) must be solved numerically since the characteristic drying-rate function $f(\Phi)$ is not known in a suitable algebraic form. However, the limiting relationships embodied in Eq. (10.7) and (10.8) do provide a basis for examining the general drying behaviour in a kiln. The differences in kiln-drying behaviour between hardwoods and the sapwood of softwoods are best illustrated by a plot of drying rate against the local board-average moisture content with position in the kiln as parameter. These graphs, in terms of non-dimensional parameters, are shown in Fig. 10.2.

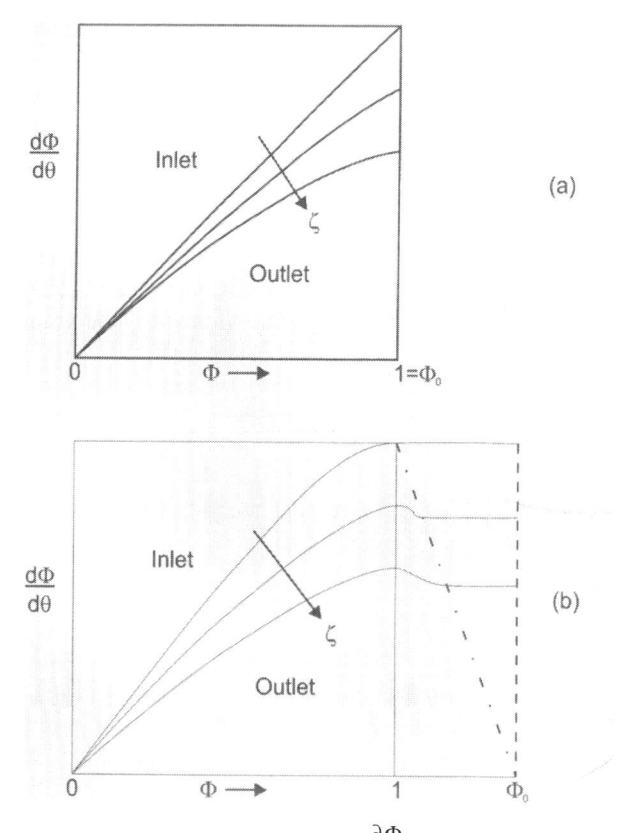

Fig. 10.2a,b. Variation of drying rate $\frac{\partial \Phi}{\partial \theta}$ as function of moisture content Φ at kiln positions ζ from the air inlet for a single-tracked kiln with $NTU \approx 1$. a Hardwood, with $f = \Phi$. b Sapwood of a softwood, with $f = \Phi^{1/2}$

These graphs neglect any moisture loss during the warm-up period. The variation of initial drying rates follow directly from Eq. (10.12) with $f(\Phi) = 1$:

$$\frac{\partial \Phi}{\partial \theta} = \Pi = 1 - \exp(-\zeta). \tag{10.15}$$

The local drying rate simply reflects the variation in external drying conditions within the kiln at the start.

Since the board-average mass-transfer coefficients also fall with distance in the airflow direction towards an asymptotic value, equal increments of ζ represent dwindling steps in distance. The drying curves for the hardwood show diminishing rates with both time and distance, with the rate curves converging to a common point at equilibrium. The drying curves for the sapwood of a softwood, however, are more complex, with rate maxima appearing within the kiln when the local moisture content is close to the critical value with the boards at the air inlet having a moisture content below it. These rate profiles follow from considering the changing humidity potential in the kiln during the course of the drying schedule, as explained by Keey (1978). Once the moisture content of the material at the air inlet has fallen below the critical point for unhindered drying, there is less convected moisture moving forward through the stack: so, at downstream positions, local drying rates can rise as long as the material there remains in the first period of drying.

The drying behaviour under normal kiln operations is more complex than that revealed in Fig. 10.2 because often there are set changes in drying conditions as the schedule proceeds and the direction of the airflow is reversed periodically to smooth board-to-board moisture-content variations. The analysis of such practical schedules is considered in the following Section 10.3.

Equation (10.12) was first derived by van Meel (1958) for any batch convective dryer. The balances over the volume element imply:

$$\frac{\partial Y_G}{\partial z} \gg \frac{\partial Y_G}{\partial t}$$

and

$$\frac{\partial X}{\partial z} \ll \frac{\partial X}{\partial t}.$$

In other words, the humidity is assumed to change more quickly with distance than with time, but the fall of moisture content with time is considered to be much greater than the moisture-content change across the kiln. The first condition holds, since the humidity at the air-inlet face of the stack is normally held at a fixed value, but the second condition is uncertain as significant moisture-content variations across the kiln appear during the course of the drying. However, the significant variation is from board to board, as each board is an entity in itself. If the volume element in the mass balance is taken over each individual board, a uniform, depth-averaged moisture content may be assumed to exist in that board to a first approximation. For practical purposes, then, the assumed inequalities may be considered to hold.

10.3 Practical Kiln Schedules

Van Meel's method was first used by Tetzlaff (1967) to examine a twin-stack kiln for drying 100×50 mm sapwood boards of *Pinus radiata* at 170 °F (77 °C) dry-bulb temperature and 150 °F (65.5 °C) wet-bulb temperature and with a kiln-air velocity of 500 ft min^{-1} (2.5 m s^{-1}). Later, more extensive calculations, for American, Australian and New Zealand schedules may be found in Ashworth's thesis (1977), but the results are plotted for unrealistic initial moisture contents and kiln widths. Aspects of this work are developed and reported by Ashworth and Keey (1979). More recent examples are provided by Keey and Pang (1994) and Nijdam and Keey (1996).

An example of the impact of a typical schedule on the course of drying is illustrated in Fig. 10.3 for the case of a three-step schedule for a softwood. This schedule speeds up the drying process by one-quarter compared to the case when the inlet humidity potential is retained at a constant value throughout the whole schedule (Keey 1972). The foreshortening of the drying time is obtained at the expense of maintaining fairly large moisture-content

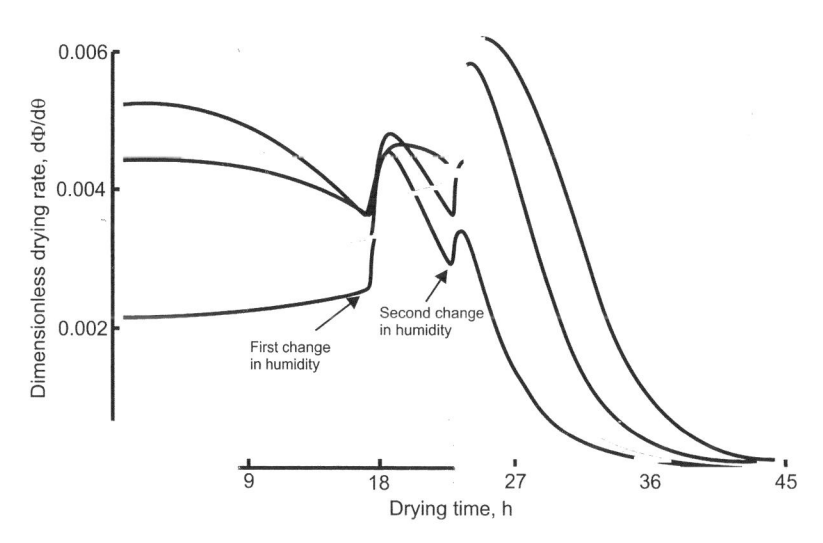

Fig. 10.3. A three-step schedule for drying a softwood. Inlet humidity potential: 0.00610 (green); 0.00878 ($\Phi = 0.264$); 0.0131 ($\Phi = 0.0868$). Initial moisture content: $\Phi_o = 1.22$. Kiln $NTU = 1.05$. (After Ashworth, 1969, and Keey, 1972)

228   Kiln Operation

variations in the boards over the kiln, caused by the stepwise boosts in the inlet humidity potential which induce temporary enhancements in the local drying rates.

Figure 10.4 shows the influence of increasing the number of flow-direction reversals on reducing the moisture-content distribution throughout the kiln in the absence of changes in the inlet-air humidity potential. Without any flow switchovers, the moisture-content profiles with time show a leaf-form, charac-

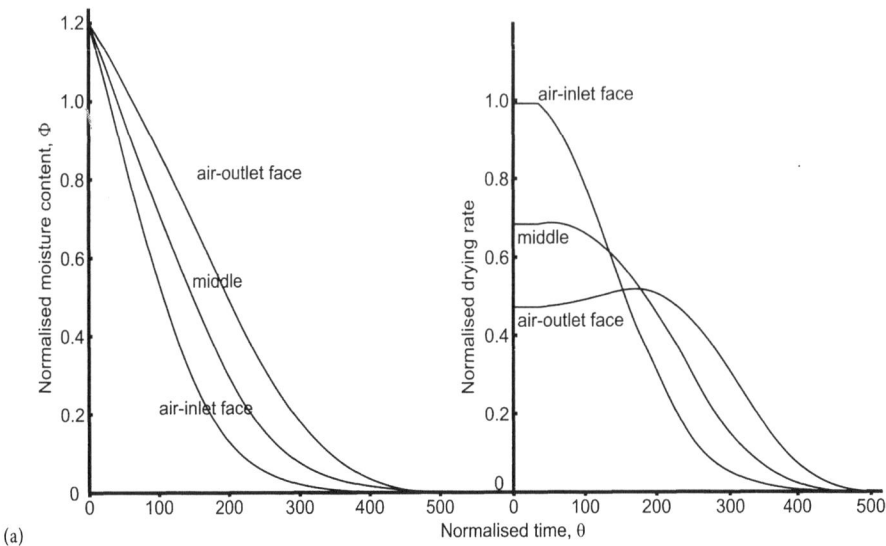

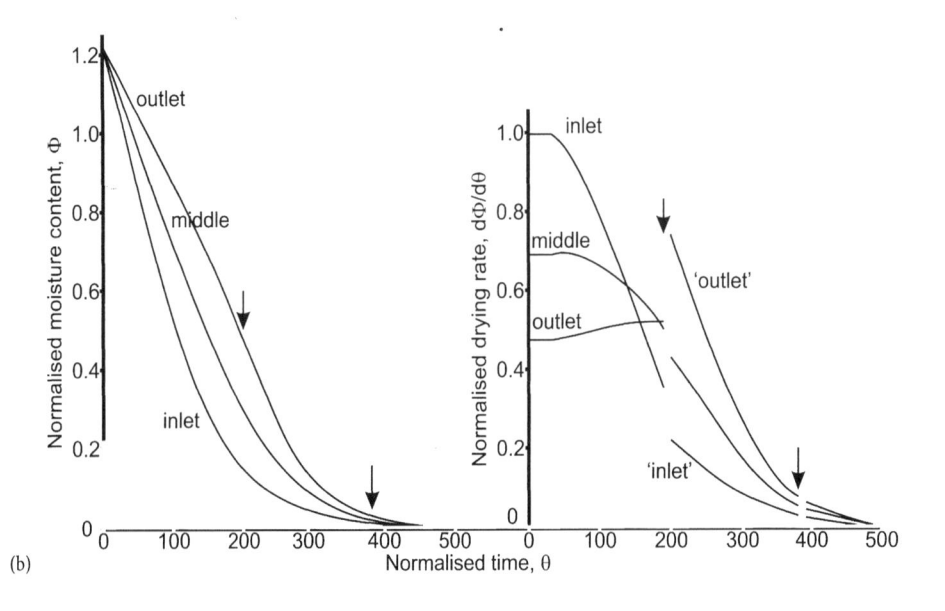

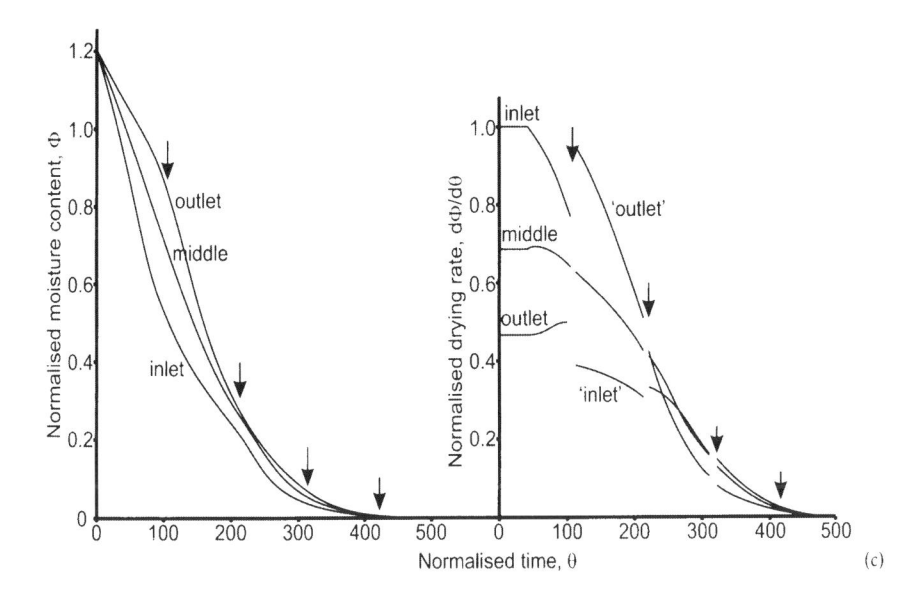

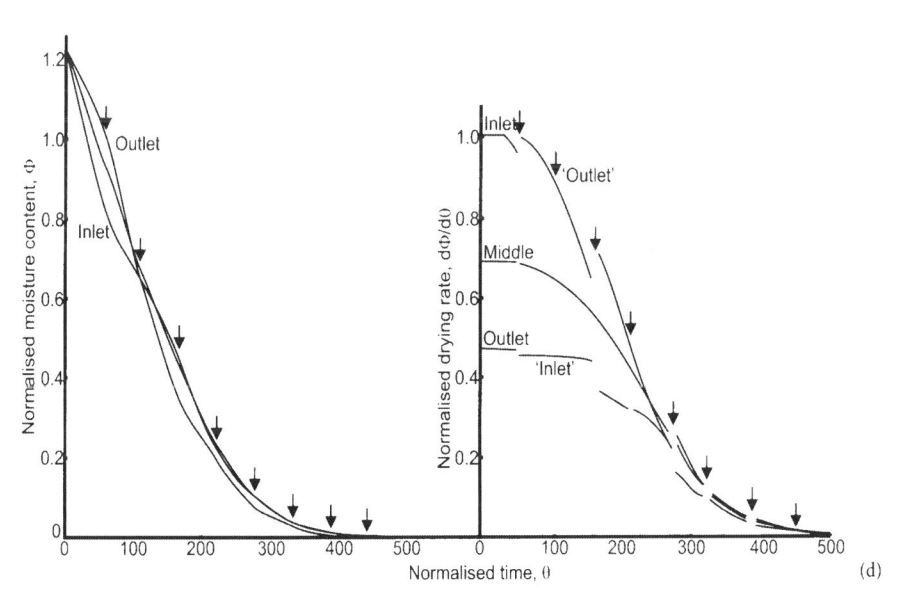

Fig. 10.4a–d. Board-average moisture contents and local drying rates as function of time and distance in the airflow direction in a twin-stack kiln. Sapwood *Pinus radiata*, 100 × 50 mm, dried at 77/65.5 °C and 2.5 m s^{-1}. **a** 0, **b** 2, **c** 4, **d** 8 reversals of airflow direction. Arrows indicate when reversals are undertaken. (After Tetzlaff 1967)

teristic of convective batch drying, with wide variations in moisture content in the kiln about half-way into the schedule. As drying proceeds, the absolute variation becomes less although the relative variance becomes greater, as shown by Keey (1992) for the mathematically analogous situation of continuous through-circulation drying on a perforated band. The plots also show a slight rise in drying rates within the kiln as the position of the critical point moves through the stack from board to board. The kiln velocity is low by the standards of the 1990s for softwood drying, and thus the kiln-wide variation in a modern kiln would be somewhat less than that indicated in Fig. 10.4.

Flow reversals, as expected, reduce the moisture-content variability over the kiln, but there is little benefit in using more than four flow reversals according to Tetzlaff's calculations. Indeed, there is some evidence to suggest that even less are needed.

Ashworth (1977) examined flow-reversal policies in some detail. He showed that there is always an irreducible minimum moisture-constant change at a given moisture content, even if an infinite number of switchovers were undertaken. If a switchover takes place at the same time as the drying schedule requires a step increase in the wet-bulb depression, then very large drying rates can result for a short period at the air-inlet side of the stack. Presumably, such drying rates would result in large, and possibly excessive moisture-content gradients near the surface of the wood. Flow reversals are more effective the narrower the width of the kiln stack, as flow-switching has a diminishing effect on the drying in the interior of the stack as it becomes wider. Reversals essentially trim the excessive over- and underdrying of the outer portions of the stack.

Pang (1994) considered the impact of various flow-reversal policies on the high-temperature drying of *Pinus radiata* at dry/wet-bulb temperatures of 120/70 °C and a kiln-air velocity of $5\,\mathrm{m\,s^{-1}}$ through a 2.4-m-wide stack. On drying heartwood without flow-switching, a maximum moisture-content difference across the stack is estimated to be $0.062\,\mathrm{kg\,kg^{-1}}$. Reversing the flow direction every 8 h gives very little benefit, but early reversals at 3 and 4 h after the start are beneficial, reducing the maximum moisture-content difference to $0.04\,\mathrm{kg\,kg^{-1}}$. Interestingly, there was essentially no difference in the moisture-content variability between making a single flow switchover after 4 h and having reversals every 4 h throughout the schedule. Similar trends were found for the drying of sapwood, except that a dual switchover, one after 2 h and the second after 6 h, reduces the maximum moisture-content difference to $0.16\,\mathrm{kg\,kg^{-1}}$ compared with a value of $0.38\,\mathrm{kg\,kg^{-1}}$ for unidirectional airflow. The greater moisture-content variability with sapwood is attributed to the relatively rapid moisture flow within the wood in the early part of the schedule, with the consequential enhanced moisture evaporation near the air-inlet side of the stack reducing the drying potential at positions downstream.

Pang's results confirm the desirability of at least one early flow reversal, and the lack of substantial benefit of later ones. Similar conclusions were obtained by Nijdam and Keey (1996) for the low-temperature drying of the same species when the dry-bulb temperature does not exceed 80 °C, for which a single reversal is made one-quarter the way into the schedule. Further, there appears to be a common minimum in the final moisture-content approach irrespective of the number of even flow reversals undertaken, but the optimum period for these reversals depended on the number to be done. This behaviour is shown clearly in Fig. 10.5, in which the standard deviation of the board-average moisture contents at the end of the schedule on drying the sapwood of a softwood is plotted against the period between reversals for various even numbers of switchovers in the schedule.

Traditionally, industry has switched the rotational direction of the fans regularly, often every 4 h, in the belief that this policy minimised kiln-wide variations in moisture content from board to board. When a number of reversals are used, changing the period between switchovers has very little effect on the variation of the final moisture contents of the boards. Wagner et al. (1996) report that there is little difference in final moisture variation whether 10 or 40 evenly-spaced reversals are used over a 60-h drying schedule for grand fir (*Abies grandis*). The evidence points to the benefits of flow reversal only in the early stage of drying as the moisture-content spread becomes larger, reflecting

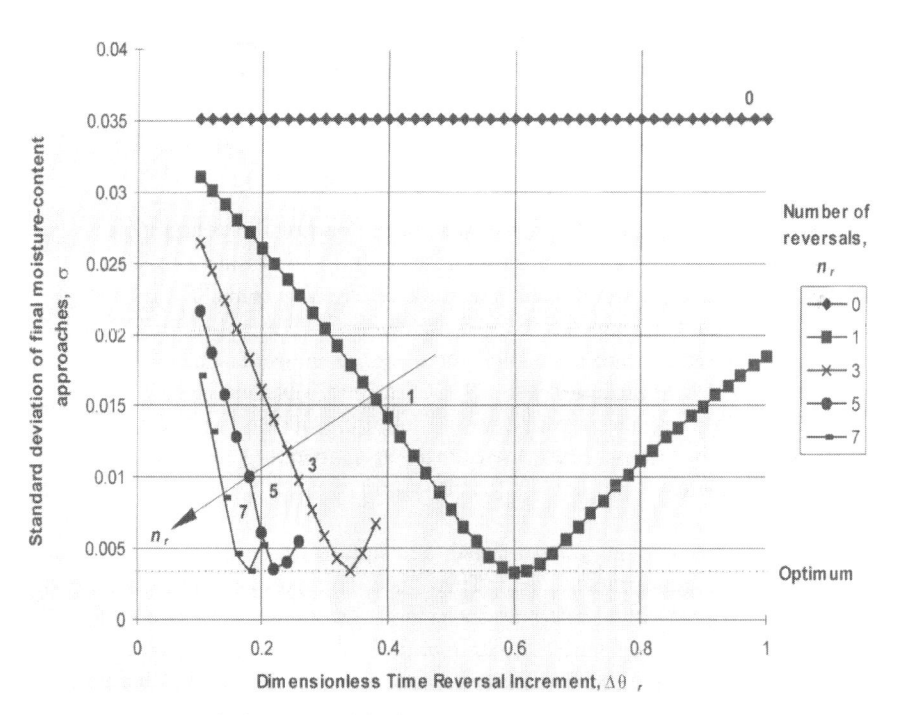

Fig. 10.5. Standard deviation of the final moisture-content approach to the specified end value $(X - X_f)/(X_o - X_f)$ as a function of the non-dimensional period between reversals θ_r and the number of reversals n. *Pinus radiata* sapwood. (Nijdam and Keey 1996)

the diminishing driving force for drying in the airflow direction, and only one is needed if done at the appropriate time. Excessive flow-switching may be counter-productive, shortening the working life of the fan system, and reducing kiln capacity if the fans are not fully bidirectional. Although manufacturers can supply bidirectional fans with less than a 10% difference between forward and reverse-flow throughput, there is some evidence from kiln audits used by Nijdam and Keey (1996) that the forward thrust of the installed fans in that case was somewhat better than the reverse thrust. The audits implied that the airflow in the reverse direction could be as much as 20% less than that in the forward flow. An attributed difference in air velocity, depending upon the airflow direction in a stack, is also noted by Culpepper (1990), who illustrates asymmetrical moisture-content profiles across the width of a stack at the end of drying in a direct-fired kiln.

Keey and Pang (1994) present profiles of the humidity potential and airstream temperature for the high-temperature drying of *Pinus radiata* at various positions in a single-track kiln, and the particular profiles for the drying of sapwood boards are reproduced in Fig. 10.6. The shape of these profiles, however, would be common to any adiabatic batch-drying process irrespective of temperature level.

The similarity in the profiles of humidity potential and of airstream temperature is to be expected, as both parameters are linearly linked through the adiabatic-saturation expression. Specifically, at the air inlet, once the wood has been warmed up to its initial steady value, the fall in air temperature T_G across the kiln is given by

$$\Theta = \frac{T_G - T_W}{T_{GO} - T_W} = \frac{T_G - T_S}{T_{GO} - T_S} = 1 - \exp(-\zeta) \tag{10.16}$$

on comparison with Eq. (10.15), for adiabatic drying when the wood's moisture content is above the critical point. In Eq. (10.16), T_S is the adiabatic-saturation and T_W the wet-bulb temperature respectively, while T_{GO} is the air-inlet temperature, assumed to be held constant.

The air temperature across the kiln is predicted to fall from about 15 °C at the beginning of the schedule to about 4 °C at the end. Thus, monitoring this temperature drop provides a rough way of following the course of drying. Towards the end of the schedule, the changes in temperature are small; the temperature difference over the kiln for the last 4 h of the process changes by only 1 °C.

Culpepper (1990), in the context of high-temperature seasoning, notes that the temperature drop across the load (TDAL) is often used as an indicator of the time to "pull the kiln". With air entering at a dry-bulb temperature of 240 °F (115 °C), a typical temperature drop at the end of drying might be 8 °F (4.4 °C). The technique must be used judiciously, however. Careful location of the thermometers is needed to sense the correct temperatures, particularly at the outlet face of the stack, where the registered temperature may be influenced by air bypassing or short-circuiting past the stack. Moreover, since fan reversals cause the temperature difference over the kiln to "flip-flop", it is

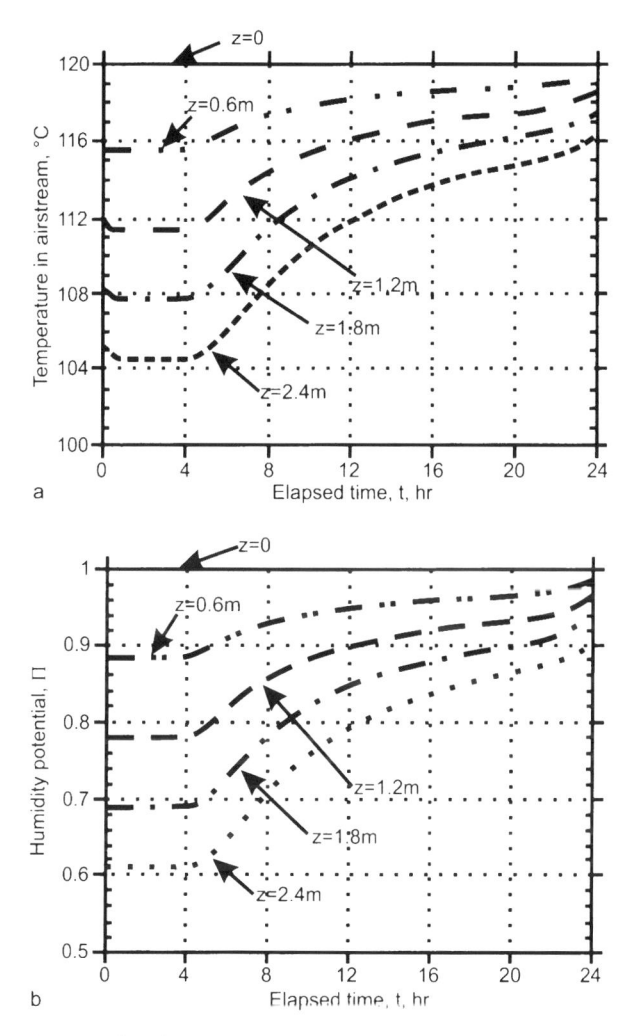

Fig. 10.6a,b. The drying of *Pinus radiata* sapwood boards, 100 × 50 mm, in a 2.4-m-wide kiln at an air velocity of 5 m s^{-1} at 120/70 °C dry/wet-bulb temperatures. **a** Humidity potential, **b** Airstream temperature. (Keey and Pang 1994)

important to use the temperature-drop reading for a control measure well after the temperatures have settled down following a switchover in the airflow direction through the stack.

10.4 General Practical Considerations

There is no attempt in this book to provide a detailed overview of kiln practice which may be found in other works. Industrial drying handbooks, such as those of Pratt and Turner (1986), Boone et al. (1988), Hilderbrand (1989), Mackay and Oliveira (1989), Simpson (1992) and Haslett (1998) have the advantage of a focus on specific species and local or national concerns that can be irrelevant elsewhere. A review of the kiln-drying of spruce in Europe will be totally different to one on the drying of southern pines in the United States, the kiln-seasoning of radiata pine in Australia and New Zealand, and the drying of the much-coveted sugi (*Cryptomeris japonica*) in Japan. However, certain general comments may be made.

10.4.1 Species-Grouped Schedules

The fundamental objective of kiln-drying is to dry at as fast a rate as possible compatible with minimal degrade, but where management lacks either experience or knowledge, or good quality-control systems are difficult to put in place, then the drying practices must be conservative. Grouping of tropical species for kiln-drying provides an example of this approach. The many hundreds of species, the difficulties in their identification and their diffuse occurrence in a tropical forest means that singling them out and drying them separately is rarely economic or feasible. Yet a single schedule that minimises degrade would be hopelessly conservative. Economic drying of mixed-species lumber must involve grouping of species according to some logical criteria. Simpson and Baah (1989) choose basic density and the initial green moisture content, as shown in Fig. 10.7, on the grounds that high-density lumber dries more slowly and is more susceptible to drying defects, while a high initial moisture content will require a long drying time. Other factors such as permeability, shrinkage and extractive content are not so easily measured, and so were not considered. Using this system, Hidayat and Simpson (1994) derived drying-time indices to form 12 species groups, whose lumber could be dried together, so that kiln-drying times for the reference schedule increased or decreased by 1.5 days on moving to adjacent species groups.

10.4.2 Species-Specific Schedules

New Zealand provides an extreme example. About 95% of the commercial forest harvest derives from plantation softwoods, of which about 90% is *Pinus radiata*. Comprehensive schedules have been developed for this species, with fast high-temperature schedules for structural grades and accelerated conventional schedules at somewhat lower temperatures for appearance grades (Haslett 1998).

In one sense, the situation in the United States is equally simple, as just 13 major species and species groups account for 82% of all kiln-dried lumber

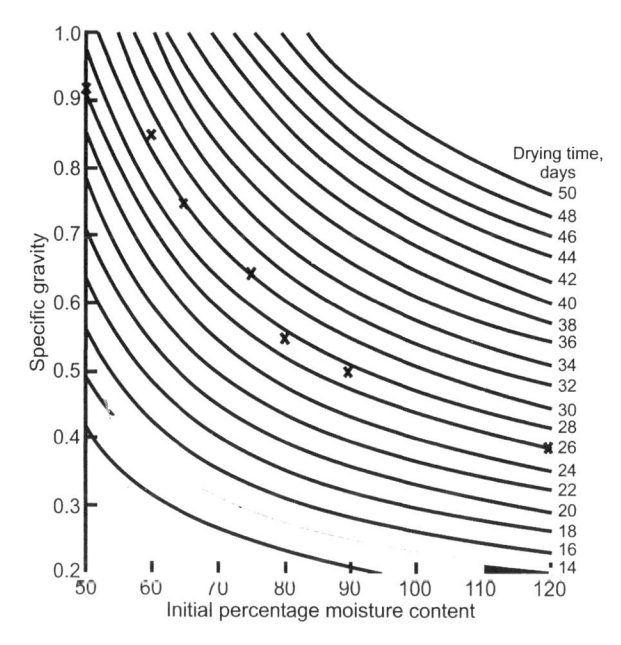

Fig. 10.7. Characteristic drying times for tropical species. Combinations of specific gravity and initial moisture content that result in equal drying times for 57-mm-thick lumber at 38 °C. Final average moisture content = 0.3 kg kg^{-1}. (After Simpson and Baah 1989)

(Rice et al. 1994). In that country, some 68.8 million m^3 are kiln-dried, about 85% of all lumber; of which 11.7 million m^3 are hardwoods and 57.1 million m^3 are softwoods. About 73% of the all the kiln-dried hardwoods comprise red and white oak, yellow poplar, maple, red alder and cherry, while 84% of the softwoods are southern yellow and ponderosa pine, Douglas-fir, western firs and western hemlock. Having large volumes of relatively few species and species groups justifies the gathering of detailed information so that species-specific schedules can be developed (Boone et al. 1988).

In the United States, almost 70% of kilns are worked with dry-bulb temperatures between 70 °C and 82 °C, with a further 20% working at higher temperatures, especially in the south (Rice et al. 1994). The bulk of the lumber dries readily: indeed, a surprising number of hardwoods, including red alder (*Alnus rubra*), aspen (*Populus balsamfera*), blackgum (*Nyssa sylvatica*) and yellow poplar (*Liriodendron tulipifera*) are being dried under high-temperature schedules, invariably with equilising and conditioning treatments.

The same report noted that the less restrictive drying conditions and moderate moisture contents for structural softwood grades mean that these woods can be dried successfully using time-based schedules, whereas most hardwoods follow more rigorous moisture-based schedules, having to be dried carefully and to a lower moisture content of 6 to 8% for interior uses such as furniture, cabinets, flooring and mouldings. Time-based schedules recognise the inherent constraints in determining the end-point condition by monitoring representative sample boards, selecting and placing sample boards within the stack and determining oven-dry weights. However, drying schedules for permeable species progress too rapidly to entertain such delays. Instead, the kiln operator may check the hot lumber intermittently during the schedule with a hand-held moisture meter using electrode probes on an extended arm to sample extensively within the stack. The recent development and introduction of impedence, dielectric and microwave-based systems capable of sampling a number of boards continuously to an accuracy of ±1% moisture content at the end of the schedule has blurred the practical distinction between time-based and moisture-based schedules for the larger kiln operators.

Rice et al. (1994) also noted that, of more than 7000 kilns surveyed in the United States, the majority (88%) were steam-fired, 7% were dehumidifer kilns and 5%, mainly in the south, were direct-fired. Less than 1% involved the use of vacuum, radio-frequency heating or other drying methods. Dehumidifying drying, which has been used for several years in Europe for drying hardwoods, is now becoming popular in America (Rosen 1995). The use of dehumidifiers in kiln-drying is considered further in Section 12.5 in Chapter 12.

10.4.3 Schedule Development

Kiln-drying involves the transfer from moisture within the wood and evaporation from its exposed surfaces, and it is crucial to keep these two processes in balance to maintain the fastest possible drying rate.

In one sense, kilns maintain relatively high humidities in order to *slow down* the rate of evaporation, particularly in the early part of the schedule when drying impermeable species. As long as the external convection controls, the fibres at the surface remain saturated, but once the internal moisture-transfer rate lags the external vapour-transfer rate, the temperature at the surface rises, its moisture content falling to maintain a moisture balance. To sustain the rate of drying, the dry-bulb temperature or the wet-bulb depression may be raised to provide a greater driving force for the overall transfer. However, a faster drying rate can lead to the development of steeper moisture-content gradients which attendant risk of checking. Modern computer-based kiln-control systems make more frequent adjustment of temperature levels possible, thus maintaining a more uniform rate of moisture loss throughout the schedule, rather than having a steep rise at the beginning of each change in a stepwise schedule. This latter policy can be dangerous if such a step change coincides with a reversal of the airflow direction in the kiln (Ashworth and Key 1979). Other possibilities with a computer-based system include schedules based on drying within a limiting strain envelope, as proposed by Langrish et al. (1997) for refractory Australian eucalypts.

It is the humidity in a kiln that is so striking to a casual visitor to a kiln site. A small wet-bulb depression means a relatively high equilibrium moisture content at kiln temperatures, with less surface shrinkage and so smaller differential strains in the early part of the schedule, especially desirable in the drying of less permeable species. In a conventional kiln schedule, the wet-bulb depression is typically about 5 to 10°C, whereas for the high-temperature drying of permeable species the wet-bulb depression can exceed 50°C as the transfer of moisture within the wood is far less rate-limiting than that in less permeable species.

10.4.4 Stacking

Whether a high-density hardwood, laminated material, or locally available wood is used for stickers, the twin overriding requirements are straightness and uniform thickness. If the stickers are too wide, poor drying and staining may result underneath them. Thus, 45 × 19 mm stickers may be acceptable for fast-drying softwoods, but would be too wide for slow-drying hardwoods, for which 30 × 30 mm might be more appropriate. With slow-drying hardwoods, stickers can be grooved, so that the contact is mainly at the edges, allowing some air movement under and along their lengths.

The placing of stickers along the stack needs to be frequent, 300 to 450 mm apart, for boards up to 30 mm thick, especially for those species or juvenile wood that are prone to warp. For thicker boards, spacings can be further apart, at 450 to 600 mm centres.

While it is much preferred to build kiln stacks of standard length with lumber of that length butted edgewise in each layer, sometimes it may be necessary to lay up the stack by placing somewhat shorter boards alternatively flush with opposite ends of the stack to ensure that the overall stack length matches that of the kiln. With random-length material, the packets must still be squared at both ends. This involves placing the shorter lengths within the body of the packet and having additional sticker rows to restrain the free ends. In this way, by-pass around the stack is minimised. Semiautomatic systems with stickering frames and machine stacking have improved considerably the quality of stacking over earlier manual methods, with stickers aligned above one another and immediately over the bearers that are placed between the individual packets in a kiln stack.

In high-temperature drying, when the lumber markedly softens at kiln temperatures, it is essential to use stack-restraining weights to prevent warp. The loading needed depends upon the susceptibilty of the wood to warp, the thickness of the lumber pieces and its width. For New Zealand *Pinus radiata*, Haslett (1998) recommends a stack loading of $1000 \, \text{kg m}^{-2}$ for juvenile stock, and $500 \, \text{kg m}^{-2}$ for other wood, but with higher-density woods grown in Chile, the stack loading can be as heavy as $1500 \, \text{kg m}^{-2}$.

The height of individual packets together with bearers, rail bogies and any weights should bring the entire stack to just below the top of the kiln, allowing the roof baffles to be held tightly against the shrinking load during the schedule. Baffles or blanks are also needed at either side of the stack ends, between bearers and under the bogies.

More uniform drying is achieved with a single-track kiln, but a twin-track kiln increases kiln capacity. Uniformity of drying with a twin-track kiln can be improved if a reheating coil is placed between the two stacks.

Problems with checking, warp and wastage are more acute with high-valued, slow-drying hardwoods, for which it is important to retard the rate of surface evaporation relative to the moisture transfer within the wood. End-sealing with moisture-resistant paints or waxes, and the use of S- or I-shaped plastic cleats, hinder splitting along the grain from excessive moisture loss along the grain.

10.4.5 Fan Speeds and Reversals

Although higher airflows seem justifiable at the start of the schedule when the moisture-vapour transfer is controlling, so-called Hi-Low flow strategies do not appear to be economically worthwhile, as noted in Chapter 9. However, higher kiln velocities to $10\,\mathrm{m\,s^{-1}}$ would be used in the drying of permeable lumber to take advantage of the benefits of enhanced convection compared with the drying of impermeable hardwoods when slow diffusion through the lumber is rate-controlling and kiln temperatures are normally less. Haslett (1998) recommends kiln velocities of 7 and $8\,\mathrm{m\,s^{-1}}$ when the dry-bulb temperatures are $120\,°C$ and $140\,°C$ respectively in the high-temperature drying of *Pinus radiata*, but lesser velocities of 4.5 to $6\,\mathrm{m\,s^{-1}}$ under conditions of accelerated conventional temperature (ACT) drying when the dry-bulb temperature is $90\,°C$. Walker (1993) notes that, with the slow drying of impermeable species, a kiln velocity as low as 1.5 to $2.0\,\mathrm{m\,s^{-1}}$ may be sufficient.

Conventionally, airflow reversals have been employed to counter the progressive humidification of the airflow through the stack with consequential loss of drying potential. As explained earlier, only a single switchover in the airflow direction is required about one-quarter the way through the drying schedule to achieve a benefit in reducing moisture-content variations across the stack, although normal commercial practice often involves reversals in the fan direction at regular intervals every 4h or so, and sometimes more frequently. Salin and Ohman (1998) observe that it is Scandinavian practice to reverse the airflow every 1 to 2 hours with species such as Norway spruce (*Picea abies*). The mills recover larger-dimension lumber from the centre of small-diameter logs, and various proportions of heartwood and sapwood may be found in the wood. Fairly frequent flow reversals are considered to mitigate variations in the ease of drying and the development of drying stresses.

10.4.6 Kiln Monitoring

Basic control of the progress of a kiln schedule relies on the wet- and dry-bulb readings. Psychrometry has long been a popular method of humidity determination, primarily because of simplicity with its use of rugged, industrial-type resistance thermometers and its superiority over other methods in monitoring conditions close to saturation.

Most single-zone, steam-heated kilns mount the split bulbs under the overhead coils, one in each plenum, on either side of the centre coils about 1.4 m off the floor (Culpepper, 1990). In kilns divided into two zones along their length, the bulbs are placed midway in each zone. Direct-fired kilns have their bulbs mounted in the outside plenums and placed midway along the kiln. Bulbs must be protected from radiation from hot coils, and be exposed to an air velocity of at least $5\,\mathrm{m\,s^{-1}}$ for the bulbs to record the psychrometric values correctly. A hardwood kiln may be worked at an air velocity through the stack less than this value, but at the same time will be controlled to maintain a high relative humidity with a correspondingly small wet-bulb depression. In practice, the error in reading may be small and tolerable in such a case.

Smaller lumber-drying installations having one or two kilns might have a simple electronic-control system with discrete programmable controls, timers, relays and a chart recorder for temperature and steaming control. Other options include a preset timer and endpoint control based on the temperature drop across the load (TDAL). A more sophisticated system for a medium-sized facility might have a low-cost, computer-based system running with windows-type software to give a visual readout of kiln conditions. The system might also incorporate a programmable logic controller (PLC) to enable the operator to change a schedule while the kiln is working and ramp setpoints up and down. Advanced computer-based kiln-control systems are available to provide centralised supervision in a larger installation comprising a number of kilns, together with ancillary plant such as steaming chambers and heating units. Culpepper (1990) reports that the installation of one process-control system incorporating a programmable logic controller (PLC) at a particular site increased the overall grade recovery from 70.7 to 81.9%, with a reduction in energy costs of 43% for steam and 10% for electrical power. The total savings represented a 1.2-year payback on the investment.

10.4.7 Volatile Emissions and Kiln Corrosion

Historically, there have been numerous reports of corrosion in kilns, the cause of which could be traced largely to highly acidic conditions in kilns operating at elevated temperatures (Barton 1972). Metals, unprotected concrete and wood, the wall linings of older kilns and stickers can all be attacked. For example, the drying of a hardwood with a high extractives content, *Nothofagus truncata*, caused sufficiently severe rusting of steel trolley wheels and other ferrous fittings in a pilot-plant kiln to require their replacement after a 2-week period. Galvanic cells caused by contact between dissimilar metals produce even greater corrosion whenever there are acidic vapours. Such corrosion is avoidable and the problem has disappeared in industry largely through the extensive use of aluminium linings, with plastic pipework replacing copper, and eliminating features that trap condensation. However, Little and Moschler (1995) observe that masonry or fibreglass construction is preferred when very acidic vapours are evolved.

Packman (1960) has noted that, of some 150 species studied, both hardwoods and softwoods, 120 have a pH between 4 and 6. Only one species has a pH which is consistently above pH 7, while some have a pH below 4. This natural acidity arises from the buffering of weak acids and their salts in the wood.

The corrosive nature of the volatile substances released during kiln-drying is very largely due to the presence of initial free acetic acid in the wood and also from acetic acid derived from the deacetylation/hydrolysis of the combined acetyl groups originally attached to the hemicelluloses. The rate of hydrolysis varies with species; it is a function of buffering action, the moisture content of the wood, and temperature. The two most corrosive species, sweet chestnut (*Castanea sativa*) and oak (*Quercus* sp.), which were studied by Arni et al. (1965a,b), have both a high tannin and phenolic content which they suggest may have a role in catalysing the autohydrolysis of the bound acetates. With air-dried lumber, the acidity remains constant, as the small loss of acetic acid by volatilisation is roughly counterbalanced by the slow rate of hydrolysis occurring at low temperatures. However, the wood retains its large reserve of acetyl groups which has the potential to be mobilised at higher temperatures and humidities. Oak storage cabinets are notorious for corroding zinc-plated metal. By contrast, in kiln-dried lumber, wood hydrolysis increases rapidly with rising temperature and humidity, releasing increasing amounts of acetic acid. Species that have a reputation for acidic volatile emissions include: sweet chestnut (*Castanea sativa*), oaks (*Quercus* spp.), Douglas-fir (*Pseudotsuga menzesii*) and western hemlock (*Tsuga heterophylla*). Western red cedar (*Thuja plicata*) has earned notoriety in the Pacific Northwest for its degradation of wood and corrosion of metals within old-style kilns. This was due to the thujaplicins which can amount to as much as 1.2% of the heartwood (Browning 1963): these can be extracted by steam distillation and, equally, are released during kiln drying to chelate subsequently with steel fittings.

In the United States, as elsewhere, plant operators are subject to environmental audit and must comply with the requirements of the Clean Air Act and its 1990 amendments, with limitations to emitted volatile organic compounds (VOCs) and condensible organic vapours (the blue haze of the aerosol) and particles less than $10\mu m$ in diameter (PM-10 emissions). The volatile organic emissions are classified into hazardous air pollutants (HAP), such as formalde-

Table 10.1. Concentrations of volatile emissions arising from two high-temperature kiln schedules for *Pinus radiata* (McDonald and Wastney 1995)

Compound	Concentration ($g\,m^{-3}$)	
	at dry/wet-bulb 120/70	temperatures (°C) 140/90
Formaldehyde	19.5	31.0
Acetic acid	21.7	38.2
Monoterpenes	34.8	66.4
Hydroxylated monoterpenes	12.6	10.7
Condenser residues (resins and fatty acids)	16.4	16.4

hyde, which have threshold-limit values for toxicity, and those substances which react in sunlight to produce ozone (Bradfield and Emery 1994). Green pines have the highest level of volatile emissions (especially α- and β-pinene and terpene hydrolysis products), with the exception of lodgepole pine (*Pinus contorta*). Some volatile substances, such as formaldehyde, come from the thermal degradation of the hemicelluloses and the lignin (von Marutzky and Roffael 1977). Apart from aspen (*Populus tremuloides*), hardwoods produce lower volatile emissions than the resinous softwoods, particularly as kiln temperatures are normally lower than for most softwoods.

McDonald and Wastney (1995) analysed the volatile emissions arising from the high-temperature drying of *Pinus radiata* (Table 10.1). A large softwood mill will have a number of kilns, each holding at least $100\,m^3$ of lumber, being dried within 24 h if the lumber is destined for structural markets.

The recorded concentration levels for formaldehyde may be compared with the New Zealand workplace-emission standard of $1.2\,g\,m^{-3}$ for total exposure in one day for an 8-h working day (NZ Department of Labour 1992). Clearly, there may be some element of concern about vapour-concentration levels near high-temperature kilns on the basis of these figures. Emission-control options are uncertain: Bradfield and Emery (1994) consider the use of biofilter/sorption and biodegradation. However, the necessity for stringent emission controls is debatable if mills are sited in remote and relatively windy locations.

10.4.8 Equalisation

Equalisation at the end of a schedule reduces the final between-board variations in moisture content. It is especially applicable in the drying of valuable, slow-drying hardwoods for which customers' specifications are more demanding than those seeking constructional grades. With moisture-based schedules, the kiln conditions at the end of drying are reset to give an equilibrium moisture content which is 2 to 3% moisture content below the desired value. This policy prevents further overdrying of the drier boards while allowing the wetter boards the opportunity to dry further towards the specified end-point. Equalisation continues until the wettest boards have reached the target moisture content. The process of equalisation is slow, however, since the rate of drying is much diminished at conditions close to equilibrium. For some permeable species, the conditioning period to be effective would be at least 24 h, and much longer with impermeable species.

However, Haslett (1998) recommends, contrary to some North American and European views, that softwood timber should be slightly overdried to ensure that the moisture-content variation within and between boards is minimised. Thorough final steaming, which is undertaken for stress relief, can raise the moisture content of the overdried portion of the load.

10.4.9 Stress Relief

If the lumber had been dried infinitesimally slowly, imperceptible moisture-content gradients would have been generated during drying, with no internal stresses from differential shrinkage. In practice, kiln-drying at economic rates causes significant moisture gradients to arise, with stress development. Schedules are aimed at tolerating some plastic strain or *set* without exceeding the ultimate tensile strength across the grain in the warm wood and thus no checking. However, the residual stresses, both tensile and compressive, must be relieved if the lumber is to be further processed. Steaming is one means of relieving these stresses.

A separate chamber supplied with a fully saturated, low-pressure steam from boiling water in a trough is favoured if the installation is large enough to justify the investment. Environmental control is problematic, as steaming chambers have no fans and there is a natural stratification of steam and heated moist air. In the case of high-temperature drying, the charge has to be cooled for 1 to 2 h depending upon the board thickness and on whether the stack is retained within the kiln with the fans running or not. With high-temperature dried lumber, the temperature of the wood must fall to between 70 and 95 °C to ensure that the lumber picks up moisture. The moisture-content profile in the lumber reverses, with the surface moisture content greater than that in the core, rising above 20% with a permeable species, and thereby the in-built stresses are relaxed. Stacks which have been dried to low moisture levels, say between 6 to 8% moisture content, readily take up steam, but lumber at moisture contents above 15% is slow to sorb moisture. Thus dry, permeable material requires only a relatively short steaming period, about 1 h for 25-mm-thick boards for example. On the other hand, large-dimension stock of a high-density hardwood may require many hours of reconditioning to be effective. Use of stack restraints mimimises any warp associated with any overdrying prior to steaming.

10.4.10 Destickering

When loads are withdrawn from the kiln, they must be covered or wrapped and left to cool without any opportunity for further drying. Further, if the stacks have been dried under restraint, the weights must be left on for a minimum of 12 hours to minimise warp.

Once the load has cooled, it should be destickered as soon as possible, wrapped and held in dry storage, to prevent any uptake of moisture, until being forwarded to the customer. Block stacking will aid further equalising between and within boards.

10.5 End-Moisture Specification

The kiln operator needs to know the customer's requirements for moisture content, If lumber is correctly dried, there will be no further shrinkage and the majority of problems associated with the instability of wood in its end-use are overcome. For example, an exporter to the United States must recognise the variation in desired moisture content for interior furniture components, depending upon whether the buyer is in the dry south-western area or in the damp, warm coastal south-east. In turn, the American exporter needs to recognise that climatic conditions in Australasia, Europe and Japan dictate the need for kiln-dried furniture stock having 10 to 12% moisture content, rather than the customary 6 to 8% required for the United States (Araman 1987). When considering all end-user requirements, Hoadley (1979) makes the telling point that it is the relative humidity of the air that is the *cause* and the equilibrium moisture content the *effect*. It is important for the kiln operator to understand and be able to measure the relative humidity necessary to maintain a specified moisture content.

The equilibrium moisture content is primarily a function of relative humidity, while the influence of temperature over normal climatic ranges is of secondary importance, as may be seen from the isotherms plotted in Fig. 2.4 in Chapter 2. At constant relative humidity, the equilibrium moisture content would not vary by about 0.3% moisture content per 10°C change in temperature over the range from 10°C to 30°C. However, the process of kiln-drying may influence the hygroscopicity of wood, particularly if it has been subjected to high temperatures, when some thermal degradation may

Table 10.2. Range in moisture content of untreated sapwood and heartwood of *Pinus radiata* during a year's exposure outside but under cover (After Kininmonth and Whitehouse 1991)

Wood	Drying method	Moisture Content ($kg\,kg^{-1}$)		
		Minimum	Mean	Maximum
Sapwood	Air-dried	0.154	0.188	0.213
	Kiln-dried at 70 °C	0.139	0.161	0.188
	Kiln-dried at 115 °C	0.131	0.152	0.177
Heartwood	Air-dried	0.148	0.180	0.204
	Kiln-dried at 70 °C	0.133	0.152	0.177
	Kiln-dried at 115 °C	0.110	0.124	0.149

have occurred. For example, Kininmonth and Whitehouse (1991) show that *Pinus radiata* dried at progressively higher kiln temperatures becomes less hygroscopic, as may be seen from Table 10.2. The rate of response to changes in environmental conditions also becomes less with material dried at a higher temperature.

Similar findings were found in earlier tests with small specimens of *Pinus radiata* and tawa (*Beilschmiedia tawa*), a hardwood, which were subject to fluctuating environments in a conditioning cabinet (Kininmonth 1976).

Both specifiers and end-users need to be aware of such differences, which are more pronounced with softwoods, as hardwoods generally tend to be dried at lower temperatures.

10.6 Handling Kiln-Dried Lumber

Inevitably, there is often some delay between kiln-drying and its final end-use, and in the intervening time the wood seeks to reach equilibrium with its surrounding environment. For this reason, mills invest in their own dry-store facilities and wrap packets in robust, protective materials in an endeavour to isolate the lumber from its external environment during transit.

There can be significant moisture pickup as a consequence of inappropriate storage and wrapping. The effectiveness of four kinds of wrapping compared with unwrapped material has been investigated by the New Zealand Forest Research Institute (1996). Packets of kiln-dried lumber were wrapped and left for 16 weeks over winter, either outside or partially protected by an open-sided covered shed. The results are set out in Table 10.3. Plastic film wrapping and shed storage proved the best option, with a moisture uptake of only 0.5% moisture content, representing a 6% gain in moisture. This compared with unwrapped material left in the open when the moisture content increased nearly fourfold over the 16 weeks.

Table 10.3. Stack-averaged percentage moisture contents before and after storage for 16 weeks over winter in the North Island of New Zealand (New Zealand Forest Research Institute 1996)

Location	Wrapping	Moisture content (%)	
		Ex-kiln	After 16 weeks
Outside storage	Unwrapped	9.1	34.4
	Paper-based	9.2	43.9
	Plastic fibre	8.9	33.5
	Woven plastic	8.5	11.5
	Plastic film	8.5	11.8
Open shed	Unwrapped	8.3	13.9
	Paper-based	8.3	10.6
	Plastic fibre	8.2	12.9
	Woven plastic	8.2	9.0
	Plastic film	8.1	8.6

Protective wrappings are needed to protect lumber during transit. If packs are exposed to the environment, woven plastic and plastic film wrappings are best. Direct heating from the sun can produce sweating, while damage to the wrap can result in pickup of moisture. Such packs are wrapped on all four sides and on top, with the lumber being free to breathe from below. If packs are protected, then paper-based and fibre wrapping is sufficient. Such covers project the supplier's commitment to quality and provide an opportunity for corporate branding.

When lumber is sold to a moisture specification, it is important that both the vendor and customer use a common measuring system. Although oven-drying and gravimetric methods provide a standard, it is important that test specimens are held in the oven for a fixed time (usually 24 h) at a specified temperature (usually 103 °C) to avoid unknown weight variations due to differences in volatile losses and any thermal degradation. Particular care is needed to undertake the test under standard conditions if the resin content of the wood is high. The residual moisture content of oven-dried material is ignored. With outside air at 20 °C and 40% relative humidity, the relative humidity in a through-draught oven at 100 °C would be 1%, which would leave a residual moisture content for most woods of about 0.05% (Keey 1992).

Commonly, moisture levels are checked with portable electrical-resistance meters. The relationship between moisture content and electrical resistivity is species-dependent. Many manufacturers take the resistivity of Douglas-fir heartwood at 12% moisture content and 80 °F (26.7 °C) as standard, and provide corrections for other species. A hardwood dried to about 8% moisture content would have a resistivity several orders of magnitude greater. A further difficulty arises when significant moisture gradients are present, as the meter picks up the path of least resistance rather than one reflecting the volume-averaged moisture content. However, the handiness of the resistance meter is widely recognised and is enhanced by a digital display and storage of data, so that its continuing use is assured despite uncertainties in the moisture-content values it records.

11 Pretreatments of Green Lumber

Pretreatments before drying seek to minimise seasoning time and degrade such as staining, warping, checking, collapse and honeycombing. In some cases, the effect of pretreatments is to modify moisture diffusivity or permeability, often with poor recognition of the differences between the two transport coefficients in influencing moisture movement. For most pretreatments, the efficacy claimed is species-specific and dependent on the intensity of the process.

11.1 Protecting Wood Prior to Drying

11.1.1 Wet Storage

11.1.1.1 Control of Microorganisms

In a number of countries, rafting is used to shift logs from forest to mill, while logs are stored in booms at mills relying on water for transport. There is the prospect of uneven colonisation of the partially submerged logs by microorganisms, with wide local variations in permeability, which is a poor environment for ensuring uniform product quality. Land storage is also practised in some countries, with the logs being placed under water sprinklers in the forest or mill at times when harvesting or transport may be disrupted or after major windblow events (Liese 1984). Such stockpiling has been extensively practiced in North America (Scheffer 1969), although more effective harvesting technology makes this much less prevalent today. In general, storage on land under sprinklers has been preferred because of limited availability of lakes and ponds, technical problems in loading and unloading, and losses due to sinkers (logs whose green density is greater than that of water), e.g. *Fagus sylvatica*. With land storage, intermittent oscillating sprays, delivering 40–50mm of water each day, are capable of maintaining moisture levels too high for aerobic organisms to grow.

A biocontrol mechanism operates in well-maintained log piles. Anaerobic bacteria, once established, within a water-saturated environment create conditions unfavourable to decay fungi. Only after an extended period beyond about 4 years do rot organisms gain a foothold (Hayward 1981): this follows the expected succession of fungi colonising wood, but the conditions in the log pile greatly retard that rate of succession (Butcher 1968; Bannerjee and Levy 1971).

Wet storage has the potential to improve the rate of drying and ease of preservative treatment in impermeable sapwood, but has no effect on heartwood. The bacteria that rapidly colonise the wood produce enzymes that are capable of degrading the pectinaceous pit-membrane material, thus rendering the wood more permeable (Suolahti and Wallen 1958; Ellwood and Ecklund 1959). The improvement in permeability is not linear with time (Bauch et al. 1970). An initial increase may be followed by a decline as the bacteria produce copious amounts of slime and only later, after many weeks, is there a further large increase in permeability as pit membranes degrade and rupture (Banks and Dearling 1973; Archer 1985). Wet storage is a necessary procedure for managing many stock inventories; the variable changes in permeability are incidental. However, wet storage has the potential to dramatically improve the sapwood permeability of hard-to-treat species, but before the process can become commercially viable the holding time will have to be reduced considerably. Furthermore, better process control is needed to minimise the uncertain, localised variability in permeability improvement. On the downside, disadvantages of wet storage are oxidative discolorations caused by phenolic wood compounds diffusing outwards (Höster 1974) and brown staining of wood close to the cambium caused by tannins diffusing inwards from the bark (Peek and Liese 1987). A greater probability of internal checking on drying is possible, as in oaks (*Quercus* spp.).

Ponding is more problematic; a small part of the log is above water and any exposed wood is very susceptible to microbial colonisation, as the conditions are ideal, reflecting the availability of oxygen, moisture and direct heating from sunlight. Permeability can be highly variable, as can be the incidence of stain and decay.

The influence of water storage with respect to kiln-drying rises from selective degradation of any unlignified pit membranes of the ray cells and of the bordered pits by colonising bacteria. The enzymatic degradation of pit membranes leads to greater porosity, a higher initial sapwood moisture content and an increase in drying rate. The improved permeability of the sapwood of refractory species like Douglas-fir (*Pseudotsuga menziesii*) and the spruces (*Picea* spp.) makes them more readily treatable. However, encrusted and lignified material resists enzyme degradation. Such material is found in the bordered pits of ray tracheids and to a degree in latewood. This effects radial penetration in roundwood of Douglas-fir (*Pseudotsuga menziesii*) but not its tangential penetration (Krahmer and Côté 1963). Allen et al. (1993) explored the use of mixed populations of bacteria (*Bacillus polymixa*, an *Enterobacter* and a *Coryneforme* sp.), which have low cellulolytic ability, together with a nutrient supplement in a basal solution of mineral salts, to optimise enzyme product and pit-membrane degradation. Again working with Douglas-fir, they suggested that, with incising to achieve deep radial penetration of bacteria, the bordered pits within the entire sapwood envelope could be degraded within a week or so. Heartwood remains impermeable. Even after storage for 5 years or more, the pits in the heartwood remain undegraded, probably because of the encrustation by polyphenolic materials.

Such biological control can protect against sapstain (Bernier et al. 1986). Once logs are removed from wet storage, the wood needs to be milled and processed quickly.

11.1.1.2 Relaxation of Growth Stresses

Whereas the literature is couched in terms of growth stresses, the experimental procedures measure growth strains in the standing tree, log or flitch and the corresponding growth stresses are only indirectly determined on isolated specimens in a tensile test. Typically, one might consider a strain of 0.5–1% or more to have a significant effect on wood properties. Growth stresses arise during deposition of lignin between microfibrils of the cell wall and during the spontaneous contraction of the newly formed microfibrils. According to Boyd (1985), the axial shrinkage accompanying lignification is the inverse analogy of shrinkage that accompanies drying, and both are crucially dependent on the microfibril angle; whereas Bamber (1978) favoured a mechanism invoking cellulose tension. Okuyama et al. (1986, 1994) offer a theoretical analysis and experimental data to unify these ideas.

In the long term, foresters have some scope to reduce the problem of severe growth stresses in key species, by selection of appropriate genotype, spacing/thinning, and log size, growth rate and age. In general, large stems present fewer problems as the peripheral growth stress is not sensitive to tree diameter, the peripheral strains remaining the same throughout the life of the tree. The longitudinal tensile stress is continuously generated in the newly formed cells of the cambium, with a balancing increase in compressive stresses around the pith. This means that the longitudinal growth-stress *gradient* is most severe in small-diameter, young, fast-grown plantation trees and decreases with stem diameter (Fig. 11.1). Large growth stresses are associated with a few well-known species of hardwood (*Fagus* spp., especially *F. sylvatica*, *Eucalyptus* spp. and *Quercus* spp.), whereas strains in conifers are relatively low but still noticeable amongst the *Pinus* spp. and *Cyptomeria japonica* (Kubler 1987). Large axial growth stresses can cause end-splitting of logs during or soon after felling, distortion during milling and severe shrinkage during drying.

In reaction-wood tissue, growth stresses can be very large in all species. They are an inevitable consequence of reaction-wood formation in leaning stems. Part of the difficulty with growth stresses is that they are so variable both between sites, within stands and within individual tress, with much of the latter variation being associated with reaction wood. Large growth stresses and resulting defects are greater in even-age stands where dominance shifts from

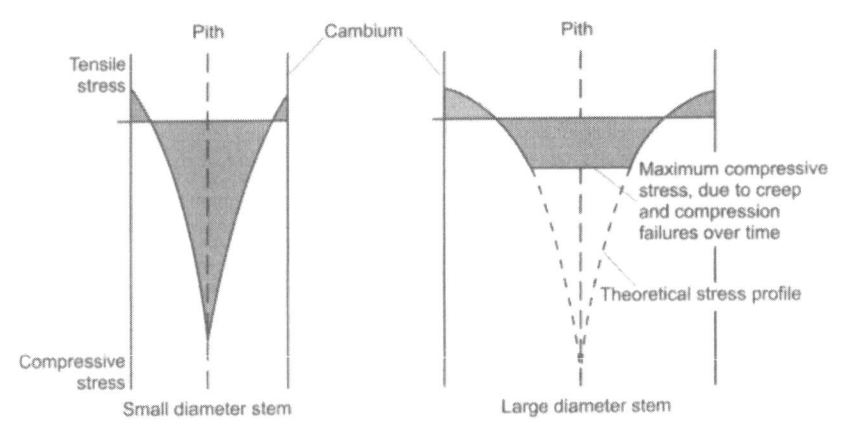

Fig. 11.1. Longitudinal growth stresses across a log. (After Boyd 1985)

tree to competing tree and there are changing stimuli at work on the live crown. Hitherto, in North America scarcely any hardwood has been used for framing because of bow and crook arising from growth stresses. On milling, the residual stresses are released, quarter-sawn boards spring/crook, while flat-sawn boards bow outwards as they are released from the log. Maeglin (1979) commented that in their trial about half of all aspen (*Populus tremuloides*) studs failed to meet industry standards for warp. Excessive downgrading of softwood studs cut from the juvenile zone is similarly influenced by instability arising from compression wood, spiral grain and high microfibril angle.

Industry has sought ways to process material subject to large growth stresses. End-splitting, if it is to occur, generally happens within a week of felling. Storage under water or under sprinklers is frequently recommended as a means of reducing growth stresses, as is ring-barking or partial defoliation some months prior to felling (Hillis 1978). It is unclear how stresses can relax in dead or drying stems whereas they are maintained over many years in living trees (Kubler 1987). Stress relaxation at elevated temperatures ($T > 80\,°C$) is understandable, as slip between macromolecules within the cell walls is possible, especially above the fibre-saturation point (Goring 1966).

11.1.2 Antisapstain Treatments

Sapstain, moulds and decay fungi breed under warm humid conditions, with many organisms preferring temperatures around 24–28 °C in temperate regions, and high temperatures in tropical countries. Moulds cause superficial discoloration, sapstain fungi penetrate throughout the entire sapwood zone, imparting a blue, green or black hue, whilst decay fungi destroy the cell walls and weaken the wood. Infected wood often displays stains in radially aligned, wedge-shaped zones throughout the sapwood, arising as a consequence of the utilisation of nutrients from foods stored in the living ray tissue. The wedge profile indicates that tangential migration is modest compared to radial penetration. Both moulds and sapstain fungi obtain nutrients from foods stored in the living ray tissue; neither utilise the cellulose and lignin and cause hardly any loss of strength. Such losses are insignificant except where wood is subject to impact (Scheffer 1973). Early studies suggesting considerable loss of toughness, such as that of Chapman and Scheffer (1940), may have been compromised by the presence of stain and rot fungi in the wood.

If logs are left in the forest or mill for longer than about 4 to 6 weeks during summer, they are likely to be infected already and no amount of chemical treatment will prevent sapstains developing in the lumber. This is especially true of pines. In Canada, apart from the west coast, harvesting has traditionally been in winter when the ground is frozen. Such logging cycles mean that considerable inventories accumulate with logs being stored for up to 10 months (Byrne 1997). Staining fungi can still grow at temperatures around freezing of water, but are arrested while the logs are actually frozen (Scheffer 1958). When these logs thaw, the stain fungi begin to grow actively again.

The greatest problems arise when the temperature is ideal and there are good drying conditions (Fig. 11.2). Recent work in New Zealand shows that radiata pine logs can be infected within 2–3 days (R. Farrell, pers. comm.). This is because sapstain fungi, such as *Ophiostoma*, *Ceratocystis* and *Sphaeropsis* spp., once established on the surface, find conditions within the logs and lumber ideal for colonisation: superficial drying accelerates sapstain penetration. There is no reliable correlation between the surface condition and the extent of internal sapstain; in other words, while sapstain can be superficial, the depth of penetration is uncertain and can permeate the entire sapwood region, given the right conditions.

The appropriate remedies are to minimise the time logs are left lying in the forest or mill yard prior to sawing and to pass the lumber through a propylactic spray or dip as it comes green off the saw. In the sawmill, the use of highly effective, broad-spectrum, long-lived fungicides, such as tetra- or pentachlorophenates often with boric acid, has declined through concern about the persistence of such chemicals in the environment and their contamination by dioxins, forcing a switch to safer but less effective chemicals, including boric acid, soda, quarternary ammonium salts and triazoles (Forintek 1985). However, by applying combinations, each having differing modes of action, it is possible to reduce the chance of tolerance developing and to achieve reasonable protection. Individual fungicides often target a limited range of fungal organisms. Hence there is the need to use products which are formulations incorporating more than one active ingredient. There are over a hundred sapstain fungi worldwide and their individual aggressiveness depends on the wood species, time of year among other factors. This has led to the development of numerous commercial antisapstain formulations. The requirement to use less aggressive chemicals emphasises the need for better quality control.

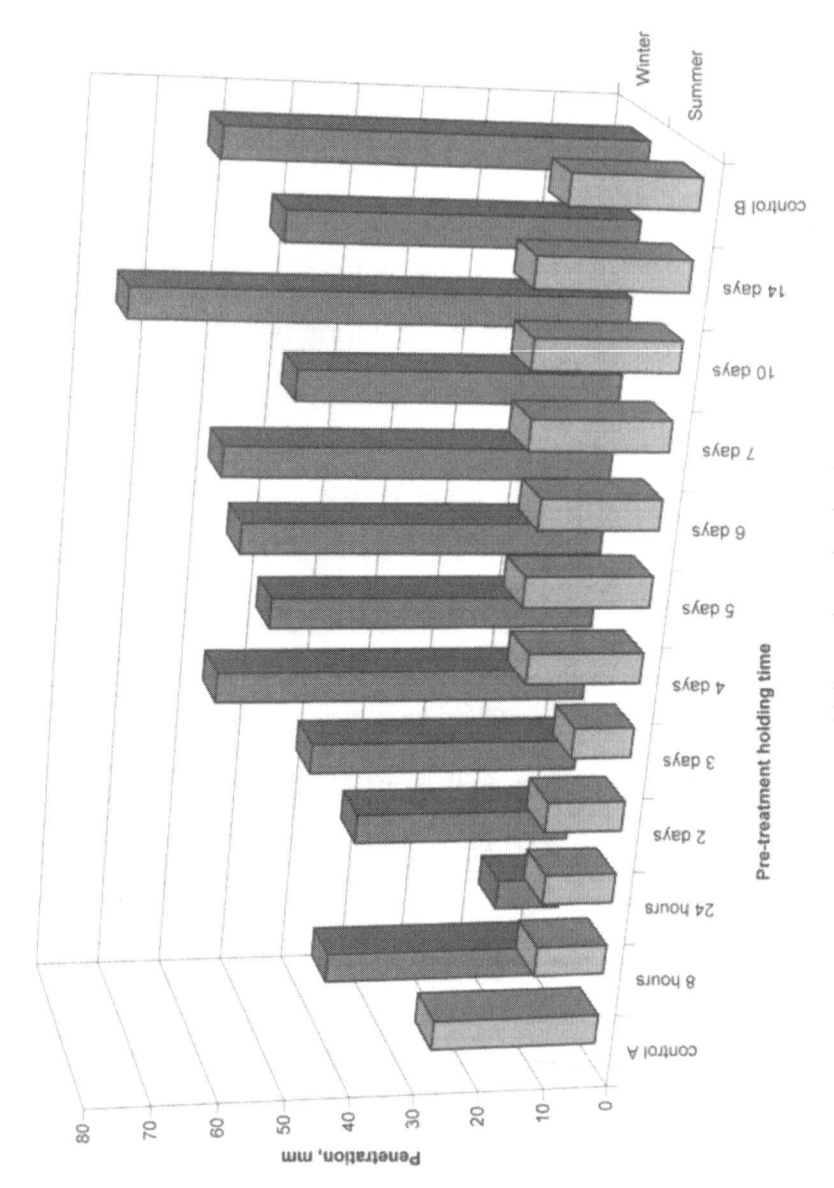

Fig. 11.2. Mean sapstain depth in roundwood billets eight weeks after felling: winter conditions 8 °C and 88% relative humidity; summer conditions 25 °C and 75% relative humidity. (After Eden et al. 1997)

Surface coverage by antisapstain treatments is superficial, offering only a thin protective envelope. With rough-sawn lumber the envelope is likely to be 1 to 2 mm at most, corresponding to the depth of the saw-damaged surface. Diffusible fungicides can offer a greater depth of protection, but greater soluble concentrations are needed to take account of the thicker envelope. In practice, the chemical must be sufficiently mobile to move into fungal cells but not so mobile that the concentration at the surface falls below the toxic threshold during the desired period. The requirements are demanding, as the antisapstain fungicidal system must control a diverse and large group of fungi and remain active for 3 to 6 months, long enough for the lumber to air-dry under relatively adverse conditions.

If the wood had become colonised while in the forest, a prophylactic spray at the mill may result in a superficial clean surface, but the interior sapwood could be badly stained by fungi. The antisapstain will then be effective only in the outer millimetre, as illustrated in Fig. 11.3, since the fungi will already have penetrated deeper.

Antisapstain treatment is essential year round in warm regions, but may be necessary only in summer months in colder climates. Traditional methods of application for sapstain control were by dipping individual pieces or whole packs of lumber. In recent years, more rigorous chemical containment has been introduced to eliminate contamination of the mill environment and groundwater from spillage and dripping timber. The treatment area requires roofing

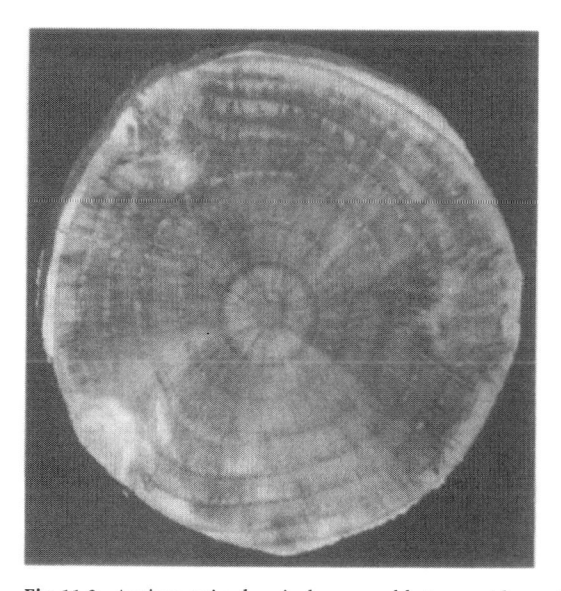

Fig. 11.3. Antisapstain chemicals are unable to provide control if the stain has already penetrated beyond the treatment envelope, which is liable to happen if timber is left too long between harvesting and milling. (Courtesy National Forest Library, Forest Research, Rotorua NZ)

and the floor must be backdraining and sealed. Where lumber was treated leaving the saws on the greenchain, the older curtain-type deluge necessitated recycling the solution. There were questions of the health of workers handling the wet timber, while the recycling of solution could lead to the stripping of the active ingredients by sawdust, the blocking of filters and accumulation of dirt in holding tanks. Modern spray booths give an even coverage independent of surface roughness. The use of low-volume spraylines delivering small charged droplets of solutions in oil means accurate transfer of metered amounts of chemical with less soaking of the wood. Many of the newer formulations are ionic solutions of water-dispersible compounds which require agitation, or emulsions. They may fix to the wood by cation exchange and can include surfactants and bonding agents that are resistant to leaching (Byrne 1992).

Future biocontrol methods may include the use of albino isolates of *Ophiostoma* species including *piliferum* strains (Cartapip™), which have shown promise in protecting red pine (*Pinus resinosa*) logs from bluestain (Behrendt et al. 1995). Kay (1997) reports considerably reduced sapstain on pine for periods as long as 100 days in field trials using pale-coloured mutant moulds such as *Trichoderma* spp. and in laboratory experiments with albino strains of *Ophiostoma* spp. Successful inoculation with any biocontrol agents, whether fungal or bacterial, involves colonisation of the surface, penetration and successful defence of its territory such as by antagonistic means or by sequestering all available nutrients. Eventually, succession by wood-decay fungi able to unlock the lignocellulosic complexes would occur, if the drying of lumber did not intervene (Wakeling 1997). Oxygenated monoterpenes, being constituents of extractives in some species, or some modification of them, offer prospects of a low-cost, natural product having considerable antifungal activity (Hill et al. 1997).

Both moulds and sapstain fungi increase the permeability of lumber, but this also allows greater uptake of moisture if the wood is rewetted during drying or at the building site, which, in turn, may favour colonisation by decay fungi and cause degrade. Once dried below a moisture content of $0.18\,kg\,kg^{-1}$, lumber is unsuited to colonisation, although existing fungi can become reactivated if rewetted. Even so, ophiostomatoid fungi will not grow on wood that has been dried and then rewetted (Seehann 1965).

There are thermophilic and heat-tolerant moulds which become active at temperatures between 35 and 57°C. These fungi, such as *Asperigullus fumigatus* and *Paecilomyces variottii*, can thrive under the controlled conditions in dehumidifier and heat-pump kilns. Whilst these fungi will not seriously degrade the lumber, they can cause unsightly discolouration. Quality control can be maintained by sterilising the kiln with a hypochlorite solution and steam sterilisation of fillets.

11.1.3 Brownstain Control

Brownstain is a visual blemish affecting lumber value for board grades, although such stain is of less concern for structural members. The discoloration is found in sapwood but not in impermeable heartwood, and often is only fully revealed after drying and dressing the boards. It is caused by the accumulation of water-soluble, mobile extractives at the wet-line or evaporative front early in the drying process. The evaporative front lies 0.5–1.0 mm below the surface at a distance corresponding to the base of the saw-damaged zone for a considerable part of the kiln schedule. The chemical and microbial action involves bacteria, feeding on wood sugars, producing enzymes which cleave various flavanoid-glucoside compounds and nitrogenous amino acids. These flavanoid fragments react together in the presence of oxygen at a neutral to mildly alkaline pH to form complex brown-coloured, tannin-like molecules. Non-microbial stains have been noted in various pines grown in the United States (*P. strobus, P. lambertiana, P. ponderosa*) where enzymes in living ray-parenchyma tissue react in similar ways (Schmidt et al. 1995).

Brownstain is especially relevant, as mills today are sawing more accurately and closer to the desired dimension. Further, the emphasis is on kiln-drying, with a market preference for light-coloured wood. While sapstain fungi can be controlled in field and mill operations, brownstain remains a problem: mills are not willing to revert to cutting oversize and planing off the discoloured zone. Brownstain in hemfir, western hemlock (*Tsuga heterophylla*) and amabilis fir (*Abies amabilis*) has been attributed to the oxidation of extractives, particularly catechin and epicatechin, to coloured chromophores (Hrutfiord et al. 1985), as well as to infection by dark-pigmented fungi (Kreber 1993).

A pale band under the stickers indicates that brownstain does not develop there. There is more limited evaporation here; immediately under the sticker mass flow within the board is as much along the grain as it is towards the surface; also the oxygen content may be limiting under the sticker.

The problem of brownstain is most acute and the stained zone is widest in moderate to high-temperature schedules when drying permeable softwoods. One solution, in the case of North American white and sugar pine (*Pinus strobus* and *P. lambertiana*), is to resort to milder schedules, especially with lower wet-bulb temperatures, initially 43/35 °C, where extractive solubility will be less, although often such schedules would normally be uneconomic. Alternatively, vacuum drying has been found to provide some control in beech (*Fagus* spp.) and oak (*Quercus* spp.) (Charrier et al. 1992; Kreber 1996). Vacuum-drying trials with pine have shown reduced incidence of brownstain at 90/60 °C and 70/60 °C, but created instead a more diffuse, grey coloration; even low partial oxygen, while again reducing brownstain, did not eliminate it (Wastney et al. 1997). Other potential solutions range from a dip-diffusional or pressure treatment with a chemical reducing agent.

11.2 Physical Methods to Improve Permeability

11.2.1 Incising

Incising is a traditional technique developed to increase the preservative uptake in impermeable roundwood and lumber (Ruddick 1987). The regularly spaced, thin cuts permit penetration of chemical to at least their roots, resulting in a uniform envelope of treated material. Incising is necessary if most North American species are to achieve the Canadian and United States standards for preservative treatment with water-borne chemicals (80% penetration of the envelope to a depth of 10 mm or more with a chemical loading of copper-chrome-arsenic salts of $6.4\,\text{kg}\,\text{m}^{-3}$). There are similar refractory species throughout the world: in Europe, the spruces (*Picea* spp.) would be amongst the more intractable. Historically, lumber was incised after drying. Incising green lumber has attracted more consideration recently and so the process becomes relevant to kiln-drying. Incising green wood results in a much improved surface appearance over dry-incised lumber, with the incisions being barely visible (Morris 1991).

11.2.2 Compression Rolling

There are two opportunities. As originally conceived by Cech and Goulet (1968), the green lumber was compressed by 5 to 20% between large, powered steel rollers. This drives the sap along the grain ahead of the nip. Furthermore, the high surface compression forces sap to move from the surface zone, either onto the surface of the board or further into the wood. The hydraulic flow of incompressible sap was thought to rupture pit membranes and render the sapwood more permeable, easier to dry and easier to treat with preservatives.

With hardwoods, it has been claimed that compression rolling can reduce the drying time for refractory species and/or ameliorate degrade during drying (Cech 1971; Cech and Pfaff 1975), with an initial claim (Goulet 1968) to savings of the order of 60% in drying time for yellow birch (*Betula alleghaniensis*). Subsequent application to several slow-drying hardwoods did not produce the hoped for improvement in drying time and behaviour (Grozdits and Chauret 1981; Günzerodt et al. 1986). These latter workers observed that physical damage in the hardwood *Nothofagus fusca* was confined to the axial sap-conducting tissue (primarily vessel walls, perforation plates and tyloses), while damage to transverse pathways (ray-to-ray walls and pits, and fibre-to-fibre pits) was limited to areas adjacent to vessels and so of little benefit for lateral flow of preservative. The effects on wood structure were dependent on the level of saturation. High levels of saturation (> 90%) led to extensive damage including delamination at ring boundaries, whereas at saturation levels of 50% the wood was less damaged but nor was the drying rate improved. Studies with softwoods have focussed almost exclusively on achieving deeper preservative penetration and higher chemical loadings. Here again results have been uncertain (Cooper 1973 vs Cech and Huffman 1970; Cech et al. 1974). Compression-rolling significantly reduced the local intensity of brownstain in the kiln-drying of *Pinus radiata*, with the colour precursors enriching less intensively a wider zone. The extent of shrinkage was increased as was, surprisingly, the drying time (Kreber and Haslett 1997).

Of more general interest is the observation that the strain is unevenly distributed, with high levels of localised strain near the surface (25–30%) and little strain (3–10%) within the majority (centre) of the boards. This offers an alternative process application, to see if mild compression might be used to provide a less superficial envelope of a chemical, such as an antisapstain formulation, than is liable to be achieved with brief dipping, curtain-coating or spraying of green lumber. Uptake could still be low, but the absorption envelope would be more uniform.

11.3 Low-Temperature Predrying

Drying in the open air can be very lengthy, ranging from months to several years, depending on lumber thickness, species and climate. As a broad generalisation, 25-mm-thick boards of American hardwoods can be dried in 50 to 200 days in the northern United States, while softwoods can be dried in 30 to 150 days, depending on the time of year. Thicker lumber dries even more slowly: doubling the board thickness more than doubles the drying time. Many hardwoods dry slowly, or very slowly; therefore a cardinal rule for drying such refractory species should be to saw or slice them to the smallest practical thickness beforehand, even if that means laminating together again after drying; although this is seldom done. Lumber dries more quickly in spring and early summer than lumber going into stacks in later summer, with drying almost ceasing during the winter at high and temperate latitudes. For example, in the upper midwest of the United States, Rietz (1972) indicates that there are only 15 effective air-drying days between 1 December and 28 February, and only 55 between mid-October and mid-April, but 165 in the remaining 6 months. Denig and Wengert (1982) have developed a simple air-drying calendar prediction based on historical weather data (relative humidity and temperature) and the initial moisture content of the wood.

With air-drying there is little control over the drying elements (wind and wind direction, temperature and sunshine, humidity and rain). The least satisfactory aspect is the exposure of lumber to weathering from alternate wetting and drying, with the surface fibres subject to severe stressing from swelling and shrinkage, aggravated perhaps by loss of permeability due to pit aspiration or tyloses. Hot, drying winds can result in excessive degrade such as surface checking; while humid, sultry weather with little air movement encourages the growth of moulds and sapstains. The increased value of lumber, high real interest rates and emphasis on just-in-time production have encouraged the trend away from air-drying and the building of greater kiln capacity.

However, even mild kiln-drying of green, impermeable hardwoods takes too long to be economic. Further, collapse and honeycombing in collapse-prone species can be mitigated by first air-drying until the cells at the centre of the boards are no longer saturated with water. In both cases, the slow air-drying period can be truncated by drying the lumber to around the fibre-saturation point before kiln-drying under more severe drying conditions.

From a practical viewpoint, collapse may be minimised by air-drying until the wood reaches about 25% moisture content and then kiln-drying; or one may kiln-dry from the green condition using a very mild, low-temperature schedule such that the collapse-susceptible temperature is never exceeded. Overall about 12% of all lumber, predominantly hardwoods, in the United States is air-dried prior to kilning (Rice et al. 1994). However, the emphasis on quality control in air-drying is resulting in greater use of covers, with shade-cloth being available in varying meshes, and drying in sheds with a wide overhang.

Terms such as forced-air drying, heated forced-air drying and low-temperature kiln-drying cover a spectrum of conditions with temperatures generally below 40°C. Vermass (1995) observed that predryers can be very large, holding up to 3500 m^3 of wood. Such systems offer drying for up to 24h a day with control over drying conditions, resulting in reduced inventory and much reduced degrade, which is especially relevant to hardwood drying for appearance grades. With such single-pass drying,. the fans are switched off when the humidity moves outside the desired band for acceptable drying.

Kiln-drying manuals observe that lumber which has been stickered for a week or longer is more prone to checking than freshly sawn material. Further, as a remedy, it is desirable to presteam before starting the actual kiln schedule. The reason for this lies in the imbalance between surface evaporation and mass transfer within the lumber at low temperatures. The rate of diffusion within the lumber at 20°C is one-tenth of that at 60°C and one-hundredth of that at 120°C, whereas the rate of evaporation is much less dependent on temperature, being determined primarily by the humidity potential as revealed in the wet-bulb depression. Therefore, at low temperatures and "favourable" drying conditions, there is great danger of steep moisture gradients developing at the surface of the lumber while the wood is waiting to go into the kiln. The stickered lumber is vulnerable to shallow surface checks which can extend into the middle of the board during kilning unless presteamed to relieve the surface stresses.

11.4 Heat Treatments

11.4.1 Steaming and Soaking in Hot Water

Traditionally, steaming has been used to sterilise wood against infection, to darken some species and to relieve initial drying stresses where stacks have been left around for some time prior to kiln-drying. Pretreating green lumber by steaming or soaking in hot water minimises the worst effects of collapse and honeycombing. The process may result in faster drying, increased permeability and some redistribution of extractives lining the cell walls (Kininmonth 1971; Alexiou et al. 1990). Such effects are in addition to the more obvious consequences, namely that preheating the wood prior to drying at a lower dry-bulb temperature provides the wood with some warmth for moisture evaporation.

Some wood hydrolysis is a consequence of high-temperature treatments that weaken the cell walls (Campbell 1961; Rosen and Laurie 1983). This reaction reduces the strength of the dried lumber. The severity of initial collapse after presteaming can be greater than that without pretreatment, despite the increased permeability of the wood (Greenhill 1938; Haslett and Kininmonth 1986). Thus steaming both increases the rate of drying (desirable) and the intensity of collapse (undesirable). The latter may be recovered subsequently by reconditioning. Unfortunately, collapse occurring at high temperatures is less completely recovered by subsequent steaming than is collapse at lower temperatures (Greenhill 1938). Moreover, recovered collapse in lumber still leaves the wood with closed, but "unhealed" checks which can reappear in service or, for example, during wood turning.

Both presteaming and high-temperature drying encourage the softening of lignin, the hemicelluloses and extractives, while thermal degradation is likely to involve acid hydrolysis from the release of acetyl groups on the hemicelluloses (Hillis 1975). The thermal softening point reflects the glass-transition temperature where the material moves from a brittle solid to a material that will stretch and flow. In green wood, the softening point for hemicelluloses is about 54–56 °C, compared with a softening temperature between 167 and 181 °C when oven-dry. Similarly for lignin, the softening points are between 72 and 128 °C and 127 and 193 °C, respectively, for the green and oven-dry condition. Cellulose, being crystalline, is not penetrated by water so its glass-transition temperature at 231–244 °C is unaffected by moisture (Goring 1966). The liberation of acetyl groups as acetic acid from the hemicelluloses weakens wood. Hardwoods with their higher acetyl content would appear more susceptible than softwoods in this regard. This progressive loss of strength is a function of moisture content, temperature and time, becoming more severe with increasing steaming temperature (Teichgräber 1966). This renders the wood more susceptible to collapse on drying. Partial hydrolysation of softwood pit membranes, deaspiration and improved permeability are features of steaming that have been reported for loblolly pine, *P. taeda* (Nicholas and Thomas 1968).

With the heartwood of *Eucalyptus pilularis*, steaming mobilises the redistribution of the continuous extractive film lining the walls of ray cells into small discrete droplets. The same effect is observed occasionally with any film lining fibres and pits. Alexiou et al. (1990) deduced that the film is unrelated to the warty layer as the latter occurs in sapwood where droplets did not form. Further, they concluded that some solubilisation of these heartwood extractives is necessary as globules are formed when steaming but not with dry heat. They attributed the 7 to 16% increase in drying rate of heartwood to improved access of water to the cell walls and enhanced diffusion. Kininmonth (1971), using both a scanning and transmission electron microscope, reached a similar conclusion for the hardwood *Nothofagus fusca*; 3 h of steaming improved the rate of diffusion, but not the permeability of the wood. The film of polyphenolic extractives, which forms a relatively uniform lining of cell lumens and in places in pit areas, was observed to blister or draw up into discrete rounded bodies, thus improving access to the cell wall without rendering the pit membranes permeable.

Hot soaking at temperatures as low as 70 °C has proved equally effective for drying the New Zealand southern beeches, *Nothofagus fusca* and *N. truncata*. Haslett and Kininmonth (1986) found that hot soaking to more than 70 °C halves the drying time for both species; that 70 °C is as effective as 100 °C; and that a short soak period is as effective as a longer period. A recommended schedule simultaneously warms both boards and water to the hot-soak temperature, holds that temperature for a further one to 4 h, depending on thickness (25 to 75 mm), before cooling the boards while still immersed (Haslett and Kininmonth 1986). However, hot soaking tends to increase collapse, so it is still necessary to steam at the end of the drying schedule to recover some of this loss of dimension. These southern beeches, like some eucalypts, are notoriously difficult to dry without excessive degrade, and the best current practice is to hot soak them before air-drying until the moisture content is below 40%, then to kiln-dry and finally to condition with steam. However, this schedule should be compared to the older traditional approach of air-drying for 1 year for every 25 mm of thickness, besides still having to tolerate considerable collapse and checking.

Finally, presteaming can be beneficial for lumber that has been filleted and then left around in the yard for more than a week or so. Presteaming adds a little moisture and plasticises the dried surface, thus reducing the likelihood of checking early in the kiln schedule.

Normally, presteaming in either a high-temperature kiln or, preferably, a steaming chamber is undertaken for between 2 and 4 hours. If kiln humidification is by sprayline, then a combination of heating coils (dry heat) and low-pressure steam injection can be used; or, in kilns lacking steam injection, the best results are by steaming from a boiling-water bath without resorting to heating coils. It is important that steam is fully saturated. Steam at 1 MPa has a temperature of 184 °C, so it can dry the lumber as it cools to 100 °C.

11.4.2 Dry Heat

Some warp and checking in lumber can be attributed to growth stresses in the original log: the steeper the gradient of longitudinal growth stress across the log diameter the greater the anticipated amount of crooking. In Japan, Tejada et al. (1977) have observed that significant stress relaxation was achievable with dry air heating of logs. Internal stresses were halved for larch (*Larix leptolepis*), cedar (*Cryptomeria japonica*), fir (*Abies sachalinensis*) and ash (*Fraxinus mandshurica*) after holding the temperature within the logs above 80 °C for 60–70 h. They were able to relieve stresses in both compression wood and normal wood, greatly reducing the stress gradient across the stem diameter. Both softening of lignin and solubilisation of the hemicelluloses are involved, with the latter being reduced to less hydrophilic polymers. These Japanese workers noted some darkening of the wood, but no significant reduction in bending strength. There are two interesting features of this study. First, the process resembles the saw-dry-rip (SDR) process which has been extensively promoted in the United States, mainly with hardwoods (Maeglin 1978; Maeglin and Boone 1983a,b; Maeglin and Adams 1990), but also with softwoods (Maeglin and Boone 1983a). In the SDR process, lightly edged, live-sawn cants and flitches are dried at high temperatures; this allows the stresses in the full width of the boards to relax in a symmetrical manner, so that lumber ripped subsequently from the dried cants and flitches remains relatively straight. Beauregard et al. (1992) suggested that the large, unrelieved longitudinal stresses found in the SRD flitches (and logs) would give rise to greater creep and stress relaxation than would occur in normal lumber. The second interesting feature of the Japanese study was the long time the logs were kept at temperatures above 80 °C to achieve stress relaxation, especially when one considers the relative brevity of the steaming period at the end of the kiln schedule.

11.5 Prefreezing

A complete, uniform freezing pretreatment attempts to minimise the severity of collapse through seeding of air bubbles in the water-saturated cell cavities. Freezing must be total to drive all the air out of solution. Once the wood is frozen, there is two-phase system of ice and air in the lumens. On thawing, these air bubbles will grow only if the release in energy due to relaxation of liquid tension exceeds the work required to create the increased surface area of the expanding bubble: this is the same argument as in Griffith's theory for brittle-crack propagation (Cottrell 1975, p. 410ff), where the incipient crack must exceed a critical value before it can grow catastrophically. Presumably, any reduction in collapse of the wood will be proportional to the proportion of cells which have been successfully air-seeded. With 25-mm stock, the phase transition from water to ice is complete within about four hours, with the freezing-point depression increasing gradually with time as the solute concentration within the residual water increases. To be effective, it appears that the temperature should be significantly below normal freezing point, lower than −7°C, with little benefit in holding at that temperature beyond 3 to 8 hours (Erickson et al. 1966). Since the adsorbed cell-wall water does not freeze at the depressed freezing point (Froix and Nelson 1975; Nanassy 1974), as determined by the solute content, then there is the possibility of enriching the water-soluble extractive content within the cell wall during prefreezing.

Haslett and Kininmonth (1986) observed no benefit of prefreezing when air-drying *Nothofagus fusca*, whereas hot soaking, which involved a disruption to the extractive film lining the cell walls, enhanced the rate of diffusion. Prefreezing, on the other hand, was found to be effective with *Sequoia sempervirens* (Erickson et al. 1966) in reducing collapse. With *Eucalyptus marginata* and *E. diversicolor*, air-drying was initially 65–50% faster than control boards without an increase in degrade, although the benefits were non-existent after 7 weeks of air-drying.

Attempts at artificially inducing bubbles in cells have failed to prevent collapse on drying. The use of sodium bicarbonate, introduced by diffusion, served only to increase collapse in *Eucalyptus regnans* (Kauman 1960). Similarly pressure impregnation for up to 17h with carbon dioxide gas at pressures of 5 to 25 MPa failed to prevent collapse in the same species (Kauman 1964), despite the release of pressure leaving the dissolved carbon dioxide in a supersaturated solution and free to form gas bubbles within the cells. Similar results were obtained with *E. delegatensis* by Jones (1993), who questioned whether nucleation of gas bubbles from the supersaturated state in wood cells was as easy as had been presumed. Also, collapse in *E. regnans* could not be prevented by applying ultrasonic energy to generate fluctuating, high local stress concentrations, so inducing cavitation and vapour bubbles (Kauman 1964).

11.6 Antishrink Chemicals

Antishrink chemicals act in two ways to reduce drying stresses. They lower the relative vapour pressure over the adsorbed solution in the cell wall, reducing moisture gradients, and they bulk the cell wall, reducing shrinkage. Water-soluble chemicals such as salts, sugars and polyethylene glycol have been used. Saturated solutions of these chemicals in the cell wall change the activity of the sap and lower the relative vapour pressure at which drying occurs. Shrinkage is reduced, depending on the amount of solute within the cell wall itself. Sodium chloride is only moderately soluble (the solubility is only 35.7 g per 100 ml of cold water) and so cannot reduce shrinkage by more than one-quarter. Further, such treated material feels damp and the salt corrodes most metallic fittings.

Only polyethylene glycol has been used to a moderate extent, and even here its use is principally in the preservation of water-logged or marine artefacts (Rowell and Barbour 1990). Kiln-drying is not involved, as the material is too weakened and susceptible to checking to allow anything but the very mildest drying. Indeed, polyethylene glycol is introduced gradually into the wood over a period of months or years depending on the size of the members being treated. Much of the cell-wall water is replaced by the polyethylene glycol, with the intention of minimising shrinkage: cross-sectional shrinkage is reduced to less than 1% when the glycol content of the solution exceeds 35%. Such artefacts may be partially decayed and the polysaccharide components may be extensively depolymerised. For hundreds of years, the slow degradation and depolymerisation of cellulose in these artefacts has resulted in the wood being far less anisotropic than when in its original condition, so that if it were dried conventionally the wood would not return to its original shape. Polyethylene glycol treatments, therefore, are designed to preserve the wood in its swollen condition (Stamm 1964; Rowell and Konkol 1987).

There is little reason for expecting to see such pretreatments being applied to lumber prior to kiln-drying. Of more interest is the potential for acetylation and related methods to block or cross-link hydroxyls within the cell wall, so preventing readsorption of moisture and wetting in dried timber and to retain colour, especially minimising the yellowing of white woods (Hon 1995).

11.7 Presurfacing

11.7.1 Problems with Surface Checking

Many small and latent fractures which act as points of weakness are created in a surface-damaged layer during sawing. On drying, stresses focus on these points of weakness, which can develop into surface checks. Checking is most strongly developed at emerging ray-tissue in flat-sawn lumber. Light planing to remove sawmarks from the surface prior to drying allows one to increase the drying rate, while reducing the susceptibility of lumber to drying defects; in the case of oaks, *Quercus alba* and *Q. falcata var. pagodaefolia*, the drying time can be reduced by about 25% (McMillen and Baltes 1972). Dressing the lumber also increases the capacity of the kiln in terms of the board length dried in a given charge by about 12%. Finally, surfacing to a uniform thickness is beneficial in controlling twist. This would be a desirable practice for warp-prone, juvenile softwood lumber in a mill which has poor size control with large sawing variation. A sawing variation of 2.5 mm in 25-mm stock is capable of resulting in a 10–15% difference in drying time, so not only are the thinner boards less restrained but they are also liable to be overdried (Simpson and Tschernitz 1980).

11.7.2 Problems with Moulds

Spores and volatile organic products from moulds on dried building lumber are of concern to health authorities (Rask-Andersen and Malmberg 1990). Mould growth is more intensive on the original undressed surface. With Scots pine (*Pinus sylvestris*), Terziev et al. (1996) attribute mould growth to the accumulation of low molecular weight sugars and nitrogenous compounds on the surface: they observed a far higher accumulation with kiln-drying at a dry-bulb temperature of 80 °C compared with that on air-drying. Their recommendation was to dress the lumber after kiln-drying rather than before.

11.8 Green Finger-Jointing and Cutting Blanks

Historically low-grade lumber with numerous defects has been uneconomic to kiln-dry as so much waste material has to be needlessly dried, to be cut out subsequently and burnt. In some remanufacturing grades, more than half of the lumber is waste, although more typically the figure would be one-third. Stacking and kiln-drying short lengths (known as *shook*) is impractical. Such material, however, provides the wood resource for green finger-jointing. Short lengths including offcuts can amount to 5% of a mill's production. Typically, this is converted to and sold as chip. Even the poorest grades with a high proportion of defects can supply shook for green finger-jointing.

Forintek in Canada has developed a process which consisted of cutting out defects in green lumber using a finger-jointer and predrying for 10–15 minutes until the moisture content of the fingers falls to about 10%, while the body of the lumber remains green. Finally, the shook is passed to a glue spreader and joint assembly (Dobie 1976). More recently, Forest Research in New Zealand has pioneered a catalysed fast-curing system (Parker 1994), with an adhesive (phenol-resorcinol-formaldehyde) and hardener applied to one set of interlocking fingers and an activator sprayed onto the opposite fingers. This *Greenweld* system can bond dry and green wood (from 12 to 180% moisture content) and cures at room temperature without radio-frequency radiation or direct heat (Fig. 11.4). With wood jointed green, the maximum strength is developed after it has been dried, although strength while green is adequate for general

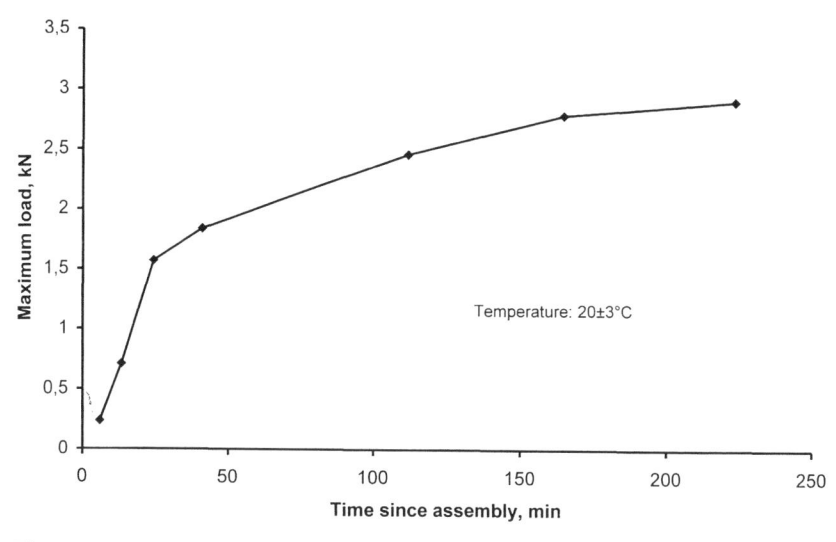

Fig. 11.4. Increase in strength of *Greenweld* finger-joints with time. (After Parker 1994)

structural uses: joints develop half of their final green strength in the first 30 minutes and full strength in 4 h. North American species including Douglas-fir (*Pseudotsuga menziesii*), southern yellow pine (*Pinus taeda*), black spruce (*Picea mariana*) as well as New Zealand *Pinus radiata* have been successfully trialled, with only some imported eucalypts and tropical hardwoods presenting major difficulties. The benefits which accrue include the production of a stable, full-length, premium-grade material, more uniform kiln-drying as there are no gaps due to short lengths within the stacks, the random distribution of jointed wood within each board, less distortion compared with lengths of low-grade lumber, and no wasted kiln production since the knotty, warp-prone part of boards are not needlessly dried. A further increase in the productive throughput of kilns is achievable by skip-planing prior to kilning.

Green finger-jointing is a partial alternative to the hardwood strategy of cutting blanks from rough-sawn, kiln-dried lumber (Araman and Hansen 1983; Hansen and Araman 1985). Short-length harvesting (<2.5 m), which minimises the effects of sweep and stem malformation in unmanaged hardwood stands, is viable because long lengths are rarely required (80% of pieces needed are less than 1.2 m and over 50% are less than 0.9 m). Small, defective hardwood logs are not cut for grade, instead they are live-sawn to give just two cants, 82.5 or 101.5 mm thick. These cants are gang-ripped to produce boards which are 25.4 or 31.7 mm thick. These boards are dried between smooth stickers, allowing them to crook with no effort being made to prevent this; by rejecting badly crooked boards future problems are minimised. However, the stacks are top-weighted to minimise cup and twist. The remaining boards are stress-free although still containing other defects. The worst defects in these standard-width boards (82.5 or 101.5 mm) are removed by cross-cutting to give one to four pieces of standard length. To simplify decision-making, only 4 out of 12 standard lengths are cut at any one time. Finally, each piece is ripped to yield a single cutting of a standard width (38, 51, 63, 76 or 89 mm). Alternatively, the material can be gang-ripped first and then cross-cut to give blanks which are slightly longer but narrower. The cuttings can be clear and defect-free or of frame-quality, admitting certain small defects. While these standard widths meet the majority of needs of furniture and kitchen-cabinetmakers, there are advantages in edge-gluing the standard-width blanks to full-width (660-mm) blanks. The purchaser can rip full-width blanks to the precise, narrower sections desired in his secondary manufacturing operation. There is one decisive marketing advantage in this approach: the buyer does not need to know or understand the vagaries of processing and grading, which differ markedly between countries. The buyer merely purchases edge-glued clear blanks of such standard dimensions as meets his stock requirements. Although furniture manufacturers use thousands of different component sizes and grades, by ordering an appropriate mix of standard blanks, trimming losses to convert to any final component should be less than 10%. This is a systematic approach to the effective use of US hardwoods with an abundance of defective trees and low-grade lumber.

The relative benefits of green finger-jointing or gang-sawing short logs depends on the latent severity of crook, cup and twist. Material with an inclination to exhibit severe crook would present more of a problem to a green finger-jointing system.

11.9 Precoating

As already noted, early work by McMillen and Baltes (1972) and others at the US Forest Products Laboratory demonstrated that faster drying (times reduced by an estimated 25%) and increased kiln capacity (up 12%) is achievable if oaks are presurfaced. More recently, Hart and Gilmore (1985) have reaffirmed that with presurfacing even the most refractory hardwoods, such as southern lowland oak (*Quercus falcata*), can be dried with much reduced surface checking. However, they observed that humidities above 90% are probably necessary during the early part of drying. They note that it is difficult to maintain such a high humidity without encountering condensation problems, and that with wide conventional stacks, such humid air quickly becomes saturated, with no drying at the outlet side of the stack. They offer an elegant, but commercially unattractive, solution. They proposed sandwiching refractory lumber between 8-mm (1/4-inch) plywood separated by 12-mm (1/2-inch) stickers. This enabled them to obtain more uniform drying across the stack and to use lower humidities in the airstream while maintaining high (unspecified) humidities at the lumber-plywood interface. Although Hart and Gilmore (1985) found the stack was virtually free of surface checking when lumber was presurfaced and

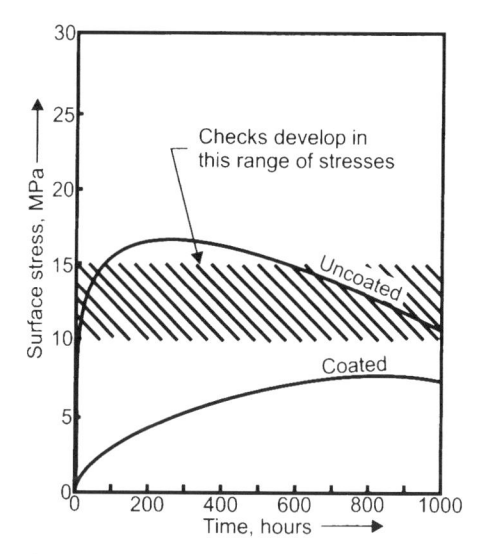

Fig. 11.5. Computed surface stresses in flat-sawn boards of Tasmanian ash (*Eucalyptus* sp.). (Schaffner and Doe 1984)

dried between plywood, as against only half of the boards without plywood inserts, they still found the wood to be susceptible to collapse. The latter was attributed to the large percentage (ca. 30%) of saturated, thin-walled ray parenchyma. However, the total schedule is complex, involving forced air-drying or dehumidifer drying to about 25% moisture content followed by kiln-drying at 60/32 °C (140/90 °F).

Checking within the first 2 to 4 h of sawing is the major cause of degrade in flat-sawn boards of *Eucalyptus regnans*, *E. delegatensis* and *E. obliqua* of the ash group of eucalypts. It is not feasible to maintain a local humidity high enough to prevent checking. Therefore, Doe and Schaffner (1982) have applied various semi-permeable coatings or films to maintain the surface moisture content above fibre saturation during the initial period of drying, thus avoiding a steep moisture gradient at the surface: the film reduces evaporation while still encouraging moisture transfer from the interior to the board surface. The film keeps the drying stresses below the critical threshold for surface checking (Fig. 11.5). The preferred system is a thick PVC film which adheres to the wet wood surface by spraying on a little liquid PVA.

11.10 Presorting

One of the more undesirable consequences of drying a mixture of heartwood and sapwood lumber, or of drying a mixture of species having varying drying rates and of drying material with slightly varying thicknesses, is that the faster-drying material within the stack is often overdried. A large variation in green moisture content and density within a kiln charge generally necessitates extended drying and conditioning. This is especially true of softwoods which are not often subjected to an extended equalisation period. Arganbright (1979), quoting from Knight and Cook (1960), noted that for *Pinus ponderosa* structural lumber the monetary loss from degrade increases steadily as the final moisture content falls.

Sorting within traditional species groups, sorting by density (Nielson and Mackay 1985; Zhang et al. 1996)) or sorting lumber into pure heartwood, pure sapwood or boards containing both heartwood and sapwood (Pang et al. 1994a) will allow optimisation of drying conditions for each kiln charge. This policy contrasts with the use of mixed charges, which risks overdrying with concomitant excessive degrade or prolonged, milder schedules to dry all the lumber under conditions that suit the slowest-drying material. The North American experience with hem-fir showed benefits by sorting into high and low basic density for western hemlock (*Tsuga heterophylla*) and amabilis fir (*Abies amabilis*) individually and for mixed hem-fir. Presorting is highly desirable considering the variation in density and green moisture content (Table 11.1). Analysis of the drying of sapwood pine boards of random density confirms the usefulness of such sorting in reducing final moisture-content variability among boards whenever there is a large range of green moisture contents in the charge (Nijdam and Keey 1999). By separating three-quarters of the boards with densities below $490\,kg\,m^{-3}$ in a range from 563 to $346\,kg\,m^{-3}$, the moisture content of the boards in the inner portion of the stack is estimated to be reduced by almost $0.02\,kg\,kg^{-1}$ to $0.12\,kg\,kg^{-1}$ at the end of the schedule. Presorting, however, does not diminish moisture-content variability due to any unevenness of the airflow through the stack.

In *Pinus radiata*, the moisture content of heartwood is in the range of 38 to 52% and of sapwood in the range of 90 to 200%: in the latter case, the value is highly density-dependant with low-density wood having a high moisture

Table 11.1. Presorting according to basic density and moisture content with hem-fir (Zhang et al. 1996)

Species: type		Basic density, $kg\,m^{-3}$			Moisture content, %		
		Min.	Av.	Max.	Min.	Av.	Max.
Hem-fir:	high density	416	447	566	25	69	123
	low density	286	376	414	21	62	136
Hemlock:	high density	435	472	563	28	59	116
	low density	316	395	434	24	64	156
Fir:	high density	349	386	496	25	49	113
	low density	229	325	349	26	58	137

content at saturation. The difference in moisture contents in heartwood and sapwood is exceptionally large and, with differing permeabilities, points to the desirability of presorting. Mixed-wood boards dry more slowly than boards composed entirely of heartwood or sapwood. When drying at high temperature (120/70 °C) to a moisture content of 6%, the drying times were calculated, and verified experimentally, to be 10 h for heartwood, 11 h for sapwood and 14 h for the mixed-boards (Pang et al. 1994a). In practice, the slowest-drying material consists largely of sapwood with only a very thin strip or zone of heartwood along one face of the lumber. Haslett (1998) states that heartwood requires less than half the drying time of sapwood. This different interpretation is likely to be the result of Haslett's sapwood including a little heartwood or the dry "intermediate zone" of the sapwood, both of which are highly impermeable, or it could relate to experience with wider kilns which would have a significantly higher moisture loading with sapwood boards. In mixed-wood boards, the capillary water in the saturated sapwood adjacent to the heartwood can either diffuse slowly through the thin heartwood barrier (no mass flow) or migrate through the entire thickness of the board to the opposite surface. Initial drying of such a heartwood-sapwood board is equivalent to a sapwood board which is twice as thick, since much of the free moisture evaporates from only one face of the lumber. Such boards are slower in drying than either pure heartwood or pure sapwood, which incidentally take about the same time to dry. Some companies are sorting timber according to heart, sap and mixed boards, and adjust their kiln schedules accordingly. Drying times are affected by both moisture content and transport coefficients. Sapwood of other pines is not necessarily as permeable as that of *Pinus radiata*; thus heartwood *P. ponderosa* might dry in 48 h while its sapwood dries in 80 to 90 h, despite the moisture contents of heartwood and sapwood being very similar for both species (ca. 0.45 kg kg^{-1} and 1.5 kg kg^{-1}, respectively; Arganbright 1979).

Drying of wetwood (see Chap. 7) poses particular problems. To give a single example, Smith and Dittman (1960) compared the drying times for 47.6-mm heartwood, sapwood and wetwood of white fir (*Abies concolor*) under a 71/63 °C schedule. Wetwood, as illustrated in Table 11.2, takes much longer to dry.

Table 11.2. Drying to 20% moisture content of 47.6-mm (1⅞-inch) white fir (*Abies concolor*) under a 71/63 °C (160/145 °F) schedule

Wetwood		Sapwood		Heartwood	
Green moisture content (kg kg^{-1})	Drying time (h)	Green moisture content (kg kg^{-1})	Drying time (h)	Green moisture content (kg kg^{-1})	Drying time (h)
1.93	158	1.45	84	0.57	42

Such differing rates of drying are species-specific. For those species with large differences in drying time, options for management (Ward and Pong 1980) include:

- Overdrying non-wetwood stock with degrade of overdried material during planing, such as splitting of cupped material.
- Sorting and redrying underdried material after a normal kiln run.
- Sorting stock prior to kiln-drying.

Undetected, underdried pockets or streaks of wetwood may result in collapse subsequent to dressing the boards, or steam pressure could rupture the boards if the latter are laminated or edge-glued using radiofrequency heating.

Commercial systems to sort lumber according to the green moisture content include measuring the temperature rise at the surface after the application of a fixed quantity of infrared energy (Van Handel 1990) or by impingement with a laser beam (Jamroz and Tremblay 1991).

Variations in density are treated somewhat differently. Ideally, one would like to know the basic density beforehand. For example assuming 90% saturation, then sapwood with a basic density of $350\,kg\,m^{-3}$ would have a moisture content of about $2\,kg\,kg^{-1}$, whereas at a basic density of $500\,kg\,m^{-3}$ the moisture content would be only $1.2\,kg\,kg^{-1}$. Encouragingly, with southern pine lumber (Taylor and Won Tek So 1990) it has proved viable to sort lumber according to green density, with denser material requiring longer to dry. As a result, there is less moisture-content variation within the dried stacks, as predicted by Nijdam and Keey (1999) for *Pinus radiata* sapwood boards of variable green density.

12 Less-Common Drying Methods

The majority of lumber-drying kilns are worked at atmospheric pressure and are direct-fired or steam-heated, sometimes using wastewood as the primary fuel. However, other heating arrangements have been advocated, including the use of solar, microwave and radiofrequency energy, or of superheated steam, besides the possibility of working under vacuum to reduce kiln temperatures with heat-sensitive wood. These less-common drying methods are reviewed in this chapter.

12.1 Solar Kilns

The use of solar radiation to deliberately dry materials is as old as civilisation. In 1984, Imré reviewed the application of solar dryers to a wide variety of drying problems, listing some general advantages of using solar energy as its "free" nature (although there are costs in collecting and using it), its availability in remote locations, the absence of monopoly of its use, and the lack of polluting effect. Listed disadvantages include its intermittent nature, dependence on time, season and weather, and the relatively low energy flux compared with conventional energy sources. One feature of the operation of some solar kilns is that the air inside the kiln can become saturated during the night with the fall in the outside-air temperature; this can lead to some condensation on the lumber to provide a diurnal conditioning which enhances product quality with impermeable wood (Langrish and Keey 1992).

Solar kilns for drying lumber may be defined as "structures which totally enclose the lumber to be dried and which derive some or all of the energy needed for drying from solar energy which is trapped for this purpose. They differ from air drying or forced air drying where the lumber is not totally enclosed." (Plumptre pers. comm.) Drying kilns which trap and use solar energy to dry lumber have been intensively researched only since Johnson (1961) built a small kiln in the late 1950s. Since then, a large number of experimental kilns have been built in a wide range of locations. Some of the experimental kilns are being used commercially.

Imré (1995) has classified solar dryers into three main groups, based on the energy sources used:

272   Less-Common Drying Methods

1. *Solar natural dryers* using ambient energy sources only.
2. *Semi-artificial solar dryers* with a fan driven by an electric motor, which ensures a continuous flow of air through the drying space.
3. *Solar-assisted artificial dryers* which may use an auxiliary conventional energy source if necessary.

Solar natural dryers include those in which the airflow is driven by natural convection (passively) and those in which fans driven by solar cells or small wind turbines give some forced convection. These dryers have mainly been used for the drying of loose vegetable materials and grains.

Most solar dryers used for lumber drying come into Imré's classification of *semi-artificial solar dryers* (*room dryers*), although some have been reported with dehumidifiers and other conventional energy assistance (*solar-assisted artificial dryers*). An example of a solar-dehumidifier kiln with forced-air recirculation is shown in Fig. 12.1.

Plumptre (1989, pers. comm.) has suggested another classification:

1. *Greenhouse kilns* where the solar collectors are within a structure which traps the heat in the same chamber as the wood. These may be further classified according to whether:
   a) the air circulation within the kiln is natural or forced by fans;
   b) venting relies on manual opening and closing of windows vents, or whether venting is also fan powered and controlled by humidity, temperature and/or external conditions such as sunlight; and

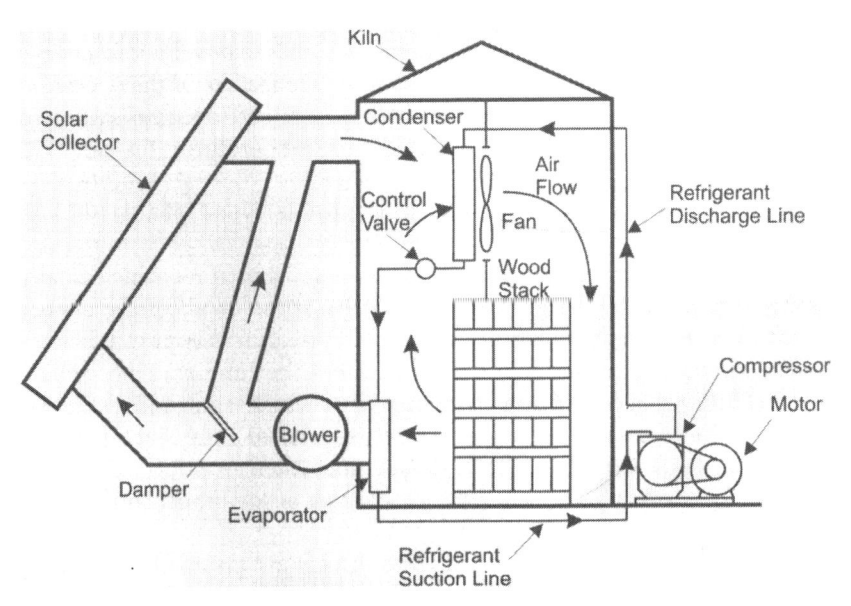

Fig. 12.1. A solar-dehumidifier kiln. (Chen et al. 1982)

c) there is insulation or glazing on the roof, walls, or floor to reduce heat loss.

Examples of greenhouse kilns include those described by Johnson (1961), Rehman and Chawla (1961), Maldonado and Peck (1962), Peck (1962), Chudnoff et al. (1966), Troxell and Mueller (1968), Casin et al. (1969), Wengert (1971), Sharma et al. (1974), Singh (1976), Bois (1977), Gough (1977), Plumptre (1979), Yang (1980) and Langrish and Keey (1992).

2. *External collector kilns* have an insulated collector in a separate structure which is connected to the kiln by insulated ducts. The collectors may be filled with either air or water. There are variants where, for example, there are box-type collectors either adjacent to the lumber chamber or within the same structure which can be isolated from the lumber chamber.

Examples of external collector kilns include those described by Read et al. (1974), McCormick and Robertson (1977), Little (1979), Lumley and Choong (1979), Chen et al. (1980) and Simpson and Tschernitz (1984).

Overall, greenhouse kilns appear to be more popular, possibly because of their greater simplicity, an important factor given that these kilns are usually operated by unskilled labour.

There are also a number of extra features that have been added to some of these designs, as follows:

1. *Heat-storage systems* including:
   a) *rock piles* beneath kilns through which hot air from the collectors is pumped during the day and from which hot air is taken at night (Read et al. 1974);
   b) *masonry partition walls or floors* (Gough 1977; Yang 1980);
   c) *heat-storage tanks* containing water (McCormick and Robertson 1977).
2. *Microprocessor control* of vents and air circulation (Steinmann et al. 1980, 1981, Steinmann 1989).
3. *Water-condensation devices* to reduce the amount of air vented and to remove water from the system as a liquid rather than losing the heat of vaporisation with the vapour (Lumley and Choong 1979).
4. *Solar-dehumidifier kilns* which are solar kilns fitted with a dehumidifier unit to reduce the humidity of the air circulated to the lumber and to return the heat obtained from the condensed water vapour to the system (Chen et al. 1982).

12.1.1 Insolation Rates and Kiln Locations

The daily world-average solar radiation or *insolation* on a horizontal surface is $3.82\,\text{kW}\,\text{h}\,\text{m}^{-2}$ (McDaniels 1984). The maximum flux is around $1\,\text{kW}\,\text{m}^{-2}$. Some data on the insolation rates in some tropical countries are given in

Table 12.1. Total daily horizontal solar insolation and sunshine hours for some tropical countries. Reprinted from Imré 1995 in Handbook of Industrial Drying 2nd edn, (ed) AS Mujumdar, by courtesy of Marcel Dekker Inc

Country	Daily average insolation ($kW\,h\,m^{-2}$)	Daily sunshine hours (h)
Cameroon	3.8–5.5	4.5–8.0
Egypt (Cairo)	6	9.6
Guatemala	5–5.3	–
India	5.8	8–10
Indonesia	4.24	–
Kenya	5.25–5.6	6–7
Malaysia	4.41	–
Mali	4.34	8.4
Mauritius	4.5	7
Mexico (Jalapa)	4.65	–
Nicaragua	5.43	–
Nigeria	3.8–7.15	5–7
Papua New Guinea	4.6–9.6	4.5–8
Philippines (Manila)	4.55	–
Sierra Leone	3.4–5.3	3–7.5
Thailand	4.25–5.66	–
Togo	4.4	5.5–7.2

Table 12.1, which shows that they are generally much higher than the world average.

Given this information, it might be expected that solar kilns would be more popular at low latitudes, where the insolation rates are highest. On the other hand, Plumptre (1989, pers. comm.), reviewing 35 solar kiln designs, notes that the location of these kilns is spread almost uniformly over the range of latitudes from 0 to 50°. This uniform distribution might be a reflection of research activity in temperate regions, however, rather than a consequence of the rate of solar drying relative to that in the open air being similar over a wide range of latitudes.

12.1.2 Absorbers

Most heat absorbers in solar kilns have used metal surfaces, painted matt black, to absorb solar energy and to transfer it to the air flowing over the absorber surface. Selective absorbers have not been widely reported, possibly on the grounds of cost. Configurations including flat, corrugated and vee surfaces have been used, together with perforated metal screens and honeycomb-patterned surfaces. Tschernitz and Simpson (1979b) have employed a bed of powdered charcoal as an absorber.

12.1.3 Glazing

The materials used for glazing have included glass and a variety of plastic films or rigid sheets:

a) *Horticultural polythene* (Plumptre 1979).
b) *Mylar* (Prins 1981).
c) *Polyvinyl fluoride* (*PVF*) (reviewed by Plumptre 1979).
d) *Polyvinyl chloride* (*PVC*) (Gough 1977).
e) *Polyester* (Peck 1962; Plumptre 1967).
f) *Resin-reinforced fibreglass* (Troxell and Mueller 1968).
g) *Polymethylmethacrylate* (Aleon 1979).

Glass is heavy, costly and durable, but it is easily broken unless it is in thick, heavy sheets. Tempered or laminated glass is stronger and, in the case of tempered glass, not much more expensive, if it is available. Optically, glass is much better than most plastics since it is transparent to most shortwave radiation and opaque to longwave radiation. Most plastics are less opaque to longwave radiation. Agricultural polythenes are very poor relative to glass, while PVC, PVF and polyesters are intermediate between polythene and glass, with some polyesters approaching the performance of glass.

All the plastics deteriorate slowly in ultraviolet light, becoming brittle. Horticultural polythene has a life of only about 1 year in the tropics (2 to 3 years in temperate regions), while the others vary up to a maximum of 5 to 6 years in the tropics.

12.1.4 Temperature Control

A thermal storage unit in the form of a rock pile was used by Read et al. (1974) to reduce (but not completely remove) the size of the diurnal temperature variations. The separate box-type collector used by Lumley and Choong (1979) enabled them to prevent night-time losses from the collectors, allowing the rock pile to act as thermal storage mass. They did this by using a differential thermostat and switched valves which connected the collector and the kiln when the collector temperature exceeded that in the kiln. Hot-water storage was also used to regulate the temperature by McCormick and Robertson (1977).

12.1.5 Humidity control

For humidity control, the majority of solar kilns reported in the literature have relied on manual operation of vents or semiautomatic operation, with the manual setting of humidistats which have operated vents according to the moisture content of the lumber. For example, in Read et al. (1974), a differential thermostat operated vents when the wet-bulb depression in the kiln fell below a selected value. This value was manually adjusted as the moisture content changed. In Troxell (1977), Rosen and Chen (1980), Simpson and Tschernitz (1984) and Langrish and Keey (1992), a hair hygrometer was used to operate venting fans when the relative humidity in the kiln fell below a selected value, with this set-point being manually adjusted as the moisture content decreased.

Many workers (including Chudnoff et al. 1966, Plumptre 1979, Tschernitz and Simpson 1979 and Langrish and Keey 1992) suggest that venting should be restricted in the initial stages of drying in order to keep the relative humidity inside the kiln high, as is usually done in conventional drying schedules. This practice, which can lead to the same or lower initial drying rates in the kiln compared with those in the open air (Gough 1977; Sharma et al. 1974; Simpson and Tschernitz 1984), has been given as a significant reason for the good quality of the lumber dried in solar kilns compared with that from open-air drying.

Humidification using water sprays was used by Chudnoff et al. (1966) and Tschernitz and Simpson (1979b). The absence of a humidification system was suggested as a reason for the severe case-hardening found in some samples of lumber dried in a solar kiln by Casin et al. (1969), who suggested that a humidification system involving heated pans of water or water spray nozzles might be useful for conditioning the lumber at the end of drying.

Recovery of the latent heat of vaporisation through condensation of outgoing water vapour was suggested by Lumley and Choong (1979), who estimated (by computer simulation) that energy losses from venting, particularly in regions of low ambient temperature or high humidity, may involve up to 25% of the solar energy input.

12.1.6 Air Circulation

Distinction needs to be made between fans used for enhancing air circulation and those used for venting. Venting fans are described in the section above (on humidity control). Maldonado and Peck (1962), Peck (1962), Gough (1977) and Langrish and Keey (1992) all used the circulation fans during daylight hours only, since operating them at night does not result in any additional solar gain. On the other hand, a continuous airflow through the stack, even with the cooler air at night, was suggested by the simulations of Taylor and Weir (1985) to be beneficial in reducing the overall drying time, in spite of the additional energy loss from the kiln walls caused by circulating air around the kiln at night.

12.1.7 Energy Losses

Plumptre (1989, pers. comm.) has reviewed different estimates of energy losses from solar kilns, and his table is reproduced in Table 12.2. Further information about the operating conditions described in Table 12.2 is given in Table 12.3.

Most definitions of kiln efficiency, as incorporated in Table 12.3, calculate this parameter in the following way:

$$Efficiency(\%) = \left\{ \frac{heat\ to\ warm\ lumber + sorptive\ and\ evaporative\ heat}{total\ heat\ input} \right\} \times 100$$

The difficulty with this definition is that it is dependent on the moisture content of the lumber (which strongly influences the drying rate), with the kiln efficiency as defined in this way decreasing to zero when the lumber reaches equilibrium. The efficiencies reported in Table 12.3 are therefore only indicative in terms of indicating differences between kiln designs, as they are influenced by the moisture content, latitude, seasonal variations, climatic variations (including temperature, humidity, cloud cover, insolation and wind speed), species of lumber (density, drying behaviour), and thickness of lumber.

Table 12.2. Heat losses from three solar kilns and one conventional kiln (Plumptre 1989, pers.comm.)

Energy Losses and uses (%)	Kiln	Tschernitz and Simpson (1979)	Troxell and Mueller (1968)	Plumptre (1979)	Conventional kiln
	Location	Madison, USA	Colorado, USA	Oxford, UK	
	Worker	Win Kyi (1983)	Wengert (1971)	Prins (1981)	Skaar (1977)
Evaporation of water from wood		21.3		4.9	45.3
Hygroscopic water (water of sorption)		0.3	16.0	0.2	0.9
Energy to heat lumber load and kiln structure		2.0		0.5	4.0
Ventilation loss		35.7	14.0	9.0	16.7
Conduction/ convection losses		13.5	40.0	33.5	33.1
Floor losses		11.0	16.0		
Radiation losses		27.2	30.0	51.9	
Total		100.0	100.0	100.0	100.0

Table 12.3. Details of the solar kilns surveyed in Table 12.2 (Plumptre 1989, pers.comm.)

Kiln	Tschernitz and Simpson (1979b)	Troxell and Mueller (1968)	Plumptre (1979)	Conventional kiln
Location	Madison, USA	Colorado, USA	Oxford, UK	
Worker	Win Kyi (1983)	Wengert (1971)	Prins (1981)	Skaar (1977)
Kiln type	External collector	Fibreglass-covered greenhouse	Polythene-covered greenhouse	
Mean moisture content of lumber during drying (%)	26.3	–	18.3	–
Capacity of kiln for lumber (m^3)	2.8	2.8	7.1	–
Time of year	June	Autumn	August	–
Latitude	43°N	41°N	52°N	–
Species	*Acer saccharum*	*Pinus contorta* /*Picea engelmannii*	*Quercus robur*	–
Approximate density of species (kg m^{-3})	650	400	690	–
Thickness of lumber (mm)	30	not stated	50	–
Average efficiency (%)	23.6	16.0	5.6	50.2

12.1.8 Economics

Imré (1984) considered that semiartificial solar dryers should be compared with conventional dryers having the same performance. Particular advantages of solar dryers in this comparison include reduced capital costs due to simple construction and the substitution of solar energy for conventional energy. Solar kilns are also frequently compared with drying in the open air, for which higher productivities (cost of land and capital) and protection against the weather and insect attack are two advantages.

The payback time (n) is the time at which the sum of the investment (C) and the annual expenses (mC) with compounded interest is equal to the total savings (S) with compounded interest gained by the use of solar energy rather than conventional energy, as expressed below:

$$C(1+r)^n + mC\frac{(1+r)^n - (1-i)^n}{r-i} = S\frac{(1+r)^n - (1-i)^n}{r-i}, \quad (12.1)$$

where r is the interest rate on capital, i is the inflation rate for annual expenses and e is the energy inflation rate. This equation may be solved for the payback time n:

$$n = \frac{\ln\left[1 - \left(\dfrac{r-i}{S/C - m}\right)\right]}{\ln\left(\dfrac{1+i}{1+r}\right)}. \tag{12.2}$$

For example, when $r = 0.055$, $m = 0.05$, $i = 0.05$ and $e = 0.09$, $S/C = 0.24$ and the annual savings are 24% of the initial cost.

The savings (S) include components due to the annual displaced (conventional) energy costs, savings gained by avoiding environmental pollution or neutralising the polluting effects of dryers which use polluting energy sources (either directly or indirectly), savings in avoiding transportation costs and savings through better quality.

The costs (C) include components due to the collector, the supporting structure, the dryer shell, installation, and the heat storage, dehumidifier and control system if present.

Imré indicated that simple structures tend to give the fastest payback times (0.5 to 3 years), particularly when

- the material to be dried needs mild, low-temperature drying;
- does not demand continuous operation; and
- the task of drying ensures the utilisation of the dryer for a long period of the year.

The first condition is often true for the drying of hardwood lumber, while small lumber producers can sometimes tolerate the lack of continuity in operation. The third point depends strongly on the geographical location.

Imré (1995) felt that "heat pumps or other supplementary energy converters are generally very costly" and that "complex and connected energy systems are complicated: their operation and maintenance costs are higher".

12.1.9 Mathematical Modelling of Performance

A lumped-parameter model of the solar-kiln design proposed by Gough (1977) was reported by Taylor and Weir (1985), who aggregated all the kiln temperatures into one parameter. Energy losses were also described by one heat-loss resistance, that between the inside kiln temperature and that of the outside ambient air. This model was used to predict the effects of varying the mass of the kiln structure (by adding a rock pile), venting rate, fan-switching policy, weather, glazing, size of kiln and load, and species. The lumber-drying kinetics were described by an empirical first-order equation, fitted to the drying of mahogany, with a density correction for other species. However, the absence of detailed consideration of heat losses from the walls makes any extrapolation of this model to different kiln geometries difficult.

In the external collector-type kiln which he used, Pallet (1988) did not model the heat-transfer performance independently of the mass-transfer behaviour, and the drying behaviour was fitted to an empirical seven-parameter equation. Hence, it is difficult to establish whether the good agreement between the model and the experimental data for the drying rate and the air temperature was due to the large number of fitting parameters or to the validity of the heat-transfer model. This heat-transfer model included a single, lumped, overall heat-transfer coefficient between the air inside the kiln and that outside it, which included all convective and radiative effects. No allowance was made for the likely variations in this coefficient due to the effects of outside wind speed (on the convective coefficients) or of the effects of internal temperatures (which varied from 20 to 55 °C) on the radiative coefficients. Since these detailed effects were not modelled, and the effects would be expected to be different if the kiln geometry was altered, the approach used by Pallet (1988) is not immediately applicable to assessing the effects of changes in geometry except in a very approximate way. Nevertheless, the air temperature above the solar collector was distinguished from that in the chamber containing the wood and that of the lumber, making this model more of a distributed-parameter one than that of Taylor and Weir (1985).

The performance of a natural-convection, induced-draft, solar kiln for processing a packed bed of agricultural products was simulated by Zahed and Elsayed (1989). No experimental results were reported, and only a parametric study was made. Surprisingly, heat losses from the equipment appear to have been neglected altogether.

12.2 Dielectric Drying

Schiffmann (1995) defines dielectric (radiofrequency) frequencies of electromagnetic radiation as covering the range 1–100 MHz, while microwave frequencies range from 300 MHz to 300 GHz. The devices used for generating microwaves are called magnetrons and klystrons.

12.2.1 Mechanisms of Heating

12.2.1.1 Dipolar Rotation

Water molecules are dipolar in nature (that is, they have an asymmetric charge centre), and they are normally randomly orientated. The rapidly changing polarity of a microwave or radiofrequency field attempts to pull these dipoles into alignment with the field. As the field changes polarity, the dipoles return to a random orientation before being pulled the other way. This buildup and decay of the field, and the resulting stress on the molecules, causes a conversion of electric field energy to stored potential energy, then to random kinetic or thermal energy. Hence, dipolar molecules such as water absorb energy in these frequency ranges. The power developed per unit volume (P_v) by this mechanism is

$$P_v = k E^2 f \varepsilon' \tan\delta = k E^2 f \varepsilon'', \qquad (12.3)$$

where k is a dielectric constant, depending on the units of measurement, E is the electric field strength (V m^{-3}), f is the frequency (s^{-1}), ε' is the relative dielectric constant or relative permeability, tan δ is the loss tangent or dissipation factor, and ε'' is the loss factor.

The field strength and the frequency are fixed by the equipment, while the dielectric constant, dissipation factor and loss factor are material-dependent. The actual electric field strength is also dependent on the location of the material within the microwave/radiofrequency cavity (Turner and Ferguson 1995c), which is one reason why domestic microwave ovens have rotating turntables so that the food is exposed to a range of microwave intensities.

12.2.1.2 Ionic Conduction

There is also a heating effect due to ionic conduction, since the ions (sodium, chloride and hydroxide) in the water inside lumber are accelerated and decelerated by the changing electric field. The collisions which occur as a result of the rapid accelerations and decelerations lead to an increase in the random kinetic or thermal energy of the material. This type of heating is not significantly dependent on either temperature or frequency, and the power developed per unit volume (P_v) from this mechanism is

$$P_v = E^2 q n \mu, \qquad (12.4)$$

where q is the amount of electrical charge on each of the ions, n is the charge density (ions/m^3) and μ is the level of mobility of the ions.

12.2.2 Interactions

12.2.2.1 Moisture Content

Schiffmann (1995) notes that the dielectric constant of water is over an order of magnitude higher than most base materials (such as cellulose in wood), while the overall dielectric constant of most materials is usually nearly proportional to moisture content up to a critical value, often around 0.2 to

0.3 kg kg^{-1}. Hence, microwave and radiofrequency methods preferentially heat and dry wetter areas in most materials, a process which tends to give more uniform final moisture contents and is one attraction of the technique.

12.2.2.2 Density

The dielectric constant of air is very low compared with that of water, so a lower density usually means lower heating rates.

12.2.2.3 Temperature

For water and other small molecules, the effect of increasing temperature is to decrease the heating rate slightly, hence leading to a self-limiting effect.

Other effects (frequency, conductivity, specific heat capacity etc) are discussed by Schiffmann (1995), but are not very relevant to the drying of lumber because the range of available frequencies (which do not interfere with radio transmissions) is small, and some properties of lumber such as specific heat capacity are not highly variable.

12.2.3 Internal Pressures

A frequent feature in the literature is the suggested use of vacuum in combination with radiofrequency and microwave-power inputs. In part, this practice stems from the difficulty in achieving convective heat input under vacuum-drying conditions where the air density is low, and here microwave/radiofrequency power input is convenient. Another reason for the use of vacuum in combination with microwave and radiofrequency drying is the need to avoid generating in the lumber high internal pressures which may develop on drying at atmospheric pressure. There is anecdotal industrial evidence that high internal pressures can be created by microwave/radiofrequency drying and that these pressures may be sufficient to damage boards of lumber (Carter Holt Harvey Forests Ltd NZ 1992, pers. comm.).

Rozsa (1994), when drying an Australian hardwood, mountain ash (*Eucalyptus regnans*), found that the effect of radiofrequency drying at absolute pressures of 1 kPa compared with 10 kPa was to increase the initial drying rate twofold at the lower pressure, and there was a rapid rise in the core temperatures at the higher pressure (10 kPa) even at low power inputs. In general, this situation might lead to damaging pressures inside the lumber. Internal pressures of up to 60 kPa above the pressure outside 50 × 100 × 1660-mm boards of birch (*Betula* sp.) were reported by Antti (1992a,b) for power inputs of

1250 W. One complication in the use of vacuum drying is the need for an expensive pressure vessel as a drying chamber, so increasing the capital cost.

Control of the core temperature below 40 °C was found by Rozsa (1994) to be necessary for good quality when drying 30-mm-thick mountain ash (*Eucalyptus regnans*). A power input of 50 W for samples which had an initial mass of 0.5 kg gave a drying time of 17 h from moisture contents of 1.1 to 0.2 kg kg^{-1}, while 100-W power input gave a drying time of 12 h. The estimated heating-energy costs were 57 and 76% of the costs of conventional hardwood drying on a small scale for the lower and higher power inputs respectively (Schaffner 1991), suggesting that the heating cost would not be excessive at these low microwave-energy inputs.

12.2.4 Drying Times Relative to Conventional Kilns

Ratios of drying times for the same drying duty using microwave and radiofrequency energy to those from conventional schedules range from 0.25 for white spruce (*Picea glauca*) to 0.03 for Douglas-fir (*Pseudotsuga* menziesii), as shown in Table 12.4.

Table 12.4. Examples of reductions in drying times with microwave and radiofrequency drying

Author	Species, size of lumber	Initial moisture content (%)	Final moisture content (%)	Relative drying times (dielectric /conventional)
Avramidis et al. (1996)	Mixture of western red cedar (*Thuja plicata*), western hemlock (*Tsuga heterophylla*) and amabilis fir (*Abies amabilis*), 71–152 mm thick	80	16	0.12
Barnes et al. (1976)	Western hemlock (*Tsuga heterophylla*), 50 mm thick	86	9–11	0.04
	Douglas-fir (*Pseudotsuga menziesii*), 50 mm thick	36	11–13	0.03
Harris and Taras (1984)	Red oak (*Quercus rubra*), 25 mm thick	57–67	7–8	0.06
Miller (1971)	White spruce (*Picea glauca*), 50 mm thick	55–65	15	0.25
Smith and Smith (1994)	Red oak (*Quercus rubra*), 57 mm square	85	8	0.25

12.2.5 Economics

Barnes et al. (1976) reported trials in a kiln with a 25-kW magnetron. Tests were carried out with 60 boards of both Douglas-fir (*Pseudotsuga menziesii*) and western hemlock (*Tsuga heterophylla*), each 2″ × 8″ × 10 ft long (0.05 × 0.4 × 3.05 m). They concluded that the cost of drying per unit volume of lumber for microwave drying is influenced mainly by the initial moisture content and the density, while for conventional drying, the cost per unit volume is mainly dependent on the length of the drying cycle required to achieve acceptable degrade levels. They felt that microwave drying is most likely to be appropriate where the wood species has a low initial moisture content, causing problems with degrade in conventional drying, and/or is relatively valuable so that capital carrying charges are significant. For example, they found that microwave drying was economical for Douglas-fir (*Pseudotsuga menziesii*), which had a low initial moisture content of 40%, but not for western hemlock (*Tsuga heterophylla*), which had a higher initial moisture content of 86%. On this basis, the authors considered that microwave drying would find most applications for high-value hardwood species.

Smith and Smith (1994) have reported on the use of radio-frequency heating in a vacuum kiln with a 23-m^3 lumber capacity for drying 57-mm-square sections of red oak (*Quercus rubra*) from a moisture content of 0.85 to 0.08 kg kg^{-1}. A 300-kW radiofrequency generator operating at a frequency of 3 MHz was used, and under absolute pressures from 20 mm Hg (2.7 kPa) to 90 mm Hg (12 kPa) inside the drying chamber. The capital cost was 90% of an estimated figure for a conventional steam-heated system with the same throughput including the ancillary boiler. The energy costs were higher, however, being almost three times higher than for conventional drying.

In a recent review of dielectric and microwave drying, Schiffmann (1995) comments that, if the power requirement is high, over 50 kW, economics favour the use of higher-power tubes in the radio-frequency range. The cost of tubes varies greatly, depending upon the output power. The least expensive tubes by far are the microwave oven tubes, the output power of which is of the order of 750 W and which may be purchased, in quantity, for a few dollars each.

Radiofrequency systems can use internal metal components, and this reduces the costs associated with their use. Microwaves are reflected from internal metal surfaces. The cost of the whole system, including the generator, tube from the generator to the dryer, applicator, control system and conveyor, is much higher. Lower unit costs are associated with higher-power equipment.

Schiffmann (1995) suggests that, as a rule of thumb, radiofrequency or microwave systems become economically attractive if the drying rate can be increased fourfold over that from conventional drying for the same product quality. Since these enhancements have been noted by most workers who have investigated lumber drying by this method, the economics appear to be promising, particularly for new kilns. The question of the economic viability of retrofitting existing dryers does not seem to have been addressed.

None of these analyses of the economic viability of microwave and radiofrequency drying has made any allowance for the reluctance of industry to adopt new technology because of the risk involved. Under these circumstances, industry might demand significantly higher rates of return on investment, and this aspect should be incorporated in future analyses for which the following questions seem relevant:

1. Is there any case for replacing existing kilns with microwave/radiofrequency units?
2. Is there any case for adding microwave/radiofrequency units to existing kilns?
3. Would it be worthwhile to consider a microwave/radiofrequency unit as a heating source when installing a new kiln, instead of, or as well as conventional heat input?

Work to date does not appear to have assessed these possibilities.

12.3 Superheated-Steam Drying

The use of superheated steam has been reported in a number of drying applications: wood pulp (Svensson 1981), lime mud (Hanson and Theliander 1993), paper (Bosse and Valentin 1988; Bond et al. 1994a,b; Poirier et al. 1994; van Deventer 1997), coal (Potter et al. 1988), beet pulp (Jensen 1992) and softwood biomass (Arnoux et al. 1994). The technique relies on the principle of drying in an environment of pure steam which is above its saturation temperature.

The superheated-steam drying of wood chips has been reported by Jensen (1996), Fyhr and Rasmusson (1997a,b,c), Jensen (1996), Johansson and Rasmusson (1997) and Johansson et al. (1997). These workers all use a multiple-mechanism transport model to predict drying rates, in conjunction with experimental measurements.

For lumber, drying with superheated steam at atmospheric pressure has been evaluated for some softwoods (*Pinus pinaster, radiata, sylvestris and nigra*) and hardwoods (*Eucalyptus globulus* and *Populus* sp.) by Fernandez-Golfin and Alvarez Noves (1994) in experimental tests using empirical schedules. The feasibility of superheated-steam drying under atmospheric conditions was felt to be demonstrated by the authors for softwood lumber of 50-mm thickness or less, particularly with lumber without knots and with low resin and extractives content.

Stubbing (1994) described the use of drying at atmospheric pressure in a steamy environment, so-called *airless drying*, with the lower density of steam relative to air being used to seal the drying chamber. Substantial energy savings (90%) relative to conventional drying are claimed. No application to lumber drying has been reported, however.

Pang (1997) extended his softwood moisture-transport model to analyse the drying behaviour of sapwood *Pinus radiata* at temperatures from 120 °C to 190 °C and heartwood at 140 °C in comparison with drying data obtained from single-board studies. The surface temperature with sapwood quickly rises above 100 °C to a plateau, which is higher as the steam temperature is raised, before rising again; whereas the core temperature remained close to 100 °C for most of the drying time for this permeable wood. Profiles of mean moisture content and local temperature are illustrated in Fig. 12.2 for tests at 140 °C and 190 °C. The particular advantage of superheated-steam drying for lumber which can tolerate high kiln temperatures is that above the so-called *inversion point* drying in steam is faster than drying in perfectly dry air at the same temperature (Yoshida and Hyodo 1970). Pang (1997) has calculated this inversion temperature to be between 185 and 205 °C, but earlier work (Keey 1972) reported values as low as 160 to 175 °C from tests on evaporation from wetted-wall columns.

12.4 Vacuum Drying

Matsumoto developed the first vacuum kiln for drying wood in 1934 (Ressel 1994). Vacuum kilns have been commercially available for at least the last 20 years, and their use is regarded as standard practice in Europe for the drying of high-quality hardwoods quickly and economically which otherwise are difficult to dry (Hilderbrand 1989). The vacuum chamber resembles a cylindrical preservative-treatment vessel, except that it is thinner-walled and is worked under vacuum rather than pressure. Ressel (1994) gave a state-of-the-art review of vacuum-drying technology, while the use of vacuum for drying oakwood has been described more recently by Audebert and Temmar (1997) and Jomaa and Baixeras (1997).

Because of the enhanced relative humidity under vacuum, rate of drying can be as rapid as that at a significantly higher temperature at atmospheric pressure. Vacuum drying has the benefits of high-temperature drying without the danger of developing defects with some susceptible species. However, these benefits come at a price. The greater specific volume of moisture vapour and air under vacuum reduces the relative space available for the lumber load inside the chamber compared with that in a conventional kiln. There may be difficulties in getting adequate air circulation through the load to obtain the heat transfer needed. Malquist and Noack (1960) note that at 20 kPa the required velocity of the drying medium in a superheated-steam atmosphere is four times that at 100 kPa to achieve the same heat-transfer rate. The poor heat transfer to the boards can be enhanced by placing heated platens between them

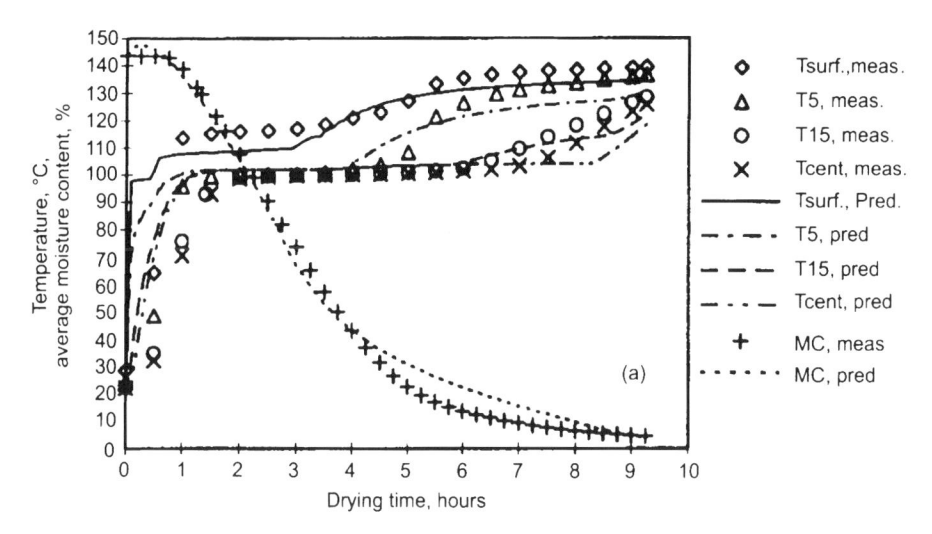

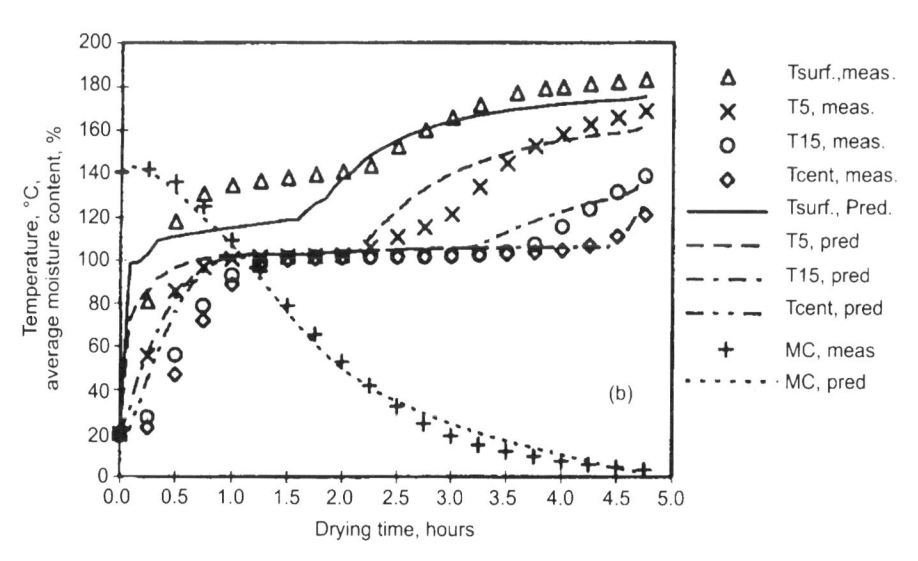

Fig. 12.2a,b. Average moisture content and local temperature in a 100 × 50-mm sapwood *Pinus radiata* board dried under superheated-steam conditions at 6.5 m s^{-1}. **a** 140°C. **b** 190°C. (Reproduced from Pang 1997 Drying Technol 15: 664–665, by courtesy of Marcel Dekker Inc)

or, in larger units, by intermittent heating with superheated steam. A third method of operation is to employ discontinuous drying, with alternating heating and drying phases under different pressures, the heating being accomplished by circulating hot air or steam at atmospheric pressure. The drying rate is controlled by the heat input. The evolved moisture vapour can be recovered by a heat pump or reemployed elsewhere by the use of vapour recompression.

Walker (1993) records one manufacturer's claim to be able to dry 50-mm-thick oak (*Quercus* sp.) from 40 to 12% moisture content in 4 days, some five to ten times faster than a conventional kiln and with half to two-thirds of the usual energy consumption. Lesser claims are made by Hilderbrand (1989), who states that drying times vary between one-half to one-third of that in conventional kilns at atmospheric pressure, depending upon the thickness of the lumber. Such differences undoubtedly reflect differences in the vaccum employed and heating arrangements. Units up to 100-m^3 capacity are available, although smaller units are commoner (Hilderbrand 1989).

Vacuum drying, when the load is stacked on bogies with aluminium platens placed between each layer, has some attractions in the drying of certain hardwoods. The technique avoids the discolorations which can sometimes arise with stickered boards of European beech (*Fagus sylvatica*) and oaks (*Quercus petrea* and *Q. robur*) when these are dried traditionally in stacks at atmospheric pressure.

Warp can be reduced by a variety of restraining devices. Simpson (1985) notes that a major constraint on drying species such as red oak (*Quercus robur*) is honeycombing, in which the lumber fails internally in tension across the grain. The checks follow the rays, which act as lines of weakness. Simpson has proposed that such species should be quarter-sawn before being dried under restraint. In contrast, flat-sawn boards can still honeycomb, as there is no restraint to checks opening along the rays.

Vacuum kilns, however, suffer from severe corrosion from released wood acids; all parts of the kiln not made from stainless steel or aluminium are likely to have a service life of less than 2 years (Ressel 1994). Screwed fittings and welded joints are preferred to gasketed flanges.

12.5 Dehumidifer Kilns

As noted in Chapter 3, drying at low temperatures, which is a feature of seasoning refractory timbers, is energy-inefficient. A dehumidifier kiln reduces the thermal energy consumption by incorporating an air-conditioning unit, which recovers heat by cooling the kiln air below its dewpoint and recycling the latent heat of condensation. Most of the moisture is removed from the kiln as liquid rather than by venting moist, warm air. Venting is normally required only for control purposes. The general arrangements of a dehumidifier kiln incorporating a solar collector for heating have been illustrated in Fig. 12.1.

Carrington et al. (1998) report values of the on-site energy use for a dehumidifier kiln drying sapwood of a softwood from 140 to 12% moisture content at a maximum dry-bulb temperature of 70 °C, which represents an upper practical limit dictated by the compressor. Their comparison of the calculated energy use with that for a gas-fired kiln is shown in Table 12.5.

Table 12.5. Kiln energy use (Carrington et al. 1998)

Energy	Dehumidifier kiln	Gas-fired kiln
Fuel, GJ m^{-3}	–	2.2
Electricity, GJ m^{-3}	0.6	0.3
Total, GJ m^{-3}	0.6	2.5

Although the capital cost of a dehumidifier kiln is normally higher than that of an equivalent direct-fired unit worked at the same temperature, the operating costs may be less unless fuel at a discounted or nil value (such as wastewood) is used. However, if very low kiln temperatures are required, costs may still favour the use of a dehumidifier kiln. A reduction in kiln temperature from 70 to 50 °C will cause the energy demand of the dehumidifier kiln to increase by approximately 7% compared with 60% for a conventional kiln venting moist air (Carrington et al. 1998).

Nevertheless, industrial experience with these dehumidifier kilns has been uneven. The smaller heating unit compared with that in a conventional kiln for the same drying duty means that the kiln temperature cannot be changed as quickly. Poor thermal insulation can severely impair the performance of the kiln. A recent case study (Carrington and Zhifa Sun 1998) revealed that a well-sealed dehumidifier kiln having an uninsulated floor took 19% longer to dry the load and used 18% more energy than a well-insulated and sealed kiln. Moreover, when the kiln with an uninsulated floor was poorly sealed, it took 80% more time to dry and used 48% more energy. Under such conditions, the benefits of providing heat recovery are effectively wasted.

References

Adesanya BA, Nanda AK, Beard JN (1988) Drying rates during high-temperature drying of yellow poplar. Drying Technol 6: 95–112

Adler E (1977) Lignin chemistry-past present and future. Wood Sci Technol 11(3): 169–218

Ahlgren L (1972) Fuktfixiering i porösa byggnadsmaterial. Moisture sorption in porous building material. Rep 36 Div Building Tech Lund Inst reported by Time 1998

Aleon D (1979) The use of solar energy in wood drying. In: Energy Aspects of the Forest Industries. Proc United Nations Economic Commission, 359–370

Alexiou PN, Wilkins AP, Harley J (1990) Effect of pre-steaming on drying rate wood anatomy and shrinkage of regrowth Eucalyptus pilularis. Wood Sci Technol 24(1): 103–110

Allen JD, Archer KJ, Walker JCF (1993) Bacterial pre-treatment of Douglas-fir to improve preservative permeability, Record 23rd Ann Conv Brit Wood Preserving and Damp-proofing Assn, Cambridge, 23: 19–27

Antti AL (1992a) Microwave drying of hardwood simultaneous measurements of pressure temperature and weight reduction. For Prod J 42(6): 49–54

Antti AL (1992b) Microwave drying of hardwood. Moisture measurements by computer tomograph. Proc 3rd IUFRO Int Wood Drying Conf Vienna Austria, 74–77

Araman PA (1987) Standard size rough dimension a potential hardwood product for the European market made from secondary-quality US hardwoods. Seminar on the valorization of secondary-quality temperate-zone hardwoods. Economic Commission for Europe Timber Committee, Nancy, France

Araman PA, Hansen BG (1983) Conventional processing of standard-size edge-glued planks for furniture and cabinet parts, a feasibility study. USDA For Serv Northeast For Exp Stn Res Pap NE-524

Archer KJ (1985) Bacterial modification of Douglas-fir roundwood permeability. PhD Thesis University of Canterbury, Christchurch, NZ

Arfvidsson J (1998) Moisture transport in porous media. Doctoral dissertation TVBH-1010 Dept building physics, Lund University, Lund, Sweden

Arganbright DM (1979) Moisture measurement problems in lumber drying. Proc symp on wood moisture content temperature and humidity relationships Oct (1979) USDA For Serv NC- For Exp Stn 62–69

Armstrong LD, Kingston RST (1960) Effect of moisture changes on creep of wood. Nature 185: 862–863

Armstrong LD, Kingston RST (1962) The effect of moisture content changes on the deformation of wood under stress. Aust J Appl Sci 13: 257–276

Arnaud G, Fohr J-P, Garnier J-P, Ricolleau C (1991) Study of the air flow in a wood drier. Drying Technol 9: 183–200

Arni PC, Cochrane GC, Gray JD (1965a) The emission of corrosive vapours by wood 1. Survey of the acid-release properties of certain freshly felled hardwoods and softwoods. J Appl Chem 15: 305–313

Arni PC, Cochrane GC, Gray JD (1965b) The emission of corrosive vapours by wood 2. The analysis of the vapours emitted by certain freshly felled hardwoods and softwoods by gas chromatography and spectrophotometry. J Appl Chem 15: 463–468

Arnoux L, Orange PA, Scott K, Langrish TAG, Keey RB, Gilmour IA (1994) Multiple-effect superheated-steam drying of woody biomass fuels for pulverised fuel combustion. Proc 9th Int Drying Symp IDS '94 Gold Coast Australia A: 157–164

References

Ashworth JC (1969) Batch drying-an analogue computer simulation of drying schedules employed when kiln-drying NZ timbers. BE Rep Chem Eng Dept Univ Canterbury NZ

Ashworth JC (1977) The mathematical simulation of batch-drying of softwood timber. PhD Thesis, Univ Canterbury, NZ, 2 vols

Ashworth JC (1980) Design of drying schedules for kiln-drying of softwood timber. In Mujumdar AS (ed) Drying '80. Hemisphere, Washington, DC 431–442

Ashworth JC, Keey RB (1979) The kiln seasoning of softwood timber boards. Chem Eng 347(8): 593–598, 607

ASTM (1996) Standard method for preparation of extractive-free wood. D1105 ASTM Annual Book of ASTM Standards Vol 410 Wood. Philadelphia, Pennsylvania

Audebert P, Temmar A (1997) Vacuum drying of oakwood moisture strains and drying process. Drying Technol 15: 2281–2302

Avramidis S, Liu F (1994) Drying characteristics of thick lumber in a laboratory radio-frequency/vacuum dryer. Drying Technol 12: 1963–1981

Avramidis S, Liu F, Neilson BJ (1994) Radio-frequency/vacuum drying of softwoods drying of thick western red cedar with constant electrode voltage. For Prod J 44(1): 41–47

Avramidis S, Zhang L, Hatzikiriakos SG (1996) Moisture transfer characteristics in wood during radio frequency/vacuum drying. Proc 5th IUFRO Int Wood Drying Conf Quebec City Canada, 125–134

Avramidis S, Zwick RL (1992) Exploratory radio frequency/vacuum drying of three BC coastal softwoods. For Prod J 42: 7–8, 17–24

Avramidis S, Zwick RL (1996) Commercial scale RF/V drying of softwood lumber. Part II. Drying characteristics and lumber quality. For Prod J 46(6): 27–36

Avramidis S, Zwick RL (1997) Commercial scale RF/V drying of softwood lumber. Part III. Energy consumption and economics. For Prod J 47(1): 48–56

Avramidis S, Zwick RL, Neilson BJ (1996) Commercial scale RF/V drying of softwood lumber. Part I Basic kiln design considerations. For Prod J 46(5): 44–51

Bailey IW, Tupper WW (1918) Size variation in tracheary cells. A comparison between the secondary xylems of vascular cryptograms gymnosperms and angiosperms. Am Acad Arts Sci Proc 54: 149–204

Bajara RA, Jones EH (1976) Flow distribution manifolds. J Fluids Engl Trans ASME 98: 654–666

Bamber RK (1978) The origin of growth stresses. Proc IUFRO Conf FORPRIDECOM, Laguna, Philippines, 7 pp. Also in Forpride Digest 81: 75–96

Banks WB, Dearling TB (1973) The water storage of Scots pine sapwood in conditions of high and low oxygen concentration. Mater Org 81: 39–49

Bannerjee AK, Levy JF (1971) Fungal succession in wood fence posts. Mater Org 61: 1–25

Barber NF (1968) A theoretical model of shrinking wood. Holzforschung 22(4): 97–103

Barber NF, Meylan BA (1964) The anisotropic shrinkage of wood-a theoretical model. Holzforschung 18(5): 146–156

Barnes D, Admiraal L, Pike RL, Mathur VNP (1976) Continuous system for the drying of lumber with microwave energy. For Prod J 26(5): 31–42

Barton GM (1972) How to prevent dry kiln corrosion. Can For Ind 92(4): 27–29

Basilico C, Genevaux JM, Martin M (1988) High temperature drying of wood semi-industrial kiln experiments. Proc 6th Int Drying Symp IDS '88 Versailles France OP299–OP335

Basilico C, Moyne C, Martin M (1982) High-temperature convective drying of softwood moisture migration mechanism. Proc 3rd Int Drying Symp Birmingham 1: 46–55

Bauch J, Liese W, Berndt H (1970) Biological investigations for the improvement of the permeability of softwoods. Holzforschung 246: 199–205

Beard JN, Rosen HN, Adesanya BA (1982) Heat transfer during the drying of lumber. Proc 3rd Int Drying Symp Birmingham 1, 110–122

Beard JN, Rosen HN, Adesanya BA (1985) Temperature distribution in lumber during impingement drying. Wood Sci Technol 19: 277–286

Beauregard R, Beaudorn M, Fortin Y, Samson M (1992) Evaluating warp from three sawing processes including saw-dry-rip to produce aspect structural lumber. For Prod J 42(6): 61–64

Behrendt CJ, Blanchette RA, Farrell RL (1995) An integrated approach using biological and chemical control to prevent blue stain in pine logs. Can J Bot 73: 613-619

Bernier R, Desrochers M, Jurasek L (1986) Antagonistic effect between Bacillus subtilis and wood staining fungi. J Inst Wood Sci 10(5): 214-216

Bier H (1986) Radiata pine in cross-grain bending. NZ J Timber Constr 21: 16-19

Biggerstaff T (1965) Drying diffusion coefficients in wood as affected by temperature. For Prod J 15(3): 127-133

Blasius H (1913) Das Ähnlichkeitsgesetz bei Reibungsvorgängen in Flüssigkeiten. The similarity law for friction processes in fluids. Forsch VDI 131

Bodig J, Jayne BA (1982) Mechanics of wood and wood composites. van Nostrand Reinhold, New York, 712pp

Bois PJ (1977) Constructing and operating a small solar-heated lumber dryer. USDA For Serv For Prod Utilization Tech Rep 7

Bond JF, Mujumdar AS, van Heiningen ARP, Douglas WJM (1994a) Drying paper by impinging jets of superheated steam Part 1 Constant drying rate in superheated steam. Can J Chem Eng 72: 446-451

Bond JF, Mujumdar AS, van Heiningen ARP, Douglas WJM (1994b) Drying paper by impinging jets of superheated steam Part 2 Comparison of steam and air as drying fluids. Can J Chem Eng 72: 452-456

Bonneau P (1991) Modélisations du séchage d'un matérial hétérogène. Application à un bois résineux. Modelling of the drying of a heterogeneous material. Application to a resinous wood. PhD Thesis, University of Bordeaux, Bordeaux, France, 245pp

Booker RE (1979) A comparison of the radial tangential and axial permeabilities of two radial pines. NZFS FRI Timber Drying Rep 39 Rotorua, NZ

Booker RE (1989) Hypothesis to explain the characteristic appearance of aspirated pits. Proc 2nd Pacific Regional Wood Anat Conf, For Prod Res develop Inst, Lagina Philippines

Booker RE (1990) Changes in transverse wood permeability during the drying of Dacrydium cupressium Lamb and Pinus radiata D Don. NZ J For Sci 202: 231-244

Booker RE (1994) Collapse or internal checking, which comes first? Proc 4th IUFRO Wood Drying Conf, Rotorua, NZ, 133-140

Booker RE (1996) New theories for liquid water flow in wood. Proc 5th Int IUFRO Wood Drying Conf, Quebec City, Canada August 1996 437-445

Booker RE, Keey RB (1997) Moisture movement in sapwood Radiata pine. NZFRI Ltd Rotorua and University of Canterbury, Christchurch, NZ, unpublished

Boone RS, Kozlik CJ, Bois PJ, Wengert EM (1988) Dry kiln schedules for commercial woods temperate and tropical. USDA For Serv For Prod Lab Gen Tech Rep FPL-GTR-57

Bosse D, Valentin P (1988) The thermal dehydration of pulp in a large scale steam dryer. Proc 6th Int Drying Symp IDS '88 Versailles France A OP337-OP344

Boyd JD (1985) The key factor in growth stress generation in trees lignification or crystallisation? Int Assoc Wood Anat Bull 6(2): 139-150

Bradfield JG, Emery JA (1994) The wood panel industry's approach to compliance with the clean air act. Proc 28th Int Particleboard/Composite Materials Symp Washington State Univ, Pullman, 37-50

Bramhall G (1976a) Fick's law and bound-water diffusion. Wood Sci 8(3): 153-161

Bramhall G (1976b) Semi-empirical method to calculate kiln-schedule modifications from some lumber species. Wood Sci 8(4): 213-222

Bramhall G (1979a) Sorption diffusion in wood. Wood Sci 12(1): 3-13

Bramhall G (1979b) Mathematical model for lumber drying I. Principles involved II. The model. Wood Sci 12(1): 14-31

Bramhall G (1995) Diffusion and the drying of wood. Wood Sci Technol 29: 209-215

Bramhall G, Wellwood RW (1976) Kiln drying of western Canadian lumber. West For Prod Lab Can For Serv Inf Rep VP-X-159

Brandao A, Perré P (1996) The flying wood-a quick test to characterise the drying behaviour of tropical woods. Proc 5th IUFRO Int Wood Drying Conf Quebec City Canada, 315-324

Brauer H (1971) Stoffaustausch einschließlich chemischer Reaktionen. Mass transfer including chemical reactions. Verlag Sauerländer, Aarau

Briggs LJ (1950) Limiting negative pressure of water. J Appl Phys 21: 721–722

British Standards Institute (1990) Specifications for portable timber ladders steps trestles and lightweight staging. BS 1129. BSI, London

Brouse D (1961) Some causes of warping in plywood and veneered products. USFA For Serv For Prod Lab Rep FPL 1952 Madison WI

Brown TD (1986) Quality control in lumber manufacturing. Miller Freeman, San Francisco, 288pp

Browning BL (1963) Methods of wood chemistry. Interscience, New York 2 vols

Brunauer S, Deming LS, Deming E, Teller E (1940) A theory of the van der Waals adsorption of gases. J Am Chem Soc 62: 1723–1732

Brunauer S, Emmett PH, Teller E (1938) Adsorption of gases in multimolecular layers. J Am Chem Soc 60: 309–319

Bryan EL (1977) Low-cost profit improvement opportunities in sawmilling. In: Proc 32nd NW Wood Prod Clinic Coeur d'Alene Wash State Univ 47–59

Burden RD (1975) Compression wood in Pinus radiata on four different sites. NZ J For Sci 5(1): 152–164

Butcher JA (1968) The ecology of fungi infection in untreated sapwood of Pinus radiata. Can J Bot 46(12): 1579–1589

Byrne T (1992) Lumber protection in the 90's. Proc of a meeting held at Forintek Canada 31 January 1992 Forintek Spec Publ No SP 33

Byrne T (1997) Sapstain control in Canada. Can New Zealand learn anything from the frozen north? In Kreber B (ed) Strategies for improving protection of logs and lumber FRI Bull 204. NZFRI Rotorua, NZ, 7–15

Cambell GS (1961) The value of presteaming for drying some collapse-susceptible Eucalypts. For Prod J 11(8): 343–347

Carlsson P, Esping B, Dahlblom O (1996) Optimization of the wood drying process. Proc 5th IUFRO Int Wood Drying Conf Quebec City Canada, 529–532

Carrington G, Zhifa Sun (1998) Control of a dehumidifier drier. Int J Energy Res, in press

Carrington G, Zhifa Sun, Bannister P (1998) Dehumidifier driers – what to expect and pitfalls to avoid. Proc Wood Technol Res Centre Workshop Christchurch NZ, 6–12

Carrington M (1996) High-temperature seasoning of softwood boards. Determination of mechanical properties at elevated temperature. ME (Chem) Thesis, University of Canterbury, NZ

Carslaw HS, Jaeger JC (1986) The conduction of heat in solids, 2nd edn. Clarendon Press, Oxford, 510pp

Casin RF, Ordinario EB, Tamayo GY (1969) Solar drying of Apitong, Narra, Red Lauan and Tanguile. Philippines Lumberman 15(4): 23–30

Cave ID (1978) Modelling moisture-related mechanical properties of wood Part II Computation of properties of a model of wood and comparison with experimental data. Wood Sci Technol 12: 127–139

Cech MY (1971) Dynamic transverse compression treatment to improve drying behaviour of yellow birch. For Prod J 21(2): 41–50

Cech MY, Goulet M (1968) Transverse compression treatment of wood to improve its drying behaviour. For Prod J 18(5): 90–91

Cech MY, Huffman DR (1970) Dynamic transverse compression treatment of spruce to improve intake of preservatives. For Prod J 20(3): 47–52

Cech MY, Pfaff F (1975) Kiln-drying of 1-inch red oak. For Prod J 25(8): 30–37

Cech MY, Pfaff F, Huffman DR (1974) CCA-retention and disproportioning in white spruce. For Prod J 24(7): 26–32

Chafe SC (1996) Drying thin sections of wetwood-infected hoop pine. Holzforschung 50(1): 55–61

Chapman AD, Scheffer TC (1940) Effect of bluestain on specific gravity and strength of southern pine. J Agric Res 61: 125–133

Charrier B, Haluk JP, Janin G (1992) Prevention of brown discolorations in European oak during kiln drying by a vacuum process. Holz Roh- Werkst 50: 433–437

Chen G, Keey RB, Walker JCF (1996a) Moisture content profiles in sapwood boards on drying. In: Strumillo C, Pakowski Z (eds) Drying '96. Hemisphere, Washington, DC, 679–687

Chen G, Keey RB, Walker JCF (1996b) Stress development and permeability variation on drying sapwood above fibre saturation. Proc 5th IUFRO Int Wood Drying Conf Quebec City Canada, 455–462

Chen G, Keey RB, Walker JCF (1997a) The drying stress and check development on high-temperature kiln-seasoning of sapwood Pinus radiata boards. Part I Moisture movement and strain model. Holz Roh- Werkst 55: 59–64

Chen G, Keey RB, Walker JCF (1997b) ibid Part II Stress development. Holz Roh- Werkst 55: 169–173

Chen G, Keey RB, Walker JCF (1997c) Stress relief for sapwood Pinus radiata boards by cooling and steam-conditioning processes. Holz Roh- Werkst 55: 351–360

Chen P, Pei DCT (1989) A mathematical model of drying processes. Int J Heat Mass Transfer 32: 297–310

Chen PYS, Helmer WA, Rosen HN (1980) Pilot plant studies of solar and solar-dehumidification lumber drying. AIChE Symp Ser 200: 316–323

Chen PYS, Helmer WA, Rosen HN (1982) Experimental solar-dehumidifier kiln for drying lumber. For Prod J 32(9): 35–41

Chen S, Whitaker S (1986) Moisture distribution during constant rate drying period for unconsolidated porous media. Failure of the diffusion theory. In: Mujumdar AS (ed) Drying '86. Hemisphere, Washington, DC, 39–48

Choong ET (1963) Movement of moisture through a softwood in the hygroscopic range. For Prod J 13: 489–498

Choong ET (1965) Diffusion coefficients of softwoods by steady-state and theoretical methods. For Prod J 15(1): 21–27

Choong ET, Chen Y, Mamit JD, Ilic J, Smith WR (1994) Moisture transport properties in hardwoods. Proc 4th IUFRO Int Wood Drying Conf Rotorua NZ, 87–94

Choong ET, Skaar C (1969) Separating internal and external resistance to moisture removal in wood drying. Wood Sci 1: 200–202

Choong ET, Skaar C (1972a) Diffusion and surface emissivity in wood drying. Wood Fiber 4(2): 80–86

Choong ET, Skaar C (1972b) Diffusivity and surface emissivity in wood drying. Wood Fiber 4(2): 89–96

Christensen GN, Kelsey KE (1958) The sorption of water vapour by the constituents of wood. Determination of sorption isotherms. Aust J Appl Sci 9: 265–282

Christensen GN, Kelsey KE (1959a) Die Geschwindigkeit der Wasserdampf-sorption durch Holz. The rate of water-vapour sorption into wood. Holz Roh- Werkst 17: 178–188

Christensen GN, Kelsey KE (1959b) The sorption of water vapour by the constituents of wood. II Heats of sorption III Swelling of lignin. Aust J Appl Sci 10: 269–293

Chudnoff M, Maldonado ED, Goytía E (1966) Solar drying of tropical hardwoods. USDA For Serv ITF (Rio Piedras Puerto Rico) For Serv Res Pap ITF-2

Cloutier A, Fortin Y (1991) Moisture content-water potential relationship of wood from saturated to dry conditions. Wood Sci Technol 25(4): 263–280

Cloutier A, Fortin Y (1993) A model of moisture movement in wood based/water potential and the determination of the effective water conductivity. Wood Sci Technol 27(2): 95–114

Cloutier A, Fortin Y (1994) Wood drying modelling based on water potential a parametric study. Proc 4th IUFRO Int Wood Drying Conf Rotorua NZ, 47–54

Cloutier A, Fortin Y, Dhatt G (1992) A wood drying finite element model based on the water potential concept. Drying Technol 10: 1151–1181

Cohan LH (1944) Hysteresis and the capillary theory of adsorption of vapors. J Am Chem Soc 66: 98–105

Collignan A, Nadeau JP, Puiggali JR (1993) Description and analysis of timber drying kinetics. Drying Technol 11: 487–506

Comstock GL (1968) Physical and structural aspects of the longitudinal permeability of wood. PhD Thesis, SUNY Coll For, Syracuse NY

Comstock GL, Côté WA (1968) Factors affecting permeability and pit aspiration in coniferous sapwood. Wood Sci Technol 2: 279-291

Cooper PA (1973) Effects of species precompression and seasoning on heartwood preservative treatability of six western conifers. For Prod J 23(7): 51-9

Cottrell AH (1975) An introduction to metallurgy. 2nd edn. Edward Arnold, London, 548pp

Cousins WJ (1974) Effect of strain rate on the transverse strength of Pinus radiata wood. Wood Sci Technol 8: 307-321

Cown DJ (1992a) Corewood (juvenile wood) in Pinus radiata-should we be concerned? NZ J For Sci 22(1): 87-95

Cown DJ (1992b) New Zealand radiata pine and Douglas-fir suitability for processing. Min For FRI Bull 168 Rotorua NZ

Cown DJ, McConchie DL (1982) Rotation age and silvicultural effects on wood properties of four stands of Pinus radiata. NZ J For Sci 12(1): 71-85

Cown DJ, McConchie DL (1983) Studies on the intrinsic properties of new-crop radiata pine II: wood characteristics of 10 trees from a 24-year-old stand grown in central North Island. FRI Bull 37 NZFRI Rotorua NZ

Cown DJ Young GD, Kimberley MO (1991) Spiral grain patterns in plantation-grown Pinus radiata. NZ J For Sci 21: 206-216

Crapiste GH, Whitaker S, Rotstein E (1988) Drying of cellular material I A mass transfer theory II Experimental and numerical results. Chem Eng Sci 43: 2919-2936

Cross AD, Jones PL, Lawton J (1982) Simultaneous energy and mass transfer in radiofrequency fields I Validation of the theoretical model. Trans IChemE 60: 67-74

Culpepper L (1990) High temperature drying-enhancing kiln operations. Miller Freeman, San Francisco, 316pp

Cunderlik I, Kudela J, Molinski W (1992) Reaction beech wood in drying process. Proc 3rd IUFRO Int Wood Drying Conf Vienna Austria, 350-353

Dahlblom O, Petersson H, Ormarsson S (1994) Numerical simulation of the development of deformation and stress in wood during drying. Proc 4th IUFRO Int Wood Drying Conf Rotorua NZ, 165-172

Dalton J (1802) Experimental essays on the constitution of mixed vapours; on evaporation. Mem Lit Philos Soc Manchester 5: 535-602

Dankwearts PV, Anolick C (1962) Mass-transfer from a grid packing to an air stream. Trans IChemE 40: 203-213

De'ev A (1999) Drying and heat treatment of wood, transl. Shubin GS loc. cit. Wood Technol Res Centre, Uni Canterbury, Christchurch, NZ

Denig J, Wengert EM (1982) Estimating air-drying moisture content losses for red oak and yellow-poplar lumber. For Prod J 32(2): 26-31

Department of Labour Te Tari Mahi (1992) Workplace exposure standards. Occup Saf Health GP Publ Wellington NZ, 84pp

Dinwoodie JM (1981) Timber, its nature and behaviour. van Nostrand Reinhold, New York, 190pp

Dixon HH, Jolly J (1895) On the ascent of sap. Philos Trans Soc Lond B186: 563-576

Dobie J (1976) Economic analysis of finger jointing green lumber by the WFPL method. West For Prod Lab Can Inf Rep VP-X-160

Doe PE, Booker JD, Innes TC, Oliver AR (1996) Optimal lumber seasoning using acoustic emission sensing and real time strain modelling. Proc 5th IUFRO Int Wood Drying Conf Quebec City Canada, 209-212

Doe PE, Oliver AR, Booker JD (1994) A non-linear strain and moisture content model of variable hardwood drying schedules. Proc 4th IUFRO Int Wood Drying Conf Rotorua NZ, 203-210

Doe PE, Schaffner RD (1982) Backsawn Tasmanian oak. Aust For Ind J 48(2): 44-49

Donaldson LA (1992) Within- and between-tree variation in microfibril angle in Pinus radiata. NZ J For Sci 22(1): 77-86

Eastlin IL, Johnson JA (1993) A surface preparation technique for enhancing grain angle measurements using reflected light. For Prod J 43(2): 61-65

Eden D, Wakeling R, Chittenden C, van der Waals J, Carpenter B (1977) Time limits for holding logs to achieve successful antisapstain treatment. In: Kreber B (ed) Strategies for improving protection of logs and lumber. FRI Bull 208 NZFRI Rotorua NZ, 55–61

Ellwood EL (1954) Properties of American beech in tension and compression perpendicular to the grain and their relation to drying. Yale Univ School For Bull 61

Ellwood EL, Ecklund BA (1959) Bacterial attack of pine logs in pond storage. For Prod J 9(9): 283–292

Erickson R, Haygreen J, Hossfeld R (1966) Drying prefrozen redwood. For Prod J 16(8): 57–65

Evans JM, Bergervoet AJ, Booker RE, Burton RJ, Keey RB, Walker JCF (1994) Vapour flow in kiln-dried Pinus radiata D Don. Proc 4th IUFRO Int Wood Drying Conf Rotorua NZ, 63–70

Feist WC, Tarkow H (1967) Polymer exclusion in wood substance: a new procedure for measuring fibre saturation points. For Prod J 17(10): 65–68

Fengel D, Wegener G (1984) Wood: chemistry, ultrastructure, reactions. de Gruyter. Berlin, 613pp

Ferguson WJ, Turner IW (1996) Control volume finite element model of mechano-sorptive creep in lumber. Numer Heat Transfer A 29: 147–164

Fernandez-Golfin JI, Alvarez Noves H (1994) Some limiting factors affecting the quality in super-heated vapour drying. Proc 4th IUFRO Int Wood Drying Conf Rotorua NZ, 404–411

Fick AE (1855) Poggendorff's Annalen der Physik. Ann Phys 94: 59–61

Filoneko GK, Lebedev PD (1960) Einleitung in die Trocknungstechnik. Introduction into drying technology. VEB, Leipzig, 107pp

Fohr J-P, Chakir A, Arnaud G, du Peuty MA (1995) Vacuum drying of oak wood. Drying Technol 13: 1675–1693

Forintek (1985) Chemical opportunities for lumber protection. Proc Joint Forintek Canada and COFI Workshop Moderator RS Smith March 1985, Vancouver BC

French EE (1923) The effect of the internal organisation of the North American hardwoods upon their more important mechanical properties, Thesis NY State College Forestry, Syracuse NY

Froix MF, Nelson R (1975) The interaction of water with cellulose from nuclear magnetic relaxation times. Macromolecules 86: 726–730

Fujita M, Takabe T, Harada H (1983) Deposition of cellulose hemicelluloses and lignin in the differentiating tracheids. In: Int Symp on Wood and Pulping Chemistry 1 Jpn Tech Assoc Pulp Pap Ind pp 20–23

Furuyama Y, Kangawa Y, Hayashi K (1994) Mechanism of free water movement in wood drying. Proc 4th IUFRO Int Wood Drying Conf Rotorua NZ, 95–101

Fyhr C, Rasmuson A (1997a) Mathematical model of a pneumatic conveying dryer. AIChE J 43: 2889–2902

Fyhr C, Rasmuson A (1997b) Some aspects of the modelling of wood chips drying in superheated steam. Int J Heat Mass Transfer 40: 2825–2842

Fyhr C, Rasmuson A (1997c) Steam drying of wood chips in pneumatic conveying dryers. Drying Technol 15: 1775–1785

Gaby LI (1972) Warping in southern pine studs. USDA For Serv Southeast For Exp Stn Res Pap SE 96

Ganowicz L, Muszynski P (1994) Drying stresses in wood and the demands of continuum mechanics. Proc 4th IUFRO Int Wood Drying Conf Rotorua NZ 211–220

Gilliwald W, Tschirnich J (1967) The effect on kiln drying of the air-layer close to the boards. Holztechnologie 81: 29–35

Gittus J (1975) Creep viscoelasticity and creep fracture in solids. Appl Sci, London, 725pp

Givnish TJ (1995) Plant stems' biomechanical adaptation for energy capture and influence on species distribution. In: Gartner BL (ed) Plant stems, physiology and functional morphology. Academic Press, New York, 3–49

Glasstone S, Kaidler KJ, Eyring H (1941) Theory of rate processes. McGraw-Hill, New York, 611pp

Goff JA (1949) Standardization of the thermodynamic properties of moist air. Trans ASHVE, 459–484

Goring DAI (1966) Thermal softening adhesive properties and glass transitions in lignin hemicellulose and cellulose. In: Bolam F (ed) Consolidation of the paper web Tech Sect Br Pap Board Makers Assoc Lond, 555–575

Gough DK (1977) The design and operation of a solar lumber kiln, Fiji timbers and their uses. Dep For Suva Fiji 67
Goulet M (1968) Méthode de séchage des bois massifs. Drying method for bulk timber. Canadian Patent 79083
Grace C (1996) Drying characteristics of Nothofagus truncata heartwood. ME (Chem) Thesis, University of Canterbury, Christchurch, NZ
Grant DJ (1979) Effect of test span on the apparent modulus of elasticity of radiata pine timber in scantling sizes. For Comm NSW Aust Res Note 39
Grant DJ, MacKenzie CE, Nicol R (1971) Structural engineering properties of cypress pine. For Comm NSW Aust Res Pap 13
Greenhill WL (1938) Collapse and its removal. CSIRO Div For Prod Tech Pap 24 Melbourne Vic
Grosvenor WM (1907) Calculations for dryer design. Trans AIChE 1: 184–202
Grozdits GA, Chauret G (1981) Influence of wood-structure on seasoning and gluing: application of wood anatomy in research and development. For Prod J 31(2): 28–33
Günzerodt H, Walker JCF, Whybrew K (1986) Compression-rolling and hot-water soaking effects on the drying and treatability of Nothofagus fusca heartwood. NZ J For Sci 16(2): 223–236
Guzenda R, Olek W (1992) Results of an analysis of energy consumption in kiln driers for sawn timber equipped with computer systems for steering. Proc 3rd IUFRO Int Wood Drying Conf Vienna Austria, 260–268
Hallock H (1965) Sawing to reduce warp in loblolly pine studs. USDA For Serv For Prod Lab Res Pap FPL 51 Madison WI
Hallock H, Malcolm FB (1972) Sawing to reduce warp in plantation red pine studs. USDA For Ser For Prod Lab Res Pap FPL 164 Madison WI
Hallström A, Wimmerstadt R (1983) Drying of porous granular materials. Chem Eng Sci 38: 1507–1516
Hammel HT (1967) Freezing of xylem sap without cavitation. Plant Physiol 5: 787–800
Hansen BG, Araman PA (1985) Low-cost opportunity for small-scale manufacture of hardwood blanks. USDA For Service Northeast For Exp Stn Res Pap NE-559
Hanson C, Theliander H (1993) Steam drying and fluidized-bed calcination of lime mud. Tappi 76(11): 181–188
Haque NW (1998) Model fitting for visco-elastic creep of Pinus radiata kiln drying. Paper accepted for Publication Wood Sci Technol
Hardtke H-J, Grimsel M, Militzer K-E (1996) Drying stresses in wood and the demands of continuum mechanics. Proc 5th IUFRO Int Wood Drying Conf Quebec City Canada, 111–115
Harris JM (1954) Heartwood formation in Pinus radiata D Don. New Phytol 533: 517–524
Harris JM (1961) The dimensional stability shrinkage and intersection point and related properties of New Zealand timbers. NZFS Tech Pap 36 NZFRI Rotorua NZ
Harris JM (1969) The use of beta rays in determining wood properties. Part 1 Measuring earlywood and latewood. NZ J Sci 12: 409–418
Harris JM (1973) The use of beta rays to examine wood density of tropical pines in Malaya. In: Selection and tree breeding to improve some tropical conifers, vol 2 Burley J, Nikles DG (eds) Commonwealth For Inst, Oxford, GB, 86–94
Harris JM (1989) Spiral grain and wave phenomena in wood formation. Springer Berlin Heidelberg New York, 214pp
Harris JM, Meylan BA (1965) The influence of microfibril angle on longitudinal and tangential shrinkage of Pinus radiata. Holzforsh 19(5): 144–153
Harris RA, Taras MA (1984) Comparison of moisture content distribution, stress distribution, and shrinkage of red oak lumber dried by a radio-frequency/vacuum drying process and a conventional kiln. For Prod J 34(1): 44–54
Hart CA (1975) The drying of wood. NC Agric Ext Serv Raleigh NC
Hart CA, Gilmore RC (1985) An air-drying technique to control surface checking in refractory hardwoods. For Prod J 35(10): 43–50
Hartley ID, Avramidis S (1994) Water clustering phenomenon in two softwoods during adsorption and desorption processes. J Inst Wood Sci 13: 467–474

Hasatani M, Itaya Y (1996) Drying-induced strain and stress: a review. Drying Technol 14: 1011–1040

Haslett AN (1998) Drying radiata pine in New Zealand. FRI Bull 206 NZFRI Rotorua NZ

Haslett AN, Kininmonth JA (1986) Pretreatments to hasten the drying of Nothofagus fusca. NZ J For Sci 16: 237–246

Haslett AN, Simpson IG, Kimberley MO (1991) Utilisation of 25-year-old Pinus radiata. Part 2 Warp of structural timber in drying. NZ J For Sci 21: 228–234

Hattori N, Kitayama S, Ando K, Kubo T, Kobayashi Y (1994) Application of laser incising to microwave drying of a low pervious softwood. Proc 4th IUFRO Int Wood Drying Conf Rotorua NZ, 279–286

Hayward PJ (1981) Sprinkler storage of windthrown Pinus radiata at Balmoral NZ. PhD Thesis, University of Canterbury, Christchurch, NZ

Hearmon RFS, Paton JM (1964) Moisture content changes and creep of wood. For Prod J 14: 357–359

Helmer WA, Rosen HN, Chen PYS, Wang SW (1980) A theoretical model for solar-dehumidification drying of wood. In: Mujumdar AS (ed) Drying '80. Hemisphere, Washington, DC, 21–28

Hidayat S, Simpson WT (1994) Use of green moisture content and basic specific gravity to group tropical woods for kiln drying. USDA For Serv For Prod Lab Res Note FPL-RN-0263 Madison WI

Hieta S, Kuga S, Usuda M (1984) Electron staining of reducing ends evidences a parallel-chain structure in Valonia cellulose. Biopolymers 23: 1807–1810

Hilderbrand R (1989) Die Schnittholztrocknung. The drying of sawn timber. Hilderbrand, Maschinenbau, Nürtingen, Germany 240pp

Hill RA, Holland PT, Rohitha BJ, Parker S, Cooney J (1997) Use of natural products in sapstain control. In: Kreber B (ed) Strategies for improving protection of logs and lumber FRI Bull 204 NZFRI Rotorua NZ: 39–42

Hillis WE (1975) The role of wood characteristics in high temperature drying. J Inst Wood Sci 7: 60–67

Hillis WE (1978) Wood quality and utilization. In: Hillis WE, Brown AG (eds) Eucalypts for wood production. CSIRO Aust, 259–289

Hilsenrath J (1955) Tables of thermal properties of gases. Natl Bur Stds Circ 564 74. US Govt Printing Office, Washington, DC

Hinds HV, Reid JS (1957) Forest trees and timbers of New Zealand. NZ Govt Printer, Wellington

Hoadley RB (1979) Effect of temperature, humidity and moisture content on solid wood products and end use. In: Symp wood moisture content-temperature and humidity relationships VDI & SU Blacksburg Virginia, USDA FRL North Central For Exp Stat: 92–96

Hoadley RB (1980) Understanding wood, a craftsman's guide to wood technology. The Taunton Press, Newtown, Connecticut, 256pp

Hon DNS (1995) Stabilization of wood color: is acetylation blocking effective? Wood Fiber Sci 27: 360–367

Höster HR (1974) Verfärbung bei Buchenholz nach Wasserlagerung. Discolourations in beechwood from water storage. Holz Roh- Werkst 32: 270–277

Hougen OA, McCauley HJ, Marshall WR (1940) Limitations of diffusion equations in drying. Trans AIChE 36: 183–210

Hrutfiord BF, Luthi R, Hanover KF (1985) Colour formation in western hemlock. J Wood Chem Technol 54: 451–640

Hunt DG (1982) Limited mechano-sorptive creep of beech wood. J Inst Wood Sci 9: 136–138

Hunt DG (1984) Creep trajectories for beech during moisture changes under load. J Mater Sci 19: 1456–1467

Hunt DG (1989) Two classical theories combined to explain anomalies in wood behaviour. J Mater Sci Lett 8: 1474–1476

Hunt DG, Gril J (1996) Evidence of a physical aging phenomenon in wood. J Mater Sci Lett 15: 80–82

Hunt DG, Shelton CF (1987) Progress in the analysis of creep in wood during concurrent moisture changes. J Mater Sci 22: 313-320

Hunt GM, Garratt GA (1967) Wood preservation, 3rd edn. McGraw-Hill, New York, 433pp

Hunter AJ (1993) On movement of water through wood-the diffusion coefficient. Wood Sci Technol 27: 401-408

Imré L (1984) Aspects of solar drying. In: Mujumdar AS (ed) Drying '84. Hemisphere, Washington, DC, 43-50

Imré L (1995) Solar drying. In: Mujumdar AS (ed) Handbook of industrial drying, vol 1. Hemisphere, Washington, DC, 373-452

Incropera FP, DeWitt DP (1990) Fundamentals of heat and mass transfer. Wiley, New York, 919pp

Innes TC (1995) Stress model of a wood fibre in relation to collapse. Wood Sci Technol 29: 363-376

Jaafar F, Michalowski S (1990) Modified BET equation for sorption/desorption isotherms. Drying Technol 8: 811-827

James WL (1975) Dielectric properties of wood and hardboard variation with temperature, frequency, moisture content and grain orientation. USDA For Serv Res Pap For Prod Lab 245 Madison WI

Jamroz W, Tremblay J (1991) Laser-based moisture sensor Lamsor. In: Proc West Dry Kiln Assoc Oregon State Univ, 37-40

Jane FW (1956) The structure of wood. A & C Black, London, 427pp

Jensen AS (1992) Pressurised drying in a fluid bed with steam. In: Mujumdar AS (ed) Drying '92. Elsevier, Amsterdam, 1593-1601

Jensen AS (1996) Pressurized steam drying of sludge bark and wood chips. Pulp Pap Can 97(7): 61-64

Johansson A, Fyhr C, Rasmuson A (1997) High-temperature convective drying of wood chips with air and superheated steam. Int J Heat Mass Transfer 40: 2843-2858

Johansson A, Rasmuson A (1997) Influence of the drying medium on high-temperature convective drying of single wood chips. Drying Technol 15: 1801-1813

Johnson CL (1961) Wind-powered solar-heated lumber dryer. South Lumberman Oct 1: 41-44

Jomaa W, Baixeras O (1997) Discontinuous vacuum drying of oak wood modelling and experimental investigations. Drying Technol 15: 2129-2144

Jones TG (1993) Carbon dioxide treatments of wood. PhD Thesis, University of Canterbury, Christchurch, NZ

Jozsa LA, Middleton GR (1994) A discussion of wood quality attributes and their practical implications. Forintek Canada Corp Vancouver Special Publ SP-34

Jung HS, Lee NH, Lee S-J (1989) Stress in drying wood in relation to moisture gradient. Proc 2nd IUFRO Int Wood Drying Conf Seattle Washington, 92-103

Kamke FA, Vanek M (1994) Comparison of wood drying models. Proc 4th IUFRO Int Wood Drying Conf Rotorua NZ, 1-21

Kanagawa Y (1989) Resin distribution in lumber dried by vacuum drying combined with radio-frequency. Proc 2nd IUFRO Int Wood Drying Conf Seattle Washington, 158-163

Kanagawa Y, Furuyama Y, Hattori Y (1992) Nondestructive measurement of moisture diffusion coefficient in wood drying. Drying Technol 10: 1231-1248

Kano T, Nakagawa S, Saito H, Oda S (1994) On the quality of larch timber Larix leptolepis Gordon. 1 Influence of some conditions about the characteristics on trees logs and squares. Govt For Exp Stat Bull 162 Tokyo

Karpenko YV, Kornejev SV, Nefyedov VN, Rasev A (1996) Microwave technology and techniques of lumber drying. Proc 5th IUFRO Int Wood Drying Conf Quebec City Canada, 541-544

Kauman WG (1960) Collapse in some Eucalyptus after treatment in inorganic salt solutions. For Prod J 10(9): 963-967

Kauman WG (1964) Cell collapse in wood. Div For Prod Reprint CSIRO Australia 566, 65pp

Kawai S, Nakato K, Sadoh T (1978) Prediction of moisture distribution in wood during drying. Mokuzai Gakkaishi 248: 520-525

Kay SJ (1997) Biological control of sapstain in New Zealand. In: Kreber B (ed) Strategies for improving protection of logs and lumber. NZFRI Bull 204: 75-79

Kayihan F (1982) Simulation of heat and mass transfer with local three-phase equilibria in wood drying. Proc 3rd Int Drying Symp Birmingham 1: 123–134

Kayihan F (1985) Stochastic modelling of lumber drying in batch kilns. In: Mujumdar AS (ed) Drying '85. Hemisphere, New York, 368–375

Kayihan F (1989) Moisture movement. Proc 2nd IUFRO Int Wood Drying Conf Seattle Washington, 255–268

Kayihan F (1993) Adaptive control of stochastic batch lumber kilns. Comput & Chem Eng 17: 265–273

Keep L-B (1998) The determination of time-dependent strains in Pinus radiata under kiln-drying conditions. ME Thesis, University of Canterbury, Christchurch, NZ

Keey RB (1972) Drying principles and practice. Pergamon, Oxford, 358pp

Keey RB (1978) Introduction to industrial drying operations. Pergamon, Oxford, 376pp

Keey RB (1992) Drying of loose and particulate materials. Hemisphere, New York, 504pp

Keey RB (1994) Heat and mass transfer in kiln drying: a review. Proc 4th IUFRO Int Wood Drying Conf Rotorua NZ, 22–44

Keey RB, Chen G (1996) Moisture movement and stress development on drying softwoods. Wood Technol Research Group Workshop Univ Canterbury, Christchurch, NZ

Keey RB, Keep L-B (1999) The modulus of elasticity of Pinus radiata below fibre saturation. Holz Roh- Werkst in press

Keey RB, Ma K (1986) On the humidity-potential coefficient. Trans IChemE Ser A 64: 119–124

Keey RB, Pang S (1994) The high-temperature drying of softwood boards a kiln-wide model. Trans IChemE 72 Ser A 741–753

Keey RB, Suzuki M (1974) On the characteristic drying curve. Int J Heat Mass Transfer 17: 1455–1464

Keey RB, Walker JCF (1988) The drying of impermeable timbers. New Zealand hard beech Proc CHEMECA '88 Sydney NSW 2: 421–435

Kelsey KE (1957) The sorption of water vapour by wood (of Araucaria klinkii). Aust J Appl Sci 8(1): 42–54

Kelsey KE (1963) The shrinkage-moisture content relationship for wood with special reference to longitudinal shrinkage. CSIRO Aust For Prod Div Prog Rept 2

Kelsey KE, Clarke LN (1956) The heat of sorption of water by wood. Aust J Appl Sci 7(2): 160–175

Kerr AJ, Goring DAI (1975) The ultrastructural arrangement of the wood cell wall. Cellul Chem Technol 9: 563–73

Kestin J, Richardson PD (1963) Heat transfer across turbulent incompressible boundary layers. Int J Heat Mass Transfer 3: 305–309

Kho PCS, Keey RB, Walker JCF (1989) Effects of minor board irregularities and airflows on the drying rate of softwood timber boards in kilns. Proc IUFRO Wood Drying Symp Seattle Washington DC, 150–157

Kho PCS, Keey RB, Walker JCF (1990) The variation of local mass-transfer coefficients in streamwise direction over a series of in-line blunt slabs. Proc CHEMECA '90 Auckland NZ, 348–355

Knight E, Cook D (1960) Degrade of ponderosa pine common lumber as related to moisture content. West Pine Assoc Res Note 4.5211

Kingston RST, Clarke LN (1961) Some aspects of rheological behaviour of wood. Aust J Appl Sci 12: 211–240

Kininmonth JA (1971) Effect of steaming on the fine structure of Nothofagus fusca. NZ J For Sci 1: 129–139

Kininmonth JA (1976) Effect of timber drying temperature on subsequent moisture and dimensional changes. NZ J For Sci 6: 101–107

Kininmonth JA, Whitehouse LJ (1991) Properties and uses of New Zealand radiata pine. NZ Min For FRI Rotorua 1: 7–10

Kollmann F, Schneider A (1961) Der Einfluss der Stromungsgeschwindigkeit auf die Heissdampftrocknung von Schnittholz. The influence of flow velocity on the superheated-steam drying of sawn timber in. Holz Roh- Werkst 19(12): 461–478

Kollmann FFP, Côté WA (1968) Principles of wood science and technology. Springer, Berlin Heidelberg New York, 2 vols 1219pp

Koponen H (1987) Moisture diffusion coefficients of wood. In: Mujumdar AS (ed) Drying '87. Hemisphere, New York, 225–232

Koponen H (1988) Moisture transfer coefficients of wood and wood-based panels. Proc 6th Int Drying Symp IDS '88 Versailles France OP99–OP106

Krahmer RL, Côté WA (1963) Changes in coniferous wood cells associated with heartwood formation. Tappi 461: 42–49

Kraus H (1980) Creep analysis. Wiley, New York, 270pp

Kreber B (1993) Advances in the understanding of hemlock brownstain. Mater Org 28: 17–37

Kreber B (1996) Elucidation of factors involved in the formation of western hemlock sap browning. Mater Org 30: 11–22

Kreber B, Haslett AN (1997) Compression-rolling reduces kiln brown stain in radiata pine sapwood. For Prod J 47: 7–8, 59–63

Krischer O, Kast W (1978) Die wissenschaftlichen Grundlagen der Trocknungstechnik. The scientific fundamentals of drying technology, 3rd edn. Springer, Berlin Heidelberg New York, 489pp

Kröll K (1978) Trockner and Trocknungsverfahren. Dryers and drying processes, 2nd edn. Springer, Berlin Heidelberg New York, 654pp

Kubler H (1987) Growth stresses in trees and related wood properties. For Prod Abstr 10(3): 61–119

Langmuir I (1918) The adsorption of gases on plain surfaces of glass mica and platinum. J Am Chem Soc 40: 1361–1403

Langrish TAG (1994) Assessing the variability of mass-transfer coefficients in stacks of timber with the aid of a numerical simulation. Proc 4th IUFRO Int Wood Drying Symp Rotorua NZ, 150–157

Langrish TAG, Bohm N (1997) An experimental assessment of driving forces for drying in hardwoods. Wood Sci Technol 31: 415–422

Langrish TAG, Brooke AS, Davis CL, Musch HE, Barton GW (1997) An improved drying schedule for Australian ironbark timber optimisation and experimental validation. Drying Technol 15: 47–70

Langrish TAG, Keey RB (1992) A solar-heated kiln for drying New Zealand hardwoods. Trans IPENZ Chem/Elect/Mech 18: 9–14

Langrish TAG, Keey RB (1996) The effects of air bypassing in timber kilns on fan power consumption. Proc CHEMECA '96 Sydney Australia 2: 103–108

Langrish TAG, Keey RB, Kho PCS, Walker JCF (1993) Time-dependent flows in arrays of timber boards flow visualisation mass-transfer measurements and numerical simulation. Chem Eng Sci 48: 2211–2223

Langrish TAG, Kho PCS, Keey RB, Walker JCF (1992) Experimental measurement and numerical simulation of local mass-transfer coefficients in timber kilns. Drying Technol 10: 753–781

Lartigue C, Puiggali JR, Quintard M (1988) Proc 6th Int Drying Symp IDS '88 Versailles France OP295–OP305

Le CV, Ly NG (1992) Multilayer adsorption of moisture in wool and its application in fabric steaming. Text Res J 62: 648–657

Lee HS (1990) Flow visualization on high-temperature drying. BE Res Rep CAPE Univ Canterbury, Christchurch, NZ

Leicester RH (1971a) A rheological model for mechano-sorptive deflections of beams. Wood Sci Technol 5: 211–220

Leicester RH (1971b) Lateral deflections of lumber beam-columns during drying. Wood Sci Technol 5: 221–231

Liese W (1984) Wet storage of windblown conifers in Germany. NZ J For 29: 119–135

Lin RT (1967) Review of the dielectric properties of wood and cellulose. For Prod J 17(7): 61–66

Lin J, Cloutier A (1996) Finite element of viscoelastic behaviour of wood during drying. Proc 5th IUFRO Int Wood Drying Conf Quebec City Canada, 213–220

Liou JK, Bruin S (1982a) An approximate method for the nonlinear diffusion problem with a power relation between the diffusion coefficient and concentration. 1 Computation of desorption times. Int J Heat Mass Transfer 25: 1209–1220

Liou JK, Bruin S (1982b) An approximate method for the nonlinear diffusion problem with a power relation between the diffusion coefficient and concentration. 2 Computation of the concentration profile. Int J Heat Mass Transfer 25: 1221–1229

List RJ (1958) Smithsonian meterological tables, 6th edn. Smithsonian Inst, Washington, DC

Little RL (1979) Ongoing research-solar-heated water dries lumber. For Prod J 29: 52–53

Little RL, Moschler WW (1995) Controlling corrosion in lumber dry kiln buildings. For Prod J 45(5): 55–58

Liu F, Avramidis S, Zwick RL (1994) Drying thick western hemlock in a laboratory radio frequency/vacuum dryer with constant and variable electrode voltage. For Prod J 44(6): 71–75

Lowery DP (1972) Vapor pressures generated in wood during drying. Wood Sci 5: 73–80

Luikov AV (1966) Heat and mass transfer in capillary-porous bodies. Pergamon, Oxford, 523pp

Luikov AV (1968) Teoriya sushki. Theory of Drying. 2nd edn, Energiya Moskva

Lumley TG, Choong ET (1979) Technical and economic characteristics of two solar kiln designs. For Prod J 29(7): 49–56

Mackay JFG, Oliveira LC (1989) Kiln operator's handbook for western Canada. Forintek Canada Corp Vancouver Spec Publ SP-31

Maeglin RR (1978) Yellow-poplar studs by S-D-R. South Lumberman 237: 2944: 58–60

Maeglin RR (1979) Could S-D-R be the answer to the aspen oversupply problem? Northern Logger Timber Proc 28(1): 24–25

Maeglin RR, Adams RD (1990) Structural lumber from aspen using the saw-dry-rip SDR process. In: Proc Aspen Symp '89 Duluth USDA For Service NC- For Exp Stn NC-140: 283–293

Maeglin RR, Boone RS (1983a) An evaluation of saw-dry-rip SDR for the manufacture of studs from small ponderosa pine logs. USDA For Prod Lab Res Pap FPL 435 Madison WI

Maeglin RR, Boone RS (1983b) Manufacture of quality yellow-poplar studs using the saw-dry-rip SDR concept. For Prod J 33(3): 10–18

Maldonado ED, Peck EC (1962) Drying by solar radiation in Puerto Rico. For Prod J 12(10): 487–488

Malquist L, Noack D (1960) Untersuchungen über die Trocknung empfindlicher Laubhölzer in reinen Heißdampf ungesättiger Wasserdampf bei Unterdruck. Experiments on the drying of sensitive hardwoods in pure hot vapour of unsaturated steam under reduced pressure. Holz Roh- Werkst 18: 171–180

Martensson A, Svensson S (1996) Application of a material model describing drying stresses in wood. Proc 5th IUFRO Int Wood Drying Conf Quebec City Canada, 93–102

Martley JF (1926) Moisture movement through wood. For Prod Res Tech Pap 2 DSIR GB London

Massey BS (1990) Mechanics of fluids, 6th edn. Van Nostrand, London, 599pp

Mauget B, Perré P (1996) Numerical simulation of drying stresses using a large displacement formulation. Proc 5th IUFRO Int Wood Drying Conf Quebec City Canada, 59–68

McCormick PO, Robertson SJ (1977) Solar industrial process heat for kiln drying lumber practical application of solar energy to wood processing. Workshop Proc Virginia Polytechnic Inst and State Univ Blacksberg Virginia, 65–68

McCurdy M, Keey RB (1998) Determination of moisture saturation properties on high-temperature drying, Proc IPENZ Conf, Wellington, NZ

McDaniels DK (1984) The sun, our future energy source, 2nd edn. J Wiley, New York, 271pp

McDonald AG, Wastney S (1995) Analysis of volatile emissions from kiln drying of radiata pine. Proc 8th Int Symp Wood Pulp Chem 3: 431–436

McDonald KA, Bendtsen BA (1986) Measuring localised slope of grain by electrical capacitance. For Prod J 36(10): 75–78

McKimmy MD (1986) The genetic potential for improving wood quality. In: Douglas fir stand management of the future, Oliver CD, Hanley DP. Johnson JA (eds) Coll For Resourc, Univ Wash, Seattle, 118–122

McMahon T (1973) Shape and size in biology. Science 179: 1201–1204

McMillen JM, Baltes RC (1972) New kiln schedule for presurfaced oak lumber. For Prod J 22(5): 19-26
Meylan BA (1968) Cause of high longitudinal shrinkage in wood. For Prod J 18(4): 75-78
Meylan BA, Probine MC (1969) Microfibril angle as a parameter in timber quality assessment. For Prod J 19(4): 30-34
Michel D, Quintard M, Puiggali JR (1987) Experimental and numerical study of pine wood drying at low temperature. In: Mujumdar AS (ed) Drying '87. Hemisphere, New York, 185-191
Miller DG (1948) Application of dielectric heating to the seasoning of wood. Proc Natl Ann Meet For Prod Res Soc 2: 235-243
Miller DG (1971) Combining radio-frequency heating with kiln-drying to provide fast drying without degrade. For Prod J 21(12): 17-21
Miller DG (1973) Further report on combining radiofrequency heating with kiln-drying. For Prod J 23(7): 31-32
Milota MR, Tschernitz JL (1990) Correlation of loblolly pine drying rates at high temperatures. Wood Fiber 22: 298-313
Milota MR, Wu Q (1994) Resolution of the stress and strain components during drying of a softwood. In: Rudolph V, Keey RB (eds) Drying '94. Proc 9th Int Drying Symp Gold Coast Australia vol B: 735-742
Mollier R (1923) Ein neues Diagramm für Dampfluftgemische. A new diagram for moist air mixtures. Z VDI 67: 869-872
Morén TJ (1989) Check formation during low temperature drying of Scots pine theoretical consideration and some experimental results. Proc 2nd IUFRO Int Wood Drying Conf Seattle Washington DC, 97-100
Morén TJ, Sehlstedt-Persson M (1992) Cupping of center boards during drying due to anisotropic shrinkage. Proc 3rd IUFRO Int Wood Drying Conf Vienna Austria, 160-164
Morris PI (1991) Improved preservative treatment of spruce-pine-fir at higher moisture contents. For Prod J 41: 11-12, 29-32
Mosbrugger V (1990) The tree habit of land plants. Springer, Berlin Heidelberg New York, 161pp
Mujumdar AS (1997) Drying fundamentals. In: Baker CGJ (ed) Industrial drying of foods. Blackie, London, 27-28
Muszynski L, Olejniczak P (1996) A simple experimental method to determine some basic parameters for mechano-sorptive creep model for wood. Proc 5th IUFRO Int Wood Drying Conf Quebec City Canada, 479-483
Nanassy AJ (1974) Water sorption in green and remoistened wood studied by the broad-line component of the wide-line NMR spectra. Wood Sci 7: 61-68
Neumann RJ (1989) Kiln drying young Eucalyptus globulis boards from green. Proc 2nd IUFRO Int Wood Drying Conf Seattle Washington, 107-115
Newman AB (1931) The drying of porous solids. Diffusion and surface emission equations. Trans AIChemE 27: 203-220
Newman RH (1998) How stiff is an individual cellulose microfibril? In: Butterfield BG (ed) Microfibril Angle in Wood. Proc IAWA/IUFRO Workshop 1997 Westport NZ, 81-93
Newton I (1701) Scala graduum caloris. Scale of degrees of heat. Philos Trans R Soc Lond 22: 824-829
New Zealand Forest Research Institute (1996) Producing quality kiln-dried timber in New Zealand, Min For/NZFRI, Rotorua, NZ 46
NHLA (1988) An introduction to grading hardwood lumber. National Hardwood Lumber Association Memphis Tennessee
Nicholas DD, Thomas RJ (1968) The influence of enzymes on the structure and permeability of loblolly pine. Proc Amer Wood Preserv Assn 64: 70-76
Nielson RW, Mackay JFG (1985) Sorting of dry and green lodgepole before kiln drying. In: Nielson RW (ed) Harvesting and processing of bettle-killed timber. Spec Publ Forintek Canada SP26, 31-34
Nijdam JJ (1998) Reducing moisture-content variation in kiln-dried timber. PhD Thesis, University of Canterbury, Christchurch, NZ

Nijdam JJ, Keey RB (1996) Influence of local variations of air velocity and flow direction reversals on the drying of stacked timber boards in a kiln. Trans IChemE 74: A: 882-892

Nijdam JJ, Keey RB (1999) The drying of non-homogeneous sapwood softwood boards in a timber kiln. Drying Technol 17, in press

Nissan AH, Kaye WG, Bell JR (1959) Mechanism of drying thick porous bodies during the falling rate period.1. The psuedo wet-bulb temperature. AIChE J 5: 103-110

NLGA (1979) Standard grading rules for Canadian lumber. National Lumber Grading Authority, Vancouver, BC

Noack D (1960) Studies on drying hardwoods in pure superheated steam (unsaturated water vapour) at low pressures. Holz Roh-Werkst 18(5): 171-180

Nordon P, Banks PJ (1973) Interacting heat and mass transfer-an Australian view. 1st Australasian Heat and Mass Transfer Conf Monash University Melbourne Vic R45-R52

Northcott PL (1957) Is spiral growth the normal growth pattern? For Chron 33: 335-352

Northway R (1989) Moisture profiles and wood temperature during very high temperature drying of Pinus radiata explain lack of degrade. Proc 2nd IUFRO Int Wood Drying Conf Seattle Washington DC, 24-28

Northway R (1996) Drying strategies for plantation-grown eucalypts. Proc 5th IUFRO Int Wood Drying Conf Quebec City Canada, 289-296

Ogura T, Ohnuma K (1955) An experimental formula for the evaporation rate of wood moisture at constant drying rate. J Jpn Wood Res Soc 1: 38-41

Okuyama T, Kawai A, Kikata Y, Yamamoto H (1986) The growth stresses in reaction wood. Proc IUFRO 18th World Congr Div 5 Ljubljana Yugoslavia, 249-260

Okuyama T, Yamamoto H, Yoshida M, Hattori Y, Archer RR (1994) Growth stresses in tension wood role of microfibrils and lignification. Ann Sci For 51: 291-300

Olek W, Guzenda R, Dudzinski J (1994) Influence of equilibrium moisture content and temperature on drying process of beech Fagus sylvatica L wood. Proc 4th IUFRO Int Wood Drying Conf Rotorua NZ, 102-106

Oliver AR (1991) A model of the behaviour of wood as it dries with special reference to eucalypt materials. Research report CM91-1. Civil and Mechanical Engineering Department, Univ Tasmania, Hobart, Australia

Oliver DR, Clarke DL (1973) Some experiments in packed-bed drying. Proc Inst Mech Eng 187: 515-521

Orman HR (1955) The response of New Zealand timbers to fluctuations in atmospheric moisture conditions. NZFS Tech Pap 8 NZFRI Rotorua NZ

Ormarsson S, Dahlblom O, Petersson H (1996) Influence of annual ring orientation on shape stability of sawn lumber. Proc 5th IUFRO Int Wood Drying Conf Quebec City Canada, 427-436

Ozawa K (1972) Features in spiral grain of karamatsu (Larix kaempferi Sarg) from a stand in Tohoku district. J Jpn For Soc 54(8): 269-274

Packman DF (1960) The acidity of wood. Holzforschung 14(6): 178-183

Palka LC (1973) Predicting the effect of specific gravity moisture content temperature and strain rate on the elastic properties of softwoods. Wood Sci Technol 7: 127-141

Pallet D (1988) Simulation and validation of a solar drying lumber model. Proc 6th Int Drying Symp IDS '88 Versailles France A OP49-OP54

Pang S (1994) High-temperature drying of Pinus radiata boards in a batch kiln. PhD Thesis, University of Canterbury, Christchurch, NZ

Pang S (1996a) Development and validation of a kiln-wide model for drying of softwood lumber. Proc 5th IUFRO Int Wood Drying Conf Quebec City Canada, 103-110

Pang S (1996b) Moisture content gradient in softwood board during drying simulation from a 2-D model and measurement. Wood Sci Technol 30: 165-178

Pang S (1997) Superheated-steam drying of softwood lumber. Drying Technol 15: 651-670

Pang S, Keey RB, Langrish TAG (1992) Modelling of temperature profiles within boards during the high-temperature drying of Pinus radiata timber. In: Mujumdar AS (ed) Drying '92. Hemisphere, New York, A: 417-433

Pang S, Keey RB, Walker JCF (1994a) Modelling of the high-temperature drying of mixed sap and heartwood boards. Proc 4th Int Wood Drying Conf Rotorua NZ 430-439

Pang S, Langrish TAG, Keey RB (1993) Heat of sorption of timber. Drying Technol 11: 1071-1080

Pang S, Langrish TAG, Keey RB (1994b) Moisture movement in softwood timber at elevated temperatures. Drying Technol 12: 1897-1914

Panshin AJ (1980) Textbook of wood technology, 4th edn. McGraw-Hill, New York, 722pp

Parker J (1994) Greenweld process for engineered wood products. Int Panel and Engineered-Wood Technology Exposition Atlanta GA 5-7 Oct 1994: 1-18

Pease DA (1998) Finger-jointing plant gives green process a restart. Wood Technol 125(2): 36-40

Peck EC (1962) Drying 4/4 red oak by solar heat. For Prod J 12(3): 103-107

Peck RE, Kauh JY (1969) Evaluation of drying schedules. AIChE J 15: 85-88

Peek RD, Liese W (1987) Braunfärbungen an lagernden Fichtenstämmen durch Gerbstoffe. Brown discolorations on stored Norway spruce stems caused by tannins. Holz -Zbl 113: 98-99, 1372

Pel L (1995) Moisture transport in porous building materials. Doct Diss Tech University, Eindhoven, Netherlands, 127pp

Peralta PN (1995) Sorption of moisture by wood within a limited range of relative humidities. Wood Fiber Sci 27(1): 13-21

Peralta PN, Joseph RG (1996) Non isothermal-radio frequency drying of wood experimental setup and preliminary results. Proc 5th IUFRO Int Wood Drying Conf Quebec City Canada, 375-384

Perré P (1987) Le séchage convectif de bois résineux choix validation et utilisation d'un modèle. The convective drying of resinous woods selection validation and use of a model. Thesè de Doctorat, l'Université Paris, France

Perré P (1994) The importance of wood anatomy to drying behaviour examples based on convective microwave and vacuum drying. Proc 4th IUFRO Int Wood Drying Conf Rotorua NZ, 55-62

Perré P (1996a) The numerical modelling of physical and mechanical phenomena involved in wood drying: an excellent tool for assisting with the study of new processes. Proc 5th IUFRO Int Wood Drying Conf Quebec City Canada, 11-38

Perré P (1996b) The concept of identity drying card. In: Rudolph V, Keey RB (eds) Drying '94. Proc 9th Int Drying Symp Gold Coast Australia A 39-50

Perré P, Fohr JP, Arnaud G (1988) A model of drying applied to softwood the effect of gaseous pressure above the boiling point. Proc 6th Int Drying Symp IDS '88 Versailles France A OP279-OP286

Perré P, Moyne C (1991) Processes related to drying Part II. Use of the same model to solve transfers both in saturated and unsaturated porous media. Drying Technol 9: 1153-1179

Perré P, Turner IW (1996) The use of macroscopic equations to simulate heat and mass transfer in porous media. In: Turner IW, Mujumdar AS (eds) Mathematical modelling and numerical techniques in drying technology. Marcel Dekker, New York, 83-156

Petty JA (1981) Fluid flow through the vessels and intervascular pits of sycamore wood. Holzforschung 35: 213-216

Plumb OA, Spolek GA, Olmstead BA (1985) Heat and mass transfer in wood during drying. Int J Heat Mass Transfer 28: 1669-1678

Plumptre RA (1967) The design and operation of a small solar seasoning kiln on the equator in Uganda. Commonw For Rev 46(4): 298-309

Plumptre RA (1979) Simple solar heated lumber dryers: design, performance and commercial viability. Commonw For Rev 58(4): 243-251

Pohlhausen E (1921) Der Wärmeaustausch zwischen festen Körpern und Flüssigkeiten mit kleiner Reibung und Wärmeleitung. The heat transfer between solid bodies and fluids with small friction and heat conduction. Angew Math Mech 1: 115-121

Poirier NA, Crotogino RH, Mujumdar AS, Douglas WJM (1994) Effect of superheated steam drying on the properties of TMP paper. J Pulp Pap Sci 20(4): J97-J102

Potter OE, Li Xi Guang, Georgakopoulos S, Mao Qi Ming (1988) Some design aspects of steam fluidized heated dryers. Proc 6th Int Drying Symp IDS '88 Versailles France A OP307-OP311

Pratt GH, Turner CHC (1986) Timber drying manual. 2nd edn. Building Research Establishment Garston HMSO, London
Prins AF (1981) Oxford solar kiln research. 1978-1979 Commonw For Rev 60(3): 187-196
Puiggali JR, Quintard M (1992) Properties and simplifying assumptions for classical drying models, In: Mujumdar AS (ed) Advances in drying, vol 5. Hemisphere Washington, DC, 109-143
Ranta-Maunus A (1975) The visco-elasticity of wood at varying moisture content. Wood Sci Technol 9: 189-205
Ranta-Maunus A (1993) Rheological behaviour of wood in directions perpendicular to grain. Mater Struct 26: 362-369
Rask-Anderson A, Malmberg P (1990) Organic dust toxic syndrome in Swedish farmers, symptoms, clinical findings and exposure in 98 cases. Am J Ind Med 17: 116-117
Rasmussen EF (1961) Dry kiln operator's manual. USDA For Serv For Prod Lab Agric Handb 188 Madison WI
Read WRW, Choda A, Copper PI (1974) A solar lumber kiln. Sol Energy 15: 309-316
Rees WH (1948) Heat of absorption of water by cellulose. J Text Inst 39: T351-T367
Rehman MA, Chawla OP (1961) Seasoning of timber using solar energy. Indian For Bull 229
Ressel JB (1994) State-of-the-art report on vacuum drying of lumber. Proc 4th IUFRO Int Wood Drying Conf Rotorua NZ, 255-262
Rice RW, Howe JL, Boone RS, Tschernitz JL (1994) Kiln drying lumber in the United States. USDA For Service For Prod Lab Gen Techn Rep FPL-GTR-81 Madison WI
Rice RW, Youngs RL (1990) The mechanism and development of creep in wood during concurrent moisture changes. Holz Roh- Werkst 48: 73-79
Richardson SD (1978) Appropriate operational scale in forest industries. Proc 8th World For Conf FAO Jakarta Doc FID II/21-18
Rietz RC (1972) A calendar for air-drying lumber in the upper midwest. USDA For Serv Res Note FPL-0224 Madison WI
Riley SG, Haslett AN (1996) Reducing air velocity during timber drying. Proc 5th IUFRO Int Wood Drying Conf Quebec City Canada, 301-308
Riley SG, Wastney S, Dakin M (1999) Investigation into the softening of radiata pine by examining instantaneous strain in compression. Proc 6th IUFRO Wood Drying Conf Stellenbosch South Africa, 287-299
Rosen HN (1978) Evaluation of drying times, drying rates, and evaporative fluxes when drying wood with impinging jets. Proc 1st Int Drying Symp Montreal Canada, 192-200
Rosen HN (1980) Empirical model for characterizing wood drying curves. Wood Sci 12: 201-206
Rosen HN (1982) Function relations and approximation techniques for characterizing wood drying curves. Wood Sci 15: 49-55
Rosen HN (1987) Recent advances in the drying of solid wood. In: Mujumdar AS (ed) Advances in drying, vol 4 Hemisphere, Washington, DC, 99-146
Rosen HN (1995) Drying of wood and wood products In: Mujumdar AS (ed) Handbook of industrial drying, 2nd edn. Marcel Dekker, New York, 899-920
Rosen HN, Chen PYS (1980) Drying lumber in a kiln with external solar collectors. AIChE Symp Ser 200: 76: 82-89
Rosen HN, Laurie SE (1983) Mechanical properties of conventionally kiln-dried and pressure steam dried yellow poplar and red oak. For Prod J 33(1): 50-52
Rosenkilde A, Arfvidsson J (1997) Measurement and evaluation of moisture transport coefficients during drying of wood. Holzforschung 51: 372-380
Rowell RM, Barbour RJ (1990) Archaeological wood: properties, chemistry and preservation. Am Chem Soc Adv Chem Ser 225 Washington DC
Rowell RM, Konkol P (1987) Treatments that enhance physical properties of wood. USDA For Serv For Prod Lab FPL GTR-55 Madison WI
Rozsa AN (1994) Dielectric vacuum drying of hardwood. Proc 4th IUFRO Int Wood Drying Conf Rotorua NZ, 271-278
Ruddick JNR (1987) Incising workshop proceedings. Forintek Canada Corp Spec Publ SP-28

Salin J-G (1989) Prediction of checking surface discoloration and final moisture content by numerical methods. Proc 2nd IUFRO Int Wood Drying Conf Seattle Washington DC, 71-78

Salin J-G (1992) Numerical prediction of checking during lumber drying and a new mechano-sorptive creep model. Holz Roh- Werkst 50: 195-200

Salin J-G (1996) Prediction of heat and mass transfer coefficents for individual boards and board surfaces. Proc 5th Int IUFRO Wood Drying Conf Quebec City Canada, 49-58

Salin J-G, Ohman G (1998) Calculation of drying behaviour in different parts of timber stack. Proc 11th Int Drying Symp IDS '98 Halkidiki Greece B: 1603-1610

Schaffner RD (1991) Drying costs-a brief introduction. In: Australian Timber Seasoning Manual, Aust Furn Res Develop Inst, Launceston Tas

Schaffner RD, Doe PE (1984) Surface check reduction in eucalypt timbers using semi-preamble coatings. 21st For Prod Res Conf Nov 1984 2: D4: 7

Scheffer TC (1958) Control of decay and stain in logs and green lumber. USDA For Serv 2107, Washington DC

Scheffer TC (1969) Protecting stored logs and pulpwood in North America. Mater Org 43: 167-199

Scheffer TC (1973) Microbiological deterioration. In: Nicholas DD (ed) Wood deterioration and its prevention by preservative treatments 1. Degradation and protection of wood. Syracuse University Press, Syracuse, NY, 31-106

Schiffmann RF (1995) Microwave and dielectric drying. In: AS Mujumdar (ed) Handbook of industrial drying, 2nd edn vol 1. Marcel Dekker, New York, 345-372

Schmidt E, Christopherson E, Highley T, Freeman M (1995) Trials of new treatments for prevention of kiln brownstain of white pine Pinus strobus. Int Res Group on Wood Preservation IRG WP 60050 Doc 95-30068

Schniewind AP (1968) Recent progress in the study of the rheology of wood. Wood Sci Technol 2: 188-206

Schniewind AP ed (1989) Concise encyclopedia of wood and wood-based materials. Pergamon Oxford, 354pp

Schniewind AP, Barrett JD (1972) Wood as a linear orthotropic viscoelastic material. Wood Sci Technol 6: 43-57

Sedgewick SA, Trevans DH (1976) An estimate of the ultimate strength of water. J Appl Phys 9: L203-L205

Seehann G (1965) Über die Wirkung einer Trocknung und Erwärmung von Nadelholz auf das Wachstum von Bläuepilzen. The effect of drying and heating of softwood on the growth of bluestain fungi. Holz Roh- Werkst 23: 341-347

Senft JF, Bendtsen BA, Galligan WL (1985) Weak wood fast-grown trees make problem lumber. J For 83: 477-484

Sharma SN, Nath P, Bali BI (1974) A solar lumber seasoning kiln. J Lumber Dev Assoc India 18(2): 10-26

Sherwood TK (1929) The drying of solids 1. Ind Eng Chem 21: 12-16

Shubin GS (1990) Sushka i templovaia orbabotka drevesiny. Drying and heat treatment of wood. Lesnaia Promyshlennost Moskva, 154pp

Siau JF (1984) Transport processes in wood. Springer Berlin Heidelberg New York, 245pp

Siau JF (1995) Wood. Influence of moisture on physical properties. Dept Wood Sci For Products Virginia Tech, 227pp

Simpson WT (1973) Predicting equilibrium moisture content of wood by mathematical models. Wood Fiber 5(1): 41-49

Simpson WT (1980) Sorption theories for wood. Wood Fiber 12: 183-195

Simpson WT (1985) Process for rapid conversion of red oak logs to dry lumber. For Prod J 35(1): 51-56

Simpson WT (1992) Dry kiln operators' manual. USDA Agric Handb 188 For Serv For Prod Lab Madison WI

Simpson WT (1993) Determination and use of moisture diffusion coefficient to characterize drying of northern red oak (Quercus rubra). Wood Sci Technol 27: 409-420

Simpson WT, Baah CK (1989) Grouping tropical wood species for kiln drying. Res Note FPL-RN-0256 USDA For Serv For Prod Lab Madison WI

Simpson WT, Rosen HN (1981) Equilibrium moisture content of wood at high temperature. Wood Fiber 13: 150–158

Simpson WT, Tschernitz JL (1980) Time, costs, and energy consumption for drying red oak lumber as affected by thickness and thickness variation. For Prod J 30(1): 23–28

Simpson WT, Tschernitz JL (1984) Solar dry kiln for tropical latitudes. For Prod J 34(5): 25–34

Singh Y (1976) Studies on a solar timber seasoning kiln. IPIRI 6: 42–44

Sjöström E (1981) Wood chemistry fundamentals and applications. Academic Press Orlando, 223pp

Skaar C (1958) Moisture movement in beech below the fiber saturation point. For Prod J 8(12): 352–357

Skaar C (1977) Energy requirements for drying lumber. In: Practical application of solar energy to wood processing. For Prod Res Soc Publ P-77-17: 56–77

Skaar C (1988) Wood-water relations. Springer Berlin Heidelberg New York 283pp

Slade GW, Hunter AJ, Sutherland JW (1996) Recent advances in the simulation of radiata pine drying. Proc 5th IUFRO Int Wood Drying Conf Quebec City Canada, 77–84

Smith HH, Dittman JR (1960) Drying rate of white fir by segregations. USDA For Serv PSW- Res Note 168

Smith JM, Van Ness HC (1975) Introduction to chemical engineering thermodynamics, 3rd edn. McGraw-Hill, New York, 698pp

Smith WB, Smith A (1994) Radio-frequency/vacuum drying of red oak energy quality value. Proc 4th IUFRO Int Wood Drying Conf Rotorua NZ, 263–270

Söderström O, Salin J-G (1993) On determination of surface emission factors in wood drying. Holzforschung 47: 391–397

Solukhin RI, Kuts PC (1983) In: Keey RB (ed) Vvedenie v Technologiyu Prom'ishlenioy Sushki. Introduction to the technology of industrial drying. Nauk i Tekhnika, Minsk 20

Sørensen A (1969) Mass-transfer coefficients on truncated slabs. Chem Eng Sci 24: 1445–1460

Spalding DB (1963) Convective mass transfer. Arnold, London, 66–72

Spalt HA (1958) The fundamentals of water sorption by wood. For Prod J 8: 288–295

Sparrow EM, Niethammer JE, Chaboki A (1982) Heat transfer and pressure drop characteristics of arrays of rectangular modules encountered in electronic equipment. Int J Heat Mass Transfer 31: 961–973

Speck TH, Vogellehner D (1988) Biophysical examinations of the bending stability of various stele types and the upright axes of early vascular land plants. Bot Acta 101: 262–268

Sperry JS, Nichols KL, Sullivan JEM, Eastlack SE (1994) Xylem embolism in ring-porous diffuse-porous and coniferous trees in northern Utah and interior Alaska. Ecology 75: 1736–1752

Sperry JS, Tyree MT (1988) Mechanism of water stress-induced xylem embolism. Plant Physiol 88: 581–587

Spolek GA, Plumb OA (1981) Capillary pressure in softwood. Wood Sci Technol 15: 189–199

Stamm AJ (1946) Passage of liquids vapors and dissolved materials through softwoods. US Dept Agric Tech Bull 929 Washington DC

Stamm AJ (1956) Diffusion of water into uncoated cellophane 1 From rates of water vapor adsorption and liquid water absorption. J Phys Chem 60: 76–83

Stamm AJ (1959) Bound water diffusion of water into wood in the fiber direction. For Prod J 9(1): 27–31

Stamm AJ (1960a) Bound water diffusion of water into wood across the fiber direction. For Prod J 10(10): 524–528

Stamm AJ (1960b) Combined bound water and water vapor diffusion into Sitka spruce. For Prod J 10(12): 644–648

Stamm AJ (1962) Wood and cellulose-liquid relationships. NC Agric Exp Stn Techn Bull 150 Part 4 Raleigh NC

Stamm AJ (1964) Wood and cellulose science. Ronald Press, New York, 543pp

Stamm AJ (1967a) Movement of fluids in wood – Part 1 Flow of fluids in wood. Wood Sci Technol 1(2): 122–141

Stamm AJ (1967b) Movement of fluids in wood – Part 2 Diffusion. Wood Sci Technol 1(3): 205–230

Stamm AJ, Loughborough WK (1935) Thermodynamics of the swelling of wood. J Phys Chem 39: 121–132

Stamm AJ, Nelson RM (1961) Comparison between measured and theoretical diffusion coefficients for southern pine. For Prod J 11(11): 536–543

Stanish MA, Schajer GS, Kayihan F (1986) A mathematical model of drying for hygroscopic porous media. AIChE J 32: 1301–1311

Steinmann DE (1989) Control of equilibrium moisture content in a solar kiln. Proc 2nd IUFRO Int Wood Drying Conf Seattle Washington, 213–221

Steinmann DE, Vermaas HF, Forrer JB (1980) Solar lumber drying kilns Part I Review of previous systems and control measures and description of an automated solar kiln. J Inst Wood Sci 8(6): 254–257

Steinmann DE, Vermaas HF, Forrer JB (1981) Solar lumber drying kilns Part II Microprocessor control of a solar kiln. J Inst Wood Sci 9(1): 27–31

Strumillo C, Kudra T (1986) Drying principles, applications and design. Gordon and Breach, New York, 448pp

Stubbing TJ (1994) Airless drying. Proc 9th Int Drying Symp IDS '94 Gold Coast Australia A: 559–566

Suolahti O, Wallen A (1958) The effect of water storage on the water absorption capacity of pine sapwood. Holz Roh- Werkst 16: 8–17

Suzuki M, Endo A, Ohtani S, Maeda S (1972) Mass transfer from a discontinuous source. Proc PACHEC '72 Kyoto Japan 3: 267–276

Svensson C (1981) Steam drying of hog fuel. Tappi 64: 153–156

Tabor D (1969) Gases liquids and solids. Penguin Books, Harmondsworth, England, 290pp

Tang Y, Pearson RG, Hart CA, Simpson WT (1994) A numerical model for heat transfer and moisture evaporation processes in hot-press drying, an integral approach. Wood Fiber Sci 26(1): 78–90

Tarkow H, Stamm AJ (1961) Diffusion through the air-filled capillaries of softwoods. For Prod J 10: 247–250, 323–324

Taylor FW, Won Tek So (1990) Sorting southern pine lumber in improve drying. For Prod J 40(4): 32–36

Taylor KJ, Weir AD (1985) Simulation of a solar lumber drier. Sol Energy 34: 3: 249–255

Teichgräber R (1966) Changes in the properties of wood during steaming. Holz Roh- Werkst 24: 548–550

Teischinger A (1992) Effect of different drying temperatures on selected wood properties. Proc 3rd IUFRO Int Wood Drying Conf Vienna Austria, 211–216

Tejeda A, Okuyama H, Yamamoto H, Yoshida M (1997) Reduction of growth stress in logs by direct heat treatment: assessment of a commercial scale operation. For Prod J 47(9): 86–92

Terziev N, Bjurman J, Boutelje JB (1996) Effect on planing on mould susceptibility of kiln- and air-dried Scots pine (Pinus sylvestris) lumber. Mater Org 30(2): 95–103

Tesoro FO, Choong ET, Kimbler OK (1974) Relative permeability and the gross pore structure of wood. Wood Fiber 6(3): 226–236

Tetzlaff AR (1967) An investigation of drying schedules when kiln-dying radiata pine timber. BE Chem Rep Univ Canterbury, Christchurch, NZ

Thijssen HAC, Coumans WJ (1984) Short-cut calculation of non-isothermal drying rates of shrinking and non-shrinking particles containing an expanding gas phase. Proc 4th IUFRO Int Drying Symp IDS '84 Kyoto Japan 1: 22–30

Thomas HR, Lewis RW, Morgan K (1980) An application of the finite element method to the drying of timber. Wood Fiber 11(4): 237–243

Time B (1998) Hygroscopic moisture transport in wood. Dr Ing Thesis 1998, NTNU, Trondheim

Timell TE (1967) Recent progress in the chemistry of wood hemicelluloses. Wood Sci Technol 1(1): 45–70

Timell TE (1986) Compression wood in gymnosperms. Springer Berlin Heidelberg New York, 2130pp
Treybal RE (1968) Mass-transfer operations, 3rd edn. McGraw-Hill, New York, 784pp
Troxell HE (1977) An application of solar energy for drying lumber in the central Rocky Mountains region. In: Practical application of solar energy to wood processing. For Prod Res Soc Publ P-77-17: 49–55
Troxell HE, Mueller LA (1968) Solar lumber drying in the central Rocky Mountain region. For Prod J 18(1): 19–24
Tschernitz JL, Simpson WT (1979a) Drying rate of northern red oak lumber as an analytical function of temperature, relative humidity and thickness. Wood Sci 114: 202–208
Tschernitz JL, Simpson WT (1979b) Solar-heated, forced-air, lumber dryer for tropical latitudes. Sol Energy 22: 563–566
Tuomi RL, Temple DM (1975) Bowing in roof joists induced by moisture gradients and slope of grain. USDA For Serv For Prod Lab Res Pap FPL 262 Madison WI
Turner IW (1990) The modelling of combined microwave and convective drying of a wet porous material. PhD Thesis, University of Queensland, St Lucia, Brisbane, Australia
Turner IW, Fergusson WJ (1995a) An unstructured mesh cell-centred control volume method for simulating heat and mass transfer in porous media. Application to softwood drying. Part I The isotropic model. Appl Math Model 19: 654–667
Turner IW, Fergusson WJ (1995b) An unstructured mesh cell-centred control volume method for simulating heat and mass transfer in porous media. Application to softwood drying. Part II The anisotropic model. Appl Math Model 19: 668–674
Turner IW, Ferguson WJ (1995c) A study of the power density distribution generated during the combined microwave and convective drying of softwood. Drying Technol 13: 1411–1430
Turner IW, Perré P (1995) A comparison of the drying simulation codes TRANSPORE and WOOD2D which are used for the modelling of two-dimensional wood drying processes. Drying Technol 13: 695–735
Tuttle F (1925) A mathematical theory of the drying of wood. J Franklin Inst 200: 609–614
Tyree MT, Dixon MA (1983) Cavitation events in Thuja occidentalis L? Ultrasonic acoustic emissions from the sapwood can be measured. Plant Physiol 72: 1094–1099
Tyree MT, Sperry JS (1988) Do woody plants operate near the point of catastrophic xylem dysfunction caused by dynamic stress? Answers from a model. Plant Physiol 88: 574–580
Ugolev BN (1976) General laws of wood deformation and rheological properties of hardwood. Wood Sci Technol 10: 169–181
Uprichard JM, Lloyd JA, Harwood VD (1975) Chemistry of beech species. Beech Res News 3 Christchurch NZ
USDA (1987) Wood handbook wood as an engineering material. USDA For Serv For Prod Lab Agric Handb 72 Madison WI
van Deventer HC (1997) Feasibility of energy efficient steam drying of paper and textile including process integration. Appl Therm Eng 178-10: 1359–1431
Van Handel R (1990) Green moisture sorting at Roseburg Forest Products. In: Proc West Dry Kiln Assoc Oregon State Univ
Van Meel DA (1958) Adiabatic convection batch drying with recirculation of air. Chem Eng Sci 9: 36–44
Verkasalo E, Ross RJ, TenWolde A, Youngs RL (1993) Properties related to drying defects in red oak wetwood. USDA For Serv For Prod Lab Res Pap FPL-RP-516 Madison WI
Vermaas HF (1995) Drying eucalypts for quality material characteristics, predrying treatments, drying methods, schedules and optimisation of drying quality. S Afr For J 17(4): 41–49
Viollez PE, Suarez C (1985) Drying of shrinking bodies. AIChE J 31: 1566–1568
Voigt H, Krischer O, Schauss H (1940) Die Feuchtigkeitsbewegung bei der Verdunstungstrocknung von Holz. The movement of moisture in the evaporative drying of wood. Holz Roh- Werkst 3: 305–321

von Marutzky R, Roffael E (1977) Über die Abspaltung von Formaldehyd bei der thermischen Behandlung von Holzspänen. Liberation of formaldehyde on the thermal processing of wood chips. Holzforschung 31: 8–12

von Wedel KW, Zobel BJ, Shelbourne CJA (1993) Prevalence and effects of knots in young loblolly pine. For Prod J 18(9): 97–103

Waananan KM, Litchfield JB, Okos MR (1993) Classification of drying models for porous solids. Drying Technol 11: 1–40

Wadsö L (1993) Studies of water vapor transport and sorption in wood. Doct Diss Rep TVBM-1013, Lund Univ, Lund

Wagner FG, Gorman TM, Folk RL, Steinhagen HP, Shaw RK (1996) Impact of kiln variables and green weight on moisture uniformity of wide grand-fir lumber. For Prod J 46(11/12): 43–46

Wakelin R (1997) Issues for successful antisapstain treatment of New Zealand radiata pine logs. In: Kreber B (ed) Strategies for improving protection of logs and lumber. FRI Bull 204 NZFRI Rotorua NZ, 31–37

Walker JCF (1993) Primary wood processing principles and practice. Chapman and Hall, London, 595pp

Walker JCF, Butterfield BG (1995) The importance of microfibril angle for the processing industries. NZ J For 40: 4,34–40

Wan J, Langrish TAG (1995) A numerical simulation for solving the diffusion equations in the drying of hardwood timber. Drying Technol 13: 783–799

Ward JC, Pong WY (1980) Wetwood in trees a timber resource problem. USDA For Serv Gen Tech Rep PNW 112, 56pp

Warrell JJ, Parmeter JR (1982) Formation and properties of wetwood in white fir. Phytopathology 72: 1209–1212

Wastney S, Bates R, Kreber B, Haslett AN (1997) The potential of vacuum drying to control kiln brown stain in radiata pine. Holzforsch Holzverwert 49: 56–58

Wengert EM (1971) Improvements in solar dry kiln design. USDA For Serv For Prod Lab Res Note FPL-0212 Madison WI

Wengert EM, Lamb FM (1994) End coating of lumber to prevent end checking. Proc 2nd IUFRO Int Wood Drying Conf Seattle Washington, 164–168

Wengert G, Denig J (1995) Lumber drying today and tomorrow. For Prod J 45(5): 22–30

Whitaker S (1977) Simultaneous heat mass and momentum transfer in porous media. Advances in heat transfer, vol 13. Academic Press, New York, 119–203

Whitaker S (1980) Heat and mass transfer in porous media. In: Mujumdar AS (ed) Advances in drying,vol 1. Hemisphere, Washington, DC, 23–61

Whitaker S (1984) Moisture transport mechanisms during the drying of granular porous media. Proc 4th Int Drying Symp Kyoto Japan, 31–42

Wiedemann HGR, Nassif NM, Gostelow JP (1989) Study of the airflow profile inside a timber drying kiln. Proc 4th Australasian Heat Mass Transfer Conf Christchurch NZ, 693–701

Wiederhold P (1995) Humidity measurements. In: Mujumdar AS (ed) Handbook of industrial drying, 2nd edn. Marcel Dekker, New York, 1313–1341

Wiley AT, Choong ET (1975) An analysis of free-water flow during drying in softwoods. Wood Sci 7(4): 310–318

Win Kyi (1983) Predicting drying times of some Burmese woods for two types of solar kilns. MSc Thesis, Virginia Polytechnic Institute and State University, Blacksberg, Virginia, 149pp

Wu QL (1989) An investigation of some problems in drying of Tasmanian eucalypt timbers. MEngSci Thesis, University of Tasmania, Hobart

Wu QL (1993) Rheological behaviour of Douglas-fir as related to the process of drying. PhD Thesis, Oregon State University, Corvalis OR

Wu QL, Milota MR (1995) Rheological behaviour of Douglas-fir perpendicular to the grain at elevated temperatures. Wood Fiber Sci 27: 285–295

Yang KC (1980) Solar kiln performance at a high latitude 48°N. For Prod J 30(3): 37–40

Ying L, Kretschmann DE, Bendtsen BA (1994) Longitudinal shrinkage in fast-grown loblolly pine plantation wood. For Prod J 44(1): 58–62

Yokota T, Tarkow H (1962) Changes in dimension on heating green wood. For Prod J 12(1): 43–45

Yoshida T, Hyodo T (1970) Evaporation of water in air humid air and superheated steam. Ind Eng Chem Proc Design Dev 92: 207–214

Young RA (1986) Structure swelling and bonding of cellulose fibers. In: Young RA, Rowell RM (eds) Cellulose structure modification and hydrolysis. Wiley, New York, 91–128

Zahed AH, Elsayed MM (1989) Mathematical modelling of a solar kiln. Sol Wind Technol 61: 19–27

Zhang YC, Oliveira L, Avramidis S (1996) Drying characteristics of hem-fir squares as affected by species and basic density presorting. For Prod J 46(2): 44–50

Zienkiewicz OC, Cormeau IC (1974) Visco-plasticity-plasticity and creep in elastic solids-a unified numerical solution approach. Int J Numer Methods Eng 8: 821–845

Zienkiewicz OC, Taylor RL (1989) The finite element method, 4th edn. McGraw-Hill, London, Vols 1 and 2

Zimmermann MH (1983) Xylem structure and the ascent of sap. Springer Berlin Heidelberg New York, 143pp

Zimmermann MH, Jeje AA (1981) Vessel-length distribution in stems of some American woody plants. Can J Bot 59(10): 1882–1892

Zobel BJ, van Buijtenen (1989) Wood variation, its causes and control. Springer Berlin Heidelberg New York, 363pp

Subject Index

Activation energy   34, 109-111, 114
- for diffusion   109, 114
- temperature effect   111
- values of   109, 114
Activity   24, 28
Adsorption   26, 35, 86
- at saturation   35
- heat of   35
- isotherm for   26
Air
- composition of   44
- enthalpy of   45-46
Air velocity   66, 69-70
Airflow
- boundary layer   58-59
- in kilns   66, 203-219
- - board irregularities, effects of   216-217
- - bypassing   218
- - maldistribution in   74, 205-206, 215-216
- - reversals in   74, 227-232
- - stack ends, effects of   217-218
- over boards   59
- smoke patterns   58, 213
Angiosperms   2
Antishrink chemicals   262-263
Arrhenius relationship   107
Asperigullus fumigatus   254
Aspiration, *see* Pits, bordered

Bacillus polymixa   248
Bacteria
- aerobic   247
- anaerobic   142
Bernoulli's theorem   207
Biot number   63, 72, 100
Boards
- air velocities between   207, 211-218
- - irregularities, effect on   216-217
- - side-gaps, effect on   213
- cupping test   200
- heartwood, *see* Heartwood
- mixed-wood, *see* Drying
- sapwood, *see* Sapwood
- softwood, *see* Softwood
Bow   149, 151, 165, 167-168

Brittleheart   151
Brownstain   129, 255

Cambium   4-5, 155
Capillary
- condensation   32
- pressure   87, 98
Cavitation   6, 9, 127-8, 134
Cell wall   12-21, 37, 80, 81, 85, 142-143, 155, 164
- bacteria, effect on   142
- in softwood tracheid   18-21
- lignin content of   21
- penetration of chemicals into   143
- primary layer in   18
- reaction wood features   164
- secondary layer in   13, 18
- sorption in   37, 85
- stresses in   158
- structure of   12-21
- water in   21, 85
Cells   4, 7, 11, 12
- epithelial   7
- in hardwoods   12
- parenchyma   4, 11
Cellulose   13-14, 18, 163, 166, 183, 281
Ceratocystis sp. 251
Checking   126
Checks   139, 267
Chemical potential   83, 85, 86, 98-100, 133
- driving force   98-100
Clausius-Clapeyron equation   38, 39
Clusters
- molecule   36
Collapse   126, 151, 157-158, 163, 166, 257, 260
Compliance   176, 185, 190-195
- mechanosorptive   190
- minima   192
- on sorption   193
- threshold   191
Compression wood   145, 151, 162, 163-165, 166
Computational fluid dynamics   59, 214, 217
Conditioning   158, *see also* Stress relief
Conductivity   87, 119

Subject Index

- hydraulic 119
- thermal, see Thermal conductivity
Contact angle 24
Corewood 155
Corniforme sp. 248
Creep 172, see also Strain
- combined 195–196
- recoverable, see also Strain, viscoelastic
Crook 148, 149, 151, 161, 165, 167–168, 250, 265
Cup 265

Dehumidifer-kilns 54
Density
- basic 23, 24, 27, 28, 119, 142, 151, 152–157, 163, 167, 234, 268–270
- – within-ring variations of 154
- – within-stand variations of 156
- – within-tree variations of 154–155
- green 66, 67, 141
- zones 141
Desorption
- isotherm for 26
Dewpoint 47
Dielectric constant 281
Diffusion
- bound-water 80, 133, 137
- driving forces for 86–100
- model 85–116
- pathways 82
- vapour 80, 137
Diffusion coefficient 85, 90, 91, 94–96, 103–113, 203
- definition of 90
- experimental measurements of 104–113
- modelling of 103–104
- moisture-content effect on 112–113
- relative 100
- temperature effect on 107–111
- values of 94, 96, 104–107
- – longitudinal 105–107
- – radial 105–107
Diffusivity, see Diffusion coefficient
Drying , see also kiln drying
- accelerated conventional temperature (ACT) 238
- airless 285
- constant external conditions of 222–223
- critical point in 203, 226
- curve, characteristic 73, 221
- dielectric 280–285
- – heating mechanisms 280–281
- – internal pressures on 282–283
- fundamental equations for 118–120
- high-temperature 124–126, 137, 151, 200–201, 203, 232–233, 235, 237–238
- – airstream temperatures in 233
- – humidity potential profile in 233
- – kiln-air velocities for 238
- – maximum strain profiles in 200
- – stress profiles in 201
- ideal time of 222–223
- intermittent 202
- mixed-wood boards 136–137, 269
- models
- – diffusion 85–116
- – empirical 67–71, 114
- – graphical-analytical 71–72
- – multiple-mechanism 117–138, 285
- – selection of 82–84
- periods 130
- radiofrequency 113, 236, 280–285
- rate
- – constant 120, 221
- – curve, normalisation of 72–79, 221–222
- – kiln-wide values 228–229
- – relative 73, 78
- regimes
- – penetration 100–103, 138
- – regular 100–103
- stress, see Stress
- superheated-steam 30, 203, 285–287
- – moisture-content profiles in 287
- – temperature profiles in 287
- vacuum 113, 236, 286–288
- variable conditions of 223–227
- wetwood 269–270

Earlywood 8, 9, 18, 19, 118, 153
Embolism 9
Emissions 239–241
Energy
- activation of 34, 109–111, 114
Enthalpy
- humid 46, 50
- of sorption 37, 135
- of vaporisation 54, 135
Entropy
- of sorption 40
- water vapour 99
Evaporation
- from wood surface 54–60, 66
- rate 55, 58, 62
- subsurface 60–63, 83
Evaporative plane, see also Zone
- receding 83, 123, 128–129, 137

Extractives   74, 143, 153, 255

Fan, *see* Kiln, airflow
Fibre saturation   32–34, 35, 125, 175
- point   32–33, 67, 81, 82, 94–96, 105, 109, 115, 133, 134
- - temperature variation of   96
Fibres   2, 11
- libriform   11
Fick's first law   88
Fick's second law   82, 90, 94
Finger-jointing, green   264–266
Formaldehyde   241
Fourier number   91, 113
Friction factor   210
Fungi
- decay   247–249, 251
- sapstain   250–252
Fungicides   251–254

Galactose   14
Glucose unit   10, 11
Grain
- slope of   146–147
- spiral   4, 146–148, 151, 168
- - angle   168
Greenwood
- moisture contents   139–141
- pretreatments   247–270
- - antisapstain   250–254
- - compression rolling   256–257
- - heating   259–261
- - incising   256
- - protective   247–255
Grosvenor chart   49
Growth rings   6
Gymnosperms   2

Hagen-Poiseuille equation   79
Handling
- kiln-dried product   244–246
Hardwoods   2, 3, 8–12, 27, 67, 74, 80, 85, 87, 106–107, 114, 125, 137–138, 139–141, 152, 155–156, 158, 162, 165, 225, 238, 240, 243, 244, 256, 258–259, 265–266, 283, 285, 286
- collapse in   158
- density of   12, 152, 155
- diffuse-porous   8, 10, 156
- diffusion coefficients for   106–107
- diffusion model   114
- drying-rate profiles   225
- greenwood moisture content   140–141
- incising, need for   256

- intervessel pitting   11
- ring-porous   9, 10, 155
- structure of   8–12
- vessels in   8–9, 80
Heartwood   5, 6, 23, 24, 29, 78, 106, 122, 128, 130, 136–7, 139–140, 142–143, 216, 223, 245, 268–269
- characteristic drying functions   78, 223
- diffusion coefficients for   106
- drying periods   130
- moisture movement in   128
Heat capacity
- definition of   45
Heat treatments   259–261
- dry heat   261
- hot-water soaking   259–260
- steaming   259–260
Hemicelluose   14–16, 151, 159, 163, 166, 183, 240
- chemical structures of   15
Hooke's law   173, 176
Honeycombing   257
Humidity
- charts   49–52
- definition   43
- relative   24, 26, 28, 29, 35, 38, 44, 48, 65, 87, 243
- - definition of   44
- - environmental changes in   40–42
Humidity-potential coefficent   55–56, 74
Hydrogen-bonding   34
Hygroscopicity   24, 26, 243
Hysteresis   24–26

Initial, cambial   4
Inversion point   286
Isotherm   24–26, 28
- adsorption   24, 41
- BET   34–37
- desorption   24, 41
- for various species   28
- Langmuir   34–35

Kelvin equation   29, 36–37
Kiln
- air-state changes in   50–52
- airflow   66, 74, 203–219 *see also* Airflow
- - fan speeds   238–239
- - maldistribution   206, 215–216
- - reversals in   227–232, 238–239
- - simulation of   208
- - vortex   207
- audits   216
- box-shaped   204–206

318   Subject Index

- control systems   236, 239
- corrosion   239-241
- dehumidifier   236, 254, 272, 273, 288-289
- double-track   204, 214, 238
- drying times   70-72
- emissions   239-241
- geometrical considerations   206-207
- heat demand, ideal   52-54
- heat-pump   254
- fans   52-53
- hydraulic model   207
- plenum   204, 206-208
- predrying   258
- pressure drops in   209-211
- single-track   204-205, 214, 225, 238
- solar   68, 271-280
- - energy losses from   277
- - external collector-type   273
- - greenhouse-type   272-273
- - insolation rates   273-274
- - modelling performance of   279-280
- temperature drop over   232-233, 239
- temperature levels in   53
- velocity distributions over   205-207, 211-212
- - between boards   207
- - coefficient of variation   206
Kiln operation   221-246, see also Kiln-seasoning
Kiln-seasoning, see also Drying
- economics   219, 278-279, 284-285
- energy use   54, 283-284, 288
- heat of sorption in   40
- Hi-Lo strategy   203
- schedules   85, 216, 233-237
- - development of   236-237
- - end-moisture values   216, 231, 243-244
- - moisture-based   235-236, 242
- - practical   227-223, 234-236
- - species-grouped   234
- - species-specific   234-236
- - time-based   235-236
Kiln stack   56, 60, 62, 204-205, 210-212, 223-229
- arrangements   204-205
- heat-transfer coefficients for   62
- mass transfer from   60, 223-229
- pressure drop across   210-211
- velocity distribution down   211-212
Kinetic-loss coefficient   207-208
Kirchhoff
- potential   83, 89

- transformation   88-89
Knothole   145
Knots   69, 139, 143-146
- dead   144
- live   144
- spike   144
Knotwood   144-145

Latewood   9, 18, 19, 118, 153
Laplace equation   127
Lignification   249
Lignin   16-17, 20-21, 129, 158-159, 166, 259
- degradation   129
- in cell wall   21
- in hardwoods   17
- in softwoods   16
- microfibrils embedded in   20
- softening of   158, 259
Lumen   4, 81, 126, 130, 137, 143, 261

Mannose   14
Mass-transfer coefficient   55, 63-64, 74, 93, 135, 225
- definition   55
- experimental values   61
- measurements   63-64
- - diffusion method   64
- - sorption method   64
- over flat plate
- - blunt-edged   58-60
- - sharp-edged   56-58
Microcavities   29, 32
Microfibrils   13, 14, 17, 18-20, 159-161, 165, 166-168
- angle   160, 161, 165, 166-168
- helix   19
- in lignin   20, 159
- swelling effect of   159-160
Microorganisms, control of   247-249
Microwave, see Drying, dielectric
Mixed wood, see also Drying, 139-149
Modulus
- of elasticity   174, 176, 178-180, see also Stiffness
- - secant value   182
- - tangent value   182
Moisture
- diffusivity   86, see also Diffusion coefficient
- equations for   129-136
- free   130-132
- - longitudinal   80
- - pathways   79-82

Subject Index 319

– – vapour   125, 128, 132–133, 137
– in air   43–45
– in cell wall   21
– in wood   23–24
– isotherm   24–26, 28
– – hysteresis   24–26
– movement (internal)   66, 79–82, 85, 105, 125, 128–136
– – bound   125, 128, 133, 137
– saturation
– – for various species   28
– sorption   34–37, 41
– – correlation of   37
– vapour pressure   29, 33, 44
Moisture content
– approach (to specification)   231–232, 242
– characteristic   73, 78
– driving force   88, 90–94, 115
– equilibrium   28–31, 34, 243
– greenwood   139–141
– irreducible   132
– kiln-wide board-average values   228–229
– maximum   23, 131
– maximum hygroscopic   30
– measurement   236, 245–246
– normalised, see characteristic
– profiles   83, 91, 95, 97, 102, 121–125, 287
– storage, before and after   245
– variability in dried product   230–232
Moisture sorption   24–31
Mollier-Ramzin chart   51
Moulds   264

NTU (Number of transfer units)   215, 222, 225, 227

Ophiostoma sp. 251, 254
Osmotic pressure   88

Paecilomyces variotii 254
Parenchyma   4, 6, 8, 11, 136
– apotracheal   11
– axial   6, 11
– ray   8, 136
Particulate emissions (PM-10)   240–241
Perforation plates
– in hardwoods   8
Permeability   80, 87, 108, 128, 130, 132–134, 248, 256–264
– liquid   132, 134, 256
– pretreatment improvements in   248, 256–264
– – compression rolling   256–257

– – incising   256
– vapour   133
Phase-transformation criterion   119
Phenylpropane unit   16
Phloem   4
Pinenes   241
Pit
– membrane   142, 248, 260
– – enzymatic degradation of   248
– – hydrolysis of   260
Pith   149, 162, 165, 167–168
Pits
– bordered   6,7, 118, 128–9, 132, 136
– – aspiration of   118, 128–9, 132, 136
– damage in pretreatment   256
Polyehylene glycol
Polyphenols   136
Potential
– chemical   83, 85, 86, 133
– humidity   55–56, 74
– Kirchhoff's flow   83
Precoating   266–267
Predrying, low-temperature   257–258
Prefreezing   261–262
Presurfacing   263–264
Pressure
– capillary   131
– gradient coefficient   120
– profiles   125–127
Psychrometer   48, 239

Quality, see Wood

Radiofrequency, see Drying, microwave
Rays   2, 5, 12
– biserate   5
   multiserate   5
Reaction wood   162–166, 249
– cell-wall features   164
Resin canals   6–8
– traumatic   8
Resin pockets   8
Resistivity, electrical   245
Reynolds number   56, 57, 214
Rigidity   159, 160

Sapstain   250–254
– depth after felling   252
Sapwood   5, 6, 24, 60, 67, 78, 106, 118, 122, 128–130, 136–7, 139–140, 143, 177, 178, 216, 223, 225, 227–229, 244, 248, 268–270
– characteristic drying functions   78, 223
– diffusion coefficients for   106
– drying periods   130

- drying-rate profiles 225
- greenwood moisture content 140
- moisture-content profiles 125
- moisture movement in 128-129
- permeability improvements in 248
- tensile stress-strain curves 177

Saturation
- adiabatic 46-48
- fibre , see Fibre
- irreducible 124, 125, 132
- of voids 131, 137
- wood 35

Saw-dry-rip SDR method 169, 261

Sawing
- strategies 166

Schmidt number 56
Sherwood number 56, 58, 102
Shook 264

Shrinkage
- effect of density on 167
- gradient
- intersection point 153
- longitudinal 160, 161
- radial 159, 199
- strain 158-159, 175-176, 180-181
- tangential 159, 161, 199
- volumetric 153, 158

Softwood 2, 3, 5-8, 27, 60, 67, 74, 80, 98, 105-106, 123-124, 127-136, 139, 147, 154-157, 162, 223, 227-233, 235, 240, 244, 257, 264, 285
- conservation equations for mass and energy 134-135
- density of 155-157
- diffusion coefficients for 105-106
- drying model 123-124, 127-136
- drying periods 130
- kiln schedules for 227-233
- structural grades 235
- structure of 5-8
- tracheids, see Tracheids

Sorption
- entropy change in 40
- environmental changes 40-42
- heat of 37-40
- - differential 38
- - finite-difference value 38
- - integral 40
- - of wood 39
- isotherms 23-26, 28, 41
- paths 41
- strain compliance 193
- theories 34-37

Sorting

- by density 268
- by green moisture content 270
- separating heart/sapwood 268

Sphaeropsis sp. 251
Spring 148, 149, 161, 250

Stack
- arrangements 237-238
- destickering 243
- top-weighting 161, 237, 243
- wrapping 202, 243-245

Stemwood 145
Stickers 203-204, 211-212, 238, 239, 243
Stiffness 151, 159, 160, 176, 178

Storage
- after drying 244-246
- wet 247-250

Strain
- behaviour of lumber 172, 199-202
- components 171, 196-197
- - solution procedures 196-197
- experimental determination 198-199
- failure 177, 178
- instantaneous 171, 176-182, 199
- - linear loading 176-181
- - non-linear loading 181
- - slow-loading 182
- - unloading 181-182
- mechanical analogues of 173-175
- - Burgers model
- - elastic element 173
- - Kelvin model 174, 183, 194, 198
- - Maxwell model 174, 189, 198
- - viscous element 174
- mechanosorptive 158, 162, 172, 185, 188-196, 198, 201
- - birch-like behaviour 189
- - spruce-like behaviour 189
- profile 180, 200
- proportional limit 176, 177, 181
- radial 157, 159, 175
- sectional 159
- shrinkage 157, 159, 172, 175-176, 180-181, 199
- tangential 159, 175
- ultimate limit 176
- viscoelastic 158, 162, 171, 183-188, 190, 195-196, 201
- - Bailey-Norton equation 183, 185-188
- - mechanical analogues 183-185
- - strain-hardening 186-187
- time-hardening 186-187

Stress
- drying 171-202, 238, 242-243

- growth   141, 169, 249–250
- profiles on drying   201
- relaxation of   249–250
- relief   202, 242–243, 262
- reversal   187
- tensile tests   177, 178
- ultimate values   178
Surface-emission coefficients   64, 92–93

Temperature
- adiabatic-saturation   47
- collapse-susceptible   157–158
- cool-limit   47
- dry-bulb   48, 120
- glass-transition   186, 259
- profiles   120–121, 287
- pseudo-wet-bulb   120
- wet-bulb   47, 63, 120
- wet-line   63
Tension wood   162–166
Terpenes   241
Thermal conductivity   118
Thermal-gradient coefficient   117, 118
Thermometers
- resistance   48
Tracheids   2, 5–6, 8, 11, 60, 80, 127–8, 147, 163
- cavitation   128
- ray   8
- vasicentric   11
Trichoderma spp. 254
Twist   149, 263, 265
Tyloses   9, 256

Vacuum, *see* Drying
Valonia sp. 13
Vapour
- diffusion, *see* Diffusion
- driving force   55, 83, 85, 94–98
Veneers, drying of   121
Vessels, *see* Hardwood
Volatile organic compounds (VOCs)   240–241

Wane   139
Warp   149, 158–162, 163, 166–169, 237–238, 243
Warping *see also* Bow, Crook, Twist
Waste   238
Water
- activity   24
- potential   99, *see also* Chemical potential
Wet-bulb
- depression   48, 66, 230, 237
- temperature   47
Wet-line   63, 255
- temperature   63
Wetwood   142, 269–270
Wood   2, 12–17, 66–67, 131–169
- acids   239–241
- characteristics of   2, 66–67
- composition of   12–17, 21
- drying kinetics   65–82
- features
- - cross-grained   146
- - gross   139–161
- - knotwood   145–146
- - intrinsic   152–169
- - stemwood   145
- hydrolysis   259
- juvenile   69, 148–151, 154, 155, 157, 264
- permeable   66–67
- quality   139–170, 254, 284–285
- - microwave drying, use of   284
- - sterilisation, use of   254
- sorption isotherms for   23–24, 28, 41
- water in   23–24

Xylan   14
Xylem   4
Xylose   14

Yellowing   263

Zone
- evaporative   133–134, 137
- moisture-flow   131

Species Index

Scientific name	Common name(s)	Reference pages
Abies spp.	firs	2, 7, 8, 112, 142, 145
A. alba	white fir	79
A. amabilis	amabilis (Pacific silver)fir	140, 255, 268, 283
A. concolor	white fir	269
A. grandis	western (grand) fir	105, 108, 126, 231
A. lasiocarpa	alpine fir	68, 96, 97, 155
A. sachlinensis	Japanese fir	261
Acer spp.	maples	9, 235
A. pseudoplatinus	sycamore	71
A. rubrum	red maple	10
A. saccharum	sugar maple	35, 278
A. sacharinum	silver maple	71, 75, 76, 77
Aesculus spp.	horse-chestnuts	9
Alnus spp.	alders	
A. rubra	red alder	71, 235
Araucaria spp.	araucaria pines	89
A. cunninghamii	hoop pine	142, 189
A. klinkii	klinki pine	25
Beilschmiedia tawa	tawa	42, 244
Betula spp.	birches	28, 105, 106, 108, 126, 182, 283
B. alleghaniensis	yellow birch	71, 141, 256
B. populifolia	poplar-leaved birch	10
Castanea spp	chestnuts	9
C. sativa	sweet chestnut	240
Catalpa speciosa	northern catalpa	143
Cedrus spp.	cedars	8
Celtis occidentalis	hackberry	106
Chamaecyparis spp.	white-cedars	
C. nootkatensis	yellow cedar (cypress)	155
C. obtusa	hinoki, white cedar	112, 122, 195
Crytomeria japonica	sugi, Japanese cedar	112, 122, 234, 271
Cupressus spp.	cypresses	
C. macrocarpa	macrocarpa cypress	5
C. sempervirens	bald cypress	71
Dicorynia paraensis	angelica	69
Eucalyptus spp.	eucalypts, gums	17, 95, 122, 175, 178, 183, 195, 236, 249, 265
E. delegatensis	alpine ash	109, 262, 267
E. diversicolor	karri	262
E. globulus	bluegum	285
E. maculata	spotted gum	96, 109
E. marginata	jarrah	251
E. muelleriana	yellow stringybark	96, 109

Scientific name	Common name(s)	Reference pages
E. obliqua	messmate (stringybark)	109, 189, 267
E. paniculata	ironbark	96, 109
E. pilularis	blackbutt	260
E. regnans	mountain ash	109, 184, 262, 267, 283, 284
Fagus spp.	beeches	249, 255
F. grandiflora	American beech	10, 24, 105, 108, 112, 141, 178
F. sylvatica	European beech	17, 70, 72, 122, 123, 182, 184, 189–191, 194, 247, 249, 288
Fraxinus spp.	ashes	2, 9, 142
F. americana	white ash	10, 106
F. mandshurica	Manchurian ash	261
Gingko biloba	gingko	162
Guiaiacum offinale	lignum vitae	23, 29
Hebe spp.	hebes	5
Juglans spp.	walnuts	
J. nigra	black walnut	71
Larix spp.	larches	122, 123, 145, 150
L. decidua	European larch	3, 6
L. leptolepsis	Japanese larch	261
L. kaempferi	Japanese larch	146, 168
L. occidentalis	western larch	140, 155
Librocedus spp.	cedars	
L. decurrens	pencil cedar	27
Liriodendron tulipifera	yellow poplar	121, 235
Liquidamber styraciflua	sweetgum	106
Nothofagus spp.	southern beeches	2, 28
N. fusca	red beech	67, 103, 256, 260, 262
N. truncata	hard beech	75, 107, 122, 124, 158, 199, 204, 240, 260
Nyssa sylvatica	blackgum, black tupelo	143, 235
Ochroma spp.	balsawoods	93
O. lagopus	balsa	23, 29
O. pyramidale	balsa	2
Picea spp.	spruces	2, 6, 7, 63, 105, 106, 108, 126, 142, 145, 150, 154, 178, 189, 195, 203, 248, 256
P. abies	European (Norway) spruce	42, 84, 184, 185, 195, 238
P. englemannii	Engelmann spruce	278
P. glauca	western white spruce	68, 140, 283
P. omorika	Serbian spruce	27
P. mariana	black spruce	20, 195, 265
P. rubens	red spruce	125
P. sitchensis	sitka spruce	30, 32, 35, 104, 108, 112, 122, 140, 155, 175
Pinus spp.	pines	2, 7, 29, 65, 106, 142, 145, 189, 195, 219, 249
P. caribaea	Caribbean pine	153
P. contorta	lodgepole pine	68, 140, 150, 155, 241, 278
P. densiflora	akamatsu, Japanese red pine	112, 122
P. jeffreyi	Jeffrey pine	160
P. lambertiana	sugar pine	255

Scientific name	Common name(s)	Reference pages
P. merkusii	Merkus pine	153
P. monticola	western white pine	142
P. nigra	Austrian pine	285
P. oocarpa	oocarpa pine	153
P. palustris	southern pine	121, 122, 131, 165, 234, 235
P. pinaster	maritime pine	114, 285
P. ponderosa	ponderosa pine	23, 140, 179, 235, 255, 269
P. radiata	radiata (Monterey) pine	24, 32, 42, 52, 67, 75, 77–79, 103, 118, 121, 124–136, 140, 146, 151, 153, 154, 159, 165, 168, 175–179, 185–186, 200–202, 203, 216, 228–234, 238, 241, 244, 265, 268–270, 285, 286
P. resinosa	red pine	254
P. strobus	yellow (eastern white) pine	27, 71, 133, 142, 255
P. sylvestris	Scots (Scotch) pine	89, 94, 105, 108, 109, 142, 178, 184, 193, 285
P. taeda	loblolly pine	54, 69, 105, 108, 122, 123, 146, 153, 265
Plagianthus spp.	ribbonwoods	
P. betulinus	lowland ribbonwood	3
Platanus occidentalis	American sycamore	106, 141
Populus spp.	poplars and aspens	9, 100, 104, 122, 142, 285
P. balsamfera	aspen	235
P. deltoides	cottonwood	71
P. tremuloides	trembling aspen	10, 99, 241, 250
Prunus spp.	cherries	235
Pseudotsuga spp.	false hemlocks	
P. menziesii	Douglas-fir	6, 7, 26, 33, 54, 105, 112, 121, 122, 140, 146, 150, 151, 153, 154, 157, 178, 189, 196, 235, 240, 245, 248, 265, 283, 284
Quercus spp.	oaks	2, 9, 12, 29, 71, 105, 240, 248, 249
Q. alba	white oak	54, 71, 142, 235, 255, 263
Q. falcata	southern red oak	263, 266
Q. petrea	durmast oak	288
Q. robur	English oak	278, 288
Q. rubra	red oak	10, 68, 71, 106, 112, 113, 141, 142, 235, 283, 284
Salix spp.	willows	9, 12
Sequoia spp.	redwoods	8
S. sempervirens	redwood	28, 71, 157, 262
Sequoiadendron giganteum	giant sequoia	1
Swietenia spp.	mahogany	279
S. mahogoni	mahogany	28
Taxodium distichum	swamp cypress	2
Taxus baccata	yew	2
Tectona spp.	teak	65
Thijopsis dolabrata	Japanese thuja	27

Scientific name	Common name(s)	Reference pages
Thuja spp.	arborvitae cedars	
T. occidentalis	white cedar	143
T. plicata	western red cedar	71, 140, 155, 240, 283
Tsuga spp.	hemlocks	7, 8, 150, 154
T. canadensis	eastern hemlock	106, 113
T. heterophylla	western hemlock	133, 140, 143, 155, 240, 255, 268, 283, 284
Ulmus spp.	elms	2
U. americana	American elm	106, 141
Vitis labrusca	vine	10

Made in the USA
San Bernardino, CA
28 February 2015